Topics in
Current Physics

42

Topics in Current Physics

Founded by Helmut K. V. Lotsch

Volumes 1–38 are listed on the back inside cover

Metallic Magnetism

Edited by H. Capellmann

With Contributions by
I. A. Campbell   H. Capellmann   G. Creuzet
P. Fulde   Y. Kakehashi   E. Kisker
V. Korenman   T. Moriya   G. Stollhoff

With 95 Figures

Springer-Verlag Berlin Heidelberg New York
London Paris Tokyo

Professor Dr. Herbert Capellmann

Institut für Theoretische Physik, Rhein.-Westf. Technische Hochschule,
Templergraben, D-5100 Aachen, Fed. Rep. of Germany
and
Institut Laue Langevin, Avenue des Martyrs
F-38042 Grenoble Cedex, France

ISBN 978-3-642-50070-1   ISBN 978-3-642-50068-8 (eBook)
DOI 10.1007/978-3-642-50068-8

Library of Congress Cataloging-in-Publication Data. Metallic magnetism. (Topics in current physics ; 42 –)
Includes bibliographies. 1. Magnetism. 2. Metals—Magnetic properties. I. Capellmann, H. (Herbert),
1942–. II. Campbell, I. A. III. Series: Topics in current physics ; 42, etc. QC754.5.M48
1986 538 85-15572

© Springer-Verlag Berlin Heidelberg 1987
Softcover reprint of the hardcover 1st edition 1987

Offset printing and bookbinding: Konrad Triltsch, Graphischer Betrieb, Würzburg.
2153/3150-543210

Preface

The magnetism of iron and other transition metals had been a subject of intensive research for a long time, but the understanding of the microscopic origin of "metallic magnetism" was quite limited until the early 1970's. During the last 10 to 15 years both theory and experiment contributed towards significant progress in this field, such that today a qualitative understanding has been achieved. The word "qualitative" indicates that the knowledge is still not complete; although many properties, the ground state as well as the finite-temperature behaviour and the phase transition from magnetic order at low temperatures to the paramagnetic state at high temperatures, can be explained in a coherent way, a quantitative description still is not fully achieved.

It is certainly appropriate to summarize the developments of the last 15 years and the present-day understanding of the field, this is the aim of this Topics volume. The form chosen is a collection of reviews, written by prominent scientists who themselves contributed decisively to the progress.

Scientists with a general interest in the field as well as specialists and active researchers in metallic magnetism should be able to profit from the two-volume treatment. The subjects not covered extensively in the present first volume (in particular neutron scattering and electronic structure properties) will make up the second volume.

Grenoble, October 1986   *H. Capellmann*

Table of Contents

List of Contributors

Campbell, I.A.
Université de Paris-Sud, Centre d'Orsay, Laboratoire de Physique des Solides,
Bâtiment 510, F-91405 Orsay, France

Capellmann, H.
Institut für Theoretische Physik, Rhein.-Westf. Technische Hochschule,
Templergraben, D-5100 Aachen, Fed. Rep. of Germany
and
Institut Laue Langevin, Avenue des Martyrs, 156X,
F-38042 Grenoble Cedex, France

Creuzet, G.
Université de Paris-Sud, Centre d'Orsay, Laboratoire de Physique des Solides,
Bâtiment 510, F-91405 Orsay, France

Fulde, P.
Max-Planck-Institut für Festkörperforschung,
D-7000 Stuttgart 1, Fed. Rep. of Germany

Kakehashi, Y.
Max-Planck-Institut für Festkörperforschung,
D-7000 Stuttgart 1, Fed. Rep. of Germany
Present address: Department of Physics, Hokkaido Institute of Technology,
Teine-Maeda, Nishi-Ku, Sapporo 006, Japan

Kisker, E.
Institut für Festkörperforschung, Kernforschungsanlage Jülich,
D-5170 Jülich, Fed. Rep. of Germany

Korenman, V.
Department of Physics and Astronomy, University of Maryland,
College Park, MD 20742, USA

Moriya, T.
Institute for Solid State Physics, The University of Tokyo,
Roppongi, Minato-ku, Tokyo 106, Japan

Stollhoff, G.
Max-Planck-Institut für Festkörperforschung,
D-7000 Stuttgart 1, Fed. Rep. of Germany

1. Introduction

H. Capellmann

1.1 Preliminary Remarks

This treatment is to review the field of metallic (or itinerant) magnetism with particular emphasis on the 3d-transition metal series containing the "classical" ferromagnetic metals iron, cobalt and nickel. In a collection of reviews our present-day understanding of the topic will be presented: Why do the materials cited above order magnetically at zero temperature and what are the microscopic processes leading to the phase transition to a paramagnetic state at high temperatures?

Although magnetism (in particular, of iron: *Ferro*magnetism) has attracted attention and research activities over many hundreds of years, an understanding of the microscopic origin of metallic magnetism evolved only in the last decade. Therefore, a few remarks might be appropriate to explain why the problem is difficult and why it took so long to establish the connection between microscopic and macroscopic properties: Even today this understanding is far from being complete.

In the transition metals Cr, Mn, Fe, Co, Ni the electrons responsible for the magnetic properties in the ferromagnetically (Fe, Co, Ni) or antiferromagnetically (Cr, Mn) ordered phases participate in the Fermi surface. These electrons are itinerant, their wave functions are delocalized, therefore no truely localized magnetic moments exist. The delocalized nature of the wave functions is due to the possibility of the electrons to move around rather freely in the crystal (free-hopping processes). It is this feature, which distinguishes itinerant magnetism from the magnetism due to localized moments, such as in the insulating transition metal oxides and in rare-earth metals. In the latter systems well localized electronic wave functions not participating in the Fermi surface give rise to localized magnetic moments. The magnetism can then be discussed concentrating on the magnetic degrees of freedom alone, governed by some magnetic Hamiltonian (like a Heisenberg Hamiltonian).

This separation of the magnetic degrees of freedom from the translational degrees of freedom is not possible in itinerant systems. The relevant d electrons are delocalized: they can move freely through the crystal. The kinetic energy associated with this electron movement is characterized by the bandwidth W, which is of the order of 5 eV. This energy is much larger than the thermal energy $k_B T_c$ leading to a phase transition from the magnetically ordered to the paramagnetic state. To achieve an understanding of the magnetic properties, one has to deal with the full many-body problem incorporating magnetic and translational degrees of freedom of the electrons responsible for the magnetism. This is the reason why it took so long until even a qualitative understanding of itinerant magnetism evolved.

1.2 Ground State and Elementary Excitations

The ground state of the ferromagnetic metals Fe, Co, Ni is characterized by a band structure, in which the usual spin degeneracy of simple metals has been lifted: Electrons described by the same wave vector but different spin quantum numbers up or down have different single particle energies, the difference

$$\Delta_k = \varepsilon_{k\uparrow} - \varepsilon_{k\downarrow} \tag{1.1}$$

being the exchange splitting. Current state-of-the-art band structure calculations [1.1] are able to account for ground state properties quite accurately, the average exchange splittings obtained for Fe, Co and Ni are of the order of 1.5, 1, and 0.6 eV, respectively. The methods, based on spin-density functional theory, can also be applied successfully to describe the antiferromagnetic ground states of Cr and Mn, although for computational reasons a number of simplifications are imposed: For Cr a commensurate spin structure and for Mn a simplified crystal structure is used in the band structure calculations [1.2].

Although the ground state of the elements on the right of the 3d series is magnetically ordered with, taking iron as an example, an average magnetic moment of 2.2 μ_B per Fe atom, no localized magnetic moments exist. Let us illustrate this important difference using a simple one-band model with an assumed ferromagnetic ground state: The number of occupied spin up levels $N_\uparrow$ is larger than the corresponding number of occupied spin down levels $N_\downarrow$:

$$N_s = \sum_k f(\varepsilon_{ks}) \; ; \tag{1.2}$$

f is the $T = 0$ Fermi function.

Let us now make a very fast Gedanken experiment measuring the spin density at some arbitrary lattice site. The meaning of "fast experiment" is that the measuring time Δt is small compared to hW^{-1} where h is the Planck's constant and W the band width

$$\Delta t < h/W. \tag{1.3}$$

This means that the energy transfer in the experiment must be large compared to the band width. The result of such an experiment is characterized by the probabilities

$$w_s = 1/N \sum_k f(\varepsilon_{ks}) \tag{1.4}$$

where N is the total number of sites.

The probability to measure exactly one electron with spin up occupying the site, is $w_\uparrow(1 - w_\downarrow)$, the probability to find exactly one electron with spin down being $w_\downarrow(1 - w_\uparrow)$. The probability of finding a total spin zero is given by the sum of the probabilities for the event of finding two electrons, which is $w_\uparrow w_\downarrow$ and for the event of finding no electrons at all, which is $(1 - w_\uparrow)(1 - w_\downarrow)$. Electron-electron correlation effects will actually suppress the probability of finding two electrons simultaneously, but this does not change the conclusions of the discussion and shall be neglected here.

The average over the probabilities for these four different events leads to the average moment $\langle \mu_i \rangle$ per atom in the ground state. One has to keep in mind, however, that the amplitude of the magnetic moment is not constant in the ground state but has very fast quantum fluctuations. This is due to the fact that the electrons are itinerant and have wave functions which are phase coherent over large distances, therefore the electron density ϱ_i and as a consequence the spin density s_i as well are not described by sharp quantum numbers even in the ground state, because the operators ϱ_i and s_i do not commute with the Hamiltonian H

$$[\varrho_i, H] \neq 0 \; ; \quad [s_1^2, H] \neq 0 \; . \tag{1.5}$$

The expression "fast quantum fluctuations" is used to indicate that these fluctuations will show up directly in fast experiments with a short time constant (and hence large energy transfer). This typical "quantum fluctuation time" t_q is of order 10^{-15} s for the 3d transition metals

$$t_q \approx h/W \approx 10^{-15} s \; ; \tag{1.6}$$

t_q is the typical time for a 3d electron to move from one lattice site to another. The existence of this typical time constant is characteristic for itinerant systems and has no equivalence in localized magnetism. This will turn out to be of considerable importance for the behaviour at finite temperatures and the unusual properties of the paramagnetic phase [1.3].

In systems having well localized magnetic moments the amplitude of the moments at sites i are constants of motion

$$[S_i^2, H_{loc}] = 0 \tag{1.7}$$

and the resulting spin dynamics are quite different from that of itinerant systems.

In the latter a "slow" measurement of the spin density with a typical time constant larger than t_q (which means typical energy transfers much smaller than W) will average over the quantum fluctuations and will directly measure the average moment per atom.

The spin-flip excitation spectrum at low temperatures consists of single particle excitations and collective excitations. The single particle excitations,

3

which correspond to transitions from a spin up to a spin down band, form a continuum (the Stoner continuum). Excitations with small or vanishing energy transfer ω are in general possible for finite momentum transfer q only; single particle excitations with zero momentum transfer (q = 0) cost a finite amount of energy ($\omega \neq 0$), the exchange splitting Δ. The fact that no single particle excitations for small q and ω exist, leads to well-defined collective excitations in that region: spin waves. For the ferromagnetic transition metals the dispersion is quadratic at small q :

$$E_{sw}(q) = Cq^2 \; ; \tag{1.8}$$

C being the spin wave stiffness constant. For large q the spin waves enter the single particle continuum and become overdamped not existing as well defined excitations anymore.

An average spin wave energy E_{sw} is of the order of 50 meV

$$E_{sw} \sim 50 \, \text{meV} \; , \tag{1.9}$$

which is two orders of magnitude smaller than the band width. The typical time constant t_{sw} associated with this energy

$$t_{sw} \approx h/E_{sw} \approx 10^{-13} \, \text{s} \tag{1.10}$$

is much slower than the fast t_q. This clear separation of timescales makes possible a visualization of the spin waves in the following way.

Averaging over fast quantum fluctuations (10^{-15} s) leads to average moments per atom. These quantum averaged moments now themselves may fluctuate slowly in time if the system is in some excited configuration. The spin wave configuration corresponds to a slow wave-like precession (10^{-13} s). In the following the expression "atomic moment" or average "moment per atom" will be used, and the quotation marks are to attract the readers' attention to the fact that these objects have a definite meaning only after averaging over the very fast quantum fluctuations.

1.3 Finite Temperature Properties: What Drives the Phase Transition?

In itinerant systems the possibilities for a phase transition from a magnetically ordered state at low temperatures to a paramagnetic state at high temperatures are more diverse than in localized systems. Several different driving mechanisms for the phase transition have been proposed in the past.

1.3.1 The Temperature Dependence of Δ

The exchange splitting Δ might be strongly temperature dependent and vanish at T_c leading to a usual paramagnetic metal above T_c. This mechanism is the basic assumption of Stoner theory [1.4]. It implies that the amplitude of the "atomic moments" decreases strongly with temperature and vanishes at T_c. This might be detected in an experiment with a typical time constant slightly faster than the (slow) magnetic fluctuation time t_{sw} (10^{-13} s for Fe) $\Delta t \lesssim t_{sw}$. This is much slower than the quantum fluctuation time (10^{-15} s) and the experiment would therefore average over those fluctuations, but fast enough to observe slow magnetic fluctuations on the scale t_{sw}. If the exchange splitting is strongly temperature dependent this intermediate timescale experiment should see the amplitude of the "atomic moment" decrease with T and vanish at T_c, "slow" magnetic fluctuations of significant intensity above T_c are absent according to Stoner theory. These predictions of Stoner theory are contrary to the true situation in transition metals, as can be seen most directly in paramagnetic neutron scattering experiments [1.5–7]. The physical reason why this mechanism is not relevant in transition metals is the high cost in energy to make the exchange splitting Δ vanish. Typically at the magnetic transition temperature the thermal energy $k_B T_c$ is an order of magnitude smaller than Δ itself, thermal energies of order $k_B T_c$ do not change the exchange splitting appreciably. If the transition temperature is calculated using Stoner theory, the result obtained is typically too large by a factor of 5 to 10, which among other failures, indicates that Stoner theory does not contain the essential ingredients for the phase transition.

1.3.2 Amplitude Fluctuations

A second mechanism invoked as a possible driving force for the phase transition is based on amplitude fluctuations [1.8–10]. In this picture an average "atomic moment" still persists above T_c, but it is argued that the amplitude should show very strong fluctuations even on the long time scale of $t_{sw} \approx 10^{-13}$ s, and that these slow amplitude fluctuations should destroy long range magnetic order above T_c.

To avoid confusion, I want to give the definition of what is meant by amplitude of the "atomic moment". It is important to realize that the "atomic moment" is an object already averaged over the very fast quantum fluctuations. Let us take a typical pure quantum state $|\kappa\rangle$ of the many body system, i.e. some vector in Hilbert space, $|\kappa\rangle$ is supposed to have a finite probability of being occupied at the temperature of interest, i.e. it contributes with finite weight to the statistical averages. This will be the case if its energy per site is low enough. The amplitude $|m_i^\kappa|$ of the "atomic moments" in the configuration $|\kappa\rangle$ (the word "configuration" will be used for a pure quantum state) is defined as the amplitude of the following quantum mechanical expectation value:

$$m_i^\kappa = g\mu_B \langle \kappa | S_i | \kappa \rangle . \tag{1.11}$$

S_i is the total spin density operator for a lattice cell at site i. The quantum mechanical expectation value ensures the average over the fast quantum fluctuations in the configuration. The amplitudes $|m_i^\kappa|$ might differ from one site

5

to another, and for a given i they might differ from one $|\kappa\rangle$ to another $|\kappa'\rangle$. Remark that this definition of the amplitude is different from $(\langle S_i^2 \rangle)^{1/2}$, which would be non zero even in a simple metal like Na, which is almost free electron like.

If amplitude fluctuations drive the phase transition an experiment on the intermediate time scale $\Delta t \stackrel{<}{\sim} t_{sw}$ should show finite "atomic moments" persisting above T_c, but differing markedly in amplitude from one lattice point to another. This mechanism has mainly been invoked for "weak itinerant ferromagnets" [1.8–10], i.e. systems with low $T_c(<50\,\mathrm{K})$ and a small moment at $T = 0(<0.5\,\mu_B)$.

1.3.3 Transverse Fluctuations

A third mechnism – and the dominant one for the ferromagnetic transition metals – is based on transverse fluctuations of the magnetization (for example, spin waves are transverse fluctuations): the "atomic moments" have almost constant amplitude, but do not point into the same direction at different lattice points. This type of fluctuation is the only one possible in localized magnetic systems. There, S_i^2 corresponds to a sharp quantum number $S(S + 1)$ with S being an integer or half integer. The amplitude is a constant of motion here, only transverse fluctuations are allowed. In itinerant systems S_i^2 does not correspond to a sharp quantum number due to the delocalized electronic wave functions leading to the fast quantum fluctuations. After averaging over these fast fluctuations, however, the amplitudes $|m_i|$ might be almost the same for all i and all configurations $|\kappa\rangle$ which carry appreciable weight in the thermal averages.

This is advocated in the "fluctuating band theory" to be the case for the ferromagnetic transition metals [1.3,11–13]. Transverse fluctuations cost low energy: In the ordered state below T_c transverse fluctuations are spin waves, their typical energy scale is $0.05\,\mathrm{eV}$, and a macroscopic occupation of spin wave excitations may lead to a phase transition destroying long-range magnetic order. Although this picture is similar to what happens in a localized magnetic system, where only transverse fluctuations are possible, we shall see, that the itinerancy of the electrons in the transition metals leads to characteristic and drastic differences to localized systems in the magnetic properties of the paramagnetic phase.

The last two mechanisms (amplitude fluctuations and transverse fluctuations) might actually be mixed and cooperate to produce the phase transition, in particular in weak itinerant systems [1.14,15].

1.4 The Paramagnetic Phase

The modern theories of itinerant magnetism [1.3,8–23] have one common feature: The dominant fluctuations leading to the phase transition to the paramagnetic state in the 3d transition metals are slow transverse fluctuations (to some extent accompanied by longitudinal fluctuations in weak itinerant ferromagnets) resulting in a paramagnetic state preserving some sort of "atomic moments" above the transition temperature T_c. There has been considerable controversy, however, concerning the detailed behaviour of the magnetic correlation functions above T_c.

In the "fluctuating band theory" [1.3,11-14] the phase above T_c in its magnetic properties is still dominated by the itinerancy of the 3d electrons resulting in "exchange split local bands" with considerable short-range magnetic order. These "local bands" can be thought of as being exchange split not with respect to a global direction but with respect to the direction of the local magnetization, which fluctuates (slowly compared to the time scale of electron hopping processes t_q) in space and time, having no long-range order above T_c.

The disordered local moment picture (DLM) [1.17,18,22,23], on the other hand, starts from the assumption that the magnetic properties above T_c are close to the properties of systems having well localized magnetic moments such as insulators and rare earth metals, being well described by, e.g. a Heisenberg Hamiltonian: Although the 3d electrons are itinerant, within the DLM picture they are thought to establish local magnetic moments which can be disordered even on the scale of nearest neighbour distances above T_c.

Before discussing the theoretical ideas it is useful for the understanding to go back to the definition and the meaning of "atomic moments" in itinerant systems: Equation (1.11) defines the "atomic moments" in some typical pure quantum state $|\kappa\rangle$ of the paramagnetic phase as a quantum mechanical expectation value of S_i. We shall discuss time dependences of the operators given by

$$S_i(t) = e^{-iHt} S_i e^{iHt} \ . \tag{1.12}$$

The expectation values $m_i(t)$ will in general be time dependent, and in the paramagnetic phase an average of $m_i(t)$ over a sufficiently long time will vanish (or alternatively the average of m_i^κ over the $|\kappa\rangle$) contributing significantly at some temperature $T > T_c$). The common feature of all modern theories [1.3,8-23] is that at temperatures above T_c, the m_i have a finite amplitude for all $|\kappa\rangle$ contributing significantly to the statistical average. Transverse fluctuations of the m_i are supposed to be responsible for the phase transition.

The controversy between the fluctuating band theory and the disordered local moment picture concerns the degree of short-range magnetic order of the m_i persisting into the paramagnetic phase.

The expression "short-range magnetic order" at this point is still quite vague and needs further specification, which is best done discussing the spin density – spin density correlation function

$$C(R_i - R_j, t) \sim \langle\langle S_i(t) S_j(0)\rangle\rangle \ , \tag{1.13}$$

where the symbol $\langle\langle \ \rangle\rangle$ indicates a thermodynamic average

$$\langle\langle A \rangle\rangle = \mathrm{Tr}\{\exp{(-\beta H)}A\}/\mathrm{Tr}\{\exp{(-\beta H)}\} \ . \tag{1.14}$$

The correlation function $C(R,t)$ will, of course, reflect the characterstic times and lengths of the system.

The basic point of the fluctuating band theory [1.3,11–14] is that the itinerancy of the magnetic 3d electrons will influence the short-range behaviour of the correlation function [1.24]. This short-range part of $C(R,t)$ depends on the specific short range forces in the system. A rough estimate of the characteristic length ℓ_e over which the itinerancy will dominate the form of $C(R,t)$ at temperature $T(\gtrsim T_c)$ is

$$\ell_e \sim v_F h/kT \ . \tag{1.15}$$

v_F is the Fermi velocity and ℓ_e is a typical length an itinerant electron can travel in a thermal fluctuation time h/kT. For the 3d transition metals ℓ_e will be large compared to typical nearest neighbour distances.

The FBT argues that over distances short compared to ℓ_e the magnetic properties in itinerant systems closely resemble ground state properties due to the fact, that the itinerant d-electrons responsible for the magnetism are phase coherent of such distances. Thus significant disorder is possible only for distances larger than ℓ_e. At these larger distances $(>\ell_e)$ the "universal" behaviour (i.e., not depending on the specific short-range forces) for the equal time correlation function $C(r)$

$$C(R) \sim 1/\, R e^{-R/\xi} \tag{1.16}$$

should be recovered, where ξ is the correlation length.

To be more specific here, I shall call ξ the "thermodynamic correlation length". ξ is temperature dependent and diverges at T_c. Close to T_c where ξ is very large the critical properties show "universal behaviour", to a large extent they do not depend on the underlying microscopic picture and the short-range forces. One has to keep in mind, however, that the form $C(r)$ given in (1.16) only describes the long-range behaviour, for distances beyond which the short-range (system dependent) forces dominate the properties of the correlation function.

The difference between the "electronic coherence length" ℓ_e and the thermodynamic correlation length ξ also shows up in their different temperature dependences. Whereas the thermodynamic correlation length ξ scales with the relative temperature $(T - T_c)/T_c$, diverging with a universal exponent for $T \rightarrow T_c$, the (system dependent) electronic coherence length ℓ_e will show no anomalies around T_c, scaling with $1/T$ as argued in (1.15). If we take iron as an example, T_c is of order $1000\,\mathrm{K}$, and the melting temperature of order $1800\,\mathrm{K}$, thus in the range above T_c, where the solid still exists, ℓ_e will not change drastically and we can speak of a typical electronic coherence length ℓ_e.

To understand the meaning of short-range magnetic order it is also necessary to discuss the time dependence of the correlation function $C(R,t)$ or alternatively its Fourier transform

$$S(Q,\omega) \sim \int dt \exp\left(-i\omega t\right) \sum \exp\left[iQ(R_i - R_j)\right]\langle\langle S_i(t) \cdot S_j(0)\rangle\rangle \ . \qquad (1.17)$$

It was argued above that for short distances (i.e., large Q) the electronic co-herence length ℓ_e makes significant thermal disorder impossible, which means that $S(Q,\omega)$ should not contain Fourier components of significant intensity in the large Q region *for energies* $\hbar\omega \lesssim k_B T_c$. This last specification concerning the frequency dependence of $S(Q,\omega)$ is of central importance: The low frequency (i.e., within thermal energy range) components of $S(Q,\omega)$ in the large Q region should have weak intensity according to the fluctuating band theory, whereas there should be considerable intensity spread over large energies of up to the band width (of order several eV for the transition metals):

When introducing the "quantum fluctuation time" t_q (of order $h/W \sim 10^{-15}$s), the typical time for the itinerant 3d electrons to cover a nearest neighbour dis-tance, it was implied that on this time scale also the local spin density should change drastically. This results in significant intensity for *large Q* [in the neigh-bourhood of the Brillouin Zone (B.Z.) boundary] and *high energy* (up to the bandwidth) Fourier components of $S(Q,\omega)$. These Fourier components are a result of the quantum fluctuations in the system, they are present at all tem-peratures (also at T = 0). "Thermal disorder" is caused by fluctuations which have energies within the range kT, whereas the high energy ($\hbar\omega \gg kT$) fluc-tuations are a result of the underlying quantum mechanical properties of the itinerant electrons.

Short-range magnetic order according to the FBT implies that for *large Q* the intensity of spin fluctuations in the energy range $\hbar\omega \lesssim kT_c$ should be small. Considerable intensity, however, should be spread over the large energy region $\hbar\omega < W$ characteristic of the quantum fluctuations.

This implies a considerable difference to properties of magnetic insulators, rare earth metals or Heusler alloys, where truely localized magnetic moments determine the magnetic properties. These systems are usually well described by a Hamiltonian of the Heisenberg type:

$$H_{loc} = 1/2 \sum J_{ij} S_i \cdot S_j \ . \qquad (1.18)$$

All energies are then determined by the exchange constants J_{ij}, e.g. spin wave energies at low T as well as the transition temperature T_c, which is typically comparable to the maximum spin wave energy [1.25]. At temperatures above T_c *all* fluctuations are within thermal energy range also for large Q, leading to significant intensity of $S(Q,\omega)$ for $\hbar\omega < kT_c$ for *all* Q, thus implying thermal disorder already over nearest-neighbour distances.

Let us point out again that for the small Q part of the spectrum itinerant systems and localized systems have qualitatively *similar* behaviour (universal behaviour independent of the short-range forces). The characteristic differences due to the itinerancy of the magnetic electrons in the 3d transition metals show up at large Q (probing the short distances).

Within the disordered local moment picture [1.17,18,22,23] on the other hand, it is argued that although the electrons carrying the magnetism of transition metals are itinerant, their magnetic properties can nevertheless be described by an effective Hamiltonian of the type H_{loc}, (1.18), enabling disorder on a nearest-neighbour scale above T_c. The argument put forward is roughly as follows: Imagine at some instant of time a magnetic configuration characterized by some specific distribution of spin up and spin down electrons (in equal amounts above T_c) within the lattice. These electrons are itinerant and are therefore able to move to different lattice sites within a characteristic time t_q. The basic assumption of the DLM picture is that over time intervals $t_{sw} \gg t_q$ the electron hopping processes take place as if the lattice sites carried labels "spin up" and "spin down", such that spin up electrons predominantly visit "spin up" sites and spin down electrons predominantly visit the "spin down" sites, thereby establishing "local moments". The labeling itself is supposed to change on a characteristic time scale $t_{sw} \gg t_q$, and above T_c to be largely uncorrelated over nearest neighbour distances ("disordered local moments"). This picture therefore implies considerable thermal disorder at short distances leading to significant intensity of the Fourier components of $S(Q,\omega)$ for all Q (also large Q close to the B.Z. boundary) within the thermal energy range $\hbar\omega \lesssim kT_c$.

The theoretical controversy (LBT versus DLM picture) has been largely resolved in the last several years mainly due to neutron scattering and photoemission experiments. Neutron scattering is able to directly measure the magnetic correlation function ($S(Q,\omega)$ is proportional to the magnetic neutron scattering cross section for momentum transfer Q and energy transfer ω). Photoemission, on the other hand, allows the study of the underlying electronic structure (essentially the one electron Green's function) which gives rise to the magnetic properties and an analysis of the photoemission current (spin, angle, and energy resolved) enables conclusions about the magnetic correlation functions [1.26–29].

Both neutron scattering and photoemission play a central role as experimental tools for the study of metallic magnetism and are represented in this book in several chapters (Chap. 3 by E. Kisker and Chap. 4 by V. Korenman, on photoemission), see also [1.30] .

Results from both types of experiments seem to be incompatible with the disordered local moment picture and support the basic assumptions of the fluctuating band theory:

Recent neutron scattering experiments using polarized neutrons and polarization analysis were able to determine the total $S(Q,\omega)$ integrated over all energy transfers smaller than typical thermal energies kT_c (of order 0.1 eV for Fe), but excluding the higher-energy quantum fluctuations. It was found that even well above T_c this integrated intensity $S_{th}(Q)$ is strongly peaked around $Q = 0$ dropping to very small values with increasing Q.

A typical Q_c over which this dropoff occurs is obtained from the observed peak in $Q^2 S_{th}(Q)$ as a functio of Q (in localized systems $Q^2 S_{th}(Q)$ monotonically increases with Q) and taking iron as an example, Q_c turns out to be of order $0.4\,\text{Å}^{-1}$, considerably smaller than $Q_{B.Z.}$ (the Q values for the B.Z. boundary), which is of order $1.7\,\text{Å}^{-1}$. Q_c depends only weakly on temperature above T_c. These findings directly confirm the basic ideas of the FBT.

Spin- and angle-resolved photoemission experiments will be presented extensively by E. Kisker in Chap. 3. The results obtained by this technique again indicate a considerable degree of short-range order in the paramagnetic phase [1.29].

1.5 Organization of the Book

This treatment contains seven reviews which are to present our current understanding of itinerant magnetism. These chapters will be separated into two volumes. The first five are part of the present volume (Chaps. 2–6), the second volume [1.30] will comprise the remaining two chapters.

Chapter 2 by T. Moriya will present "A Unified Picture of Magnetism", aiming to cover the whole range from weak itinerant magnetism all the way to localized magnetism. Although this last field lies outside the scope of the book, it is certainly useful and necessary to discuss the connections of the topics treated to adjacent fields.

Chapter 3 by E. Kisker will be on "3d-Metallic Magnetism and Spin Resolved Photoemission". This experimental technique is the most important one for the study of electronic structure properties and is able to give essential information on the character of the delocalized electrons carrying the magnetism. Development and progress in this area during recent years were very important for the understanding of itinerant magnetism.

Chapter 4 by V. Korenman is closely related to the preceding one: It reviews the "Local Band Theory" with particular emphasis on the consequences of the theory on photoemission. The local band theory starts from some given band structure describing the ground state at zero temperature and discusses how finite temperature properties, the transition to and the properties of the paramagnetic phase can be understood.

Chapter 5 by P. Fulde, Y. Kakehashi and G. Stollhoff reviews "Electron Correlations in Transition Metals". Here the main objective is an understanding of the correlated ground state in the 3d-transition metal series based on an evaluation of the ground-state wave function, which allows a description of equal time correlation functions. An extension to include excitations and finite temperature effects is also included.

The volume closes with a brief chapter on "Magnetovolume Effects" by I.A. Campbell and G. Creuzet. The occurrence of magnetism in the 3d-series is accompanied by strong magnetovolume effects. Their temperature variation gives access to information about the electronic properties.

Two essential topics are not contained in the present volume; they will make up a second volume [1.30].

First there will be a chapter on "Neutron Scattering in Transition Metal Magnetism" by K.R.A. Ziebeck. Neutron scattering is an ideal experimental tool to directly study magnetic correlation functions, which to a large extend are the main objective in the understanding of magnetism. Properties like magnetization distributions and excitations in the ordered state at low temperatures, their behaviour while approaching the transition temperature T_C, the properties of the critical region (universal to a large degree), and the behaviour of the paramagnetic phase above T_C have all been studied extensively. Their review will therefore be extensive and will take up more than half of the entire volume.

The second chapter will contain a description of "Electronic Structure Properties in Metallic Magnetism" by B.L. Gyorffy. Modern state-of-the-art band structure calculations allow a quite accurate account of ground-state properties of transition metals. The extension of these first principle techniques to describe finite temperature effects is still an open field, this chapter will review progress and attempts being made in this direction.

Progress in understanding of metallic magnetism was considerable in the last ten to fifteen years, to which both experiment and theory contributed significantly. It is the aim of this 2-volume treatment to comprise our current understanding made possible by these developments.

References

1.1   C.S. Wang, J. Calloway: Phys. Rev. B15, 298 (1977)
1.2   J. Kübler: J. Magn. Magn. Mat. 20, 107, 277 (1980)
1.3   H. Capellmann: J. Magn. Magn. Mat. 28, 250 (1982)
1.4   E.P.Wohlfarth: Rev. Mod. Phys. 25, 211 (1953)
1.5   J. Déportes, D.Givord, K.R.A. Ziebeck: J. Appl. Phys. 52, 2074 (1981)
1.6   P.J. Brown, J. Déportes, D. Givord, K.R.A. Ziebeck: J. Appl. Phys. 53, 1973 (1982)
1.7   P.J. Brown, H. Capellmann, J. Déportes, D.Givord, K.R.A. Ziebeck: J. Magn. Magn. Mat. 30, 243, 335 (1982); and 31-34, 295 (1983)
1.8   T. Moriya, A. Kawabata: J. Phys. Soc. Jpn. 34, 639 (1973); and 35, 609 (1973)
1.9   T. Moriya: J. Magn. Magn. Mat. 14, 1 (1979); and J. Phys. Soc. Jpn. 51, 420 (1982)
1.10   K.K. Murata, S. Doniach: Phys. Rev. Lett. 29, 285 (1972)
1.11   H. Capellmann: J. Phys. F4, 1466 (1974); Solid State Commun. 30, 7 (1979); and Z. Phys. B34, 29 (1979)
1.12   V. Korenman, J.L. Murray, R.E. Prange: Phys. Rev. B16, 4032 (1977)
1.13   V. Korenman, R.E. Prange: Phys. Rev. B19, 4691 (1979)
1.14   J.B. Sokoloff: Phys. Rev. Lett. 31, 1417 (1973); J. Phys. F5, 528, 1946 (1975)
1.15   G.G. Lonzarich: J. Magn. Magn. Mat. 45, 43 (1984)
1.16   J.A. Hertz, M.A. Klenin: Phys. Rev. B10, 1084 (1974)
1.17   H. Hasegawa: J. Phys. Soc. Jpn. 46, 1504 (1979); J. Phys. Soc. Jpn. 49, 178, 963 (1980)
1.18   J. Hubbard: Phys. Rev. B19, 2626 (1979); B20, 4584 (1979); B23, 5974 (1981)
1.19   M. Cyrot: J. Magn. Magn. Mat. 45, 9 (1984)
1.20   M.V. You, V. Heine, A.J. Holden, P.J. Lin-Chung: Phys. Rev. Lett. 44, 1282 (1980)
1.21   P.J. Lin Chung, A.J. Holden; Phys. Rev. B23, 3414 (1981)
1.22   B.S. Shastry, D.M. Edwards, A.P. Young: J. Phys. C14, L665 (1981)
1.23   J. Staunton, B.L. Gyorffy, A.J. Pindor, G.M. Stocks, H. Winter: J. Magn. Magn. Mat. 45, 15 (1984)

1.24   H. Capellmann, V. Vieira: Solid State Commun. **43**, 747 (1982)
1.25   H.A. Mook: J. Magn. Magn. Mat. **31–34**, 250 (1983)
1.26   V. Korenman, R.E. Prange: Phys. Rev. Lett **44**, 1291 (1980)
1.27   H. Capellmann, R.E. Prange: Phys. Rev. B**23**, 4709 (1981)
1.28   P.Durham, J. Staunton, B.L. Gyorffy: J. Magn. Magn. Mat. **45**, 38 (1984)
1.29   E. Haines, R. Clauberg, R. Feder: Phys. Rev. Lett. **54**, 932 (1985)
1.30   H. Capellmann (ed.): *Metallic Magnetism II*, Topics Cur. Phys. (Springer, Berlin, Heidelberg)

2. A Unified Picture of Magnetism

T. Moriya
With 17 Figures

2.1 Background

A remarkable event in recent investigations of metallic magnetism is the success of the theory of weak itinerant ferromagnetism where spatially extended coupled modes of spin fluctuations play predominant roles. The success of the theory was even quantitative and established a new picture of magnetism in the opposite limit to the long-familiar picture of a system consisting of local magnetic moments.

As a natural consequence of this development, attempts have been made to generalize the theory of spin fluctuations so as to interpolate between the two mutually opposite extremes, the local-moment limit and the weakly ferromagnetic (extended-moment) limit. Such a theory should naturally lead to a unified picture of magnetism and be applied to substances in the intermediate regime covering a major part of itinerant-electron magnets including Fe, Co and Ni.

These recent developments in the field of magnetism are briefly summarized.

2.1.1 Historical Development

Recent investigations on metallic magnetism have brought forth significant advances in our understanding of this long-standing controversial problem [2.1]. After the success of the band theory in describing the ground state properties of metallic magnets, the central issue has been to clarify how the magnetic ordering in the ground state is destroyed with increasing temperatur or what kind of thermal excitations dominate the thermodynamical properties of metallic magnets.

The same problem has been solved long ago for insulator magnets where the electrons responsible for magnetism are localized on atoms owing to the Mott strong-correlation mechanism and local (atomic) magnetic moments are well defined [2.2]; magnetism and the other physical properties are described precisely by using the Heisenberg model with necessary generalizations to include magnetic anisotropies, etc. The thermal excitations in the Heisenberg model have well been investigated. In particular, above the Curie or magnetic-ordering temperature T_C we have randomly oriented local moments changing their directions with the average frequency of $\sim k_B T_C / h$, where k_B and h are the Boltzmann and Planck constants, respectively.

Similar situations exist for most of the rare-earth metals where the magnetic 4f electrons are well localized and form local magnetic moments. The coupling between the local moments is through the indirect mechanism mediated by the conduction electrons [2.3].

In magnetic transition metals, alloys and intermetallic compounds, on the other hand, the d electrons responsible for magnetism are not localized but itinerant. This fact has been generally accepted for more than two decades, after a long controversy over localized vs. itinerant nature of magentic d electrons themselves. The real problem since then has been to describe the finite-temperature properties of metallic magnets on the basis of the itinerant-electron model. A simple extension of the mean-field band theory to finite temperatures, or the Stoner theory, is not satisfactory for this purpose, since it neglects the vitally important exchange enhancement effect of thermal spin excitations, thus predicting too high values for the Curie temperature and failing to account for the Curie-Weiss magnetic susceptibility above T_C observed for almost all the ferromagnets. It is evident that we have to take into account the effects of electron correlation beyond the mean-field theory or the Hartree-Fock approximation (HFA). In physical terms we need to consider the effects of spin and charge density fluctuations which are neglected in HFA. Two types of approaches have been developed for this problem since the 1960s.

Because of the overwhelming success of the Heisenberg model or the local-moment model in magnetism in general and, in particular, the long-dominated explanation of the Curie-Weiss susceptibility based on the local moment model, one line of intensive investigations have been to find a way of reconciling between the itinerant-electron model and the local-moment picture. Thus the study of the local moment in metals has been the subject of intensive investigations not only in dilute-alloy problems, but in metallic magnetism in general [2.4, 5].

The other line of investigations has been the direct extension of the band magnetism to include exchange-enhanced spin-density fluctuations. Starting from the Slater-Herring theory of spin waves on the itinerant-electron model, the theories of general spin fluctuations have been developed in the 1960s on the basis of the dynamical mean-field or the random-phase approximation (RPA) [2.6]. This line of approach was quite successful in describing the spin-wave excitations from the ground state. However, the success was limited within the low-temperature limit. The drawbacks of the Stoner theory at finite temperatures have not been removed until the 1970s.

In the early 1970s remarkable progress was made by the success of the self-consistent renormalization (SCR) theory of spin fluctuations, which goes one step beyond the HF-RPA theory and renormalizes the coupled modes of spin fluctuations in a self-consistent fashion [2.1]. This theory, which applies best to weakly ferro- and antiferromagnetic metals, has given a new mechanism for the Curie-Weiss magnetic susceptibility entirely different from the conventional local-moment mechanism. The nature of spin fluctuations in weakly ferromagnetic metals, as described by the SCR theory, differs entirely from that of

16

the local moment systems. The long-wavelength modes of spin fluctuations are dominating and therefore the spin correlation is always extended in real space even much above T_C. This extended moment picture may be regarded as a limiting case opposite to the local moment limit.

Owing to the subsequent theoretical and experimental investigations it has turned out that the SCR theory is a quantitatively correct theory for weakly ferromagnetic metals just in the same sense that the magnetic insulators are correctly described by the generalized Heisenberg model. This means that for the first time we have had a quantitatively correct theory of itinerant-electron magnetism that describes overall properties of a certain class of metallic magnets. At the same time a new class of magnets with extended moments have been established. The nature of spin density fluctuations in this limit is just opposite to the well-established local moment limit. There is every reason to believe that most of the metallic magnets are in the intermediate regime between these two extremes and therefore it is necessary to find a unified picture of magnetism interpolating between these two limits. Theoretical attempts in this direction have been the subject of recent investigations. Although the theories developed so far are limited within the adiabatic approximation for spin fluctuations, a number of physical properties of metallic magnets have been elucidated qualitatively or semi-quantitatively. It is clear that the dynamical theory of spin fluctuations in the intermediate regime is to be developed for real and quantitative understanding of the metallic magnetism as a whole.

2.1.2 Outline

The purpose of the present chapter is to give a brief survey of the above-mentioned developments in the field of magnetism which took place in recent years.

We first give in Sect. 2.2 a brief sketch of the historical developments in the theory of itinerant-electron magnetism leading to the self-consistent renormalization theory of spin fluctuations or the self-consistent theory of the coupled modes of spin fluctuations. A simple discussion of this theory is given in Sect. 2.3. Since the local-moment theories of magnetism have been familiar for a long time we will not discuss them here but treat in Sect. 2.4 an attempt at a unified theory of magnetism in the above-mentioned context. In Sect. 2.5 we discuss the dynamical properties of spin fluctuations in various magnets under the light of the unified picture referring to the results of recent neutron scattering experiments.

Throughout this chapter we will not go into mathematical details but place emphasis on physical pictures. No intentional effort is made for exhaustive coverage of important topics or references. It is hoped that the readers get a good idea about recent developments in this area of research, in particular, to what extent the problem is understood and what is still to be clarified. Those who are interested in further details and references are advised to refer to [2.1].

2.2 Theory of Magnetism Based on the Band Model

2.2.1 The Ground State

The electronic structure in the ground state of a magnetic transition metal is now widely believed to be described very well by recent state-of-the-art band calculations on the basis of the exchange correlation potential expressed in terms of local charge and spin densities [2.7]. This applies to ferromagnetic as well as antiferromagnetic metals. Although there are several indications suggesting the importance of many body effects in the ground state, the general qualitative conclusions are not altered by these effects; the band theory is at least a very good starting point for the theory of itinerant-electron magnetism. It may be worth while to note that even the antiferromagnetic ground state of a magnetic insulator compound is expected to be described reasonably well by the band theory, except for the effect of zero point spin fluctuations.

Here we will not go into the ground state problem any further but discuss the thermal-excitations and the finite-temperature properties of magnets from a general point of view.

2.2.2 Magnetic Excitations from the Ground State

Let us first consider the magnetic excitations from a ferromagnetic ground state consisting of the Fermi spheres with the radii $k_\uparrow$ and $k_\downarrow$ for the electrons with up and down spins, respectively.

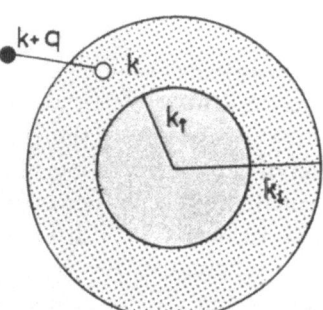

Fig. 2.1. Fermi spheres for $\downarrow$ and $\uparrow$ spin electrons and an electron-hole pair excitation

According to the one-electron picture or the Stoner theory the magnetic excitations are the spin-flip excitations of electrons across the Fermi surface. Such an excitation is shown in Fig. 2.1; a down-spin electron with the wave vector k is excited into an up-spin state with $k + q$. This Stoner excitation may also be regarded as an electron-hole pair excitation with $k\downarrow$ and $k + q\uparrow$ and the excitation energy is given by

$$\Delta E_0(\boldsymbol{k}, \boldsymbol{q}) = \varepsilon_\uparrow(\boldsymbol{k} + \boldsymbol{q}) - \varepsilon_\downarrow(\boldsymbol{k}) = \varepsilon_{\mathbf{k}+\mathbf{q}} - \varepsilon_{\mathbf{k}} + 2\Delta \ , \tag{2.1}$$

where ε_k is the band-energy dispersion and 2Δ is the exchange splitting of the up and down spin bands, for brevity, assumed to be independent of k. The intensity of the Stoner excitations with the momentum and energy transfers q and ω are given by the imaginary part of

$$\chi_0^{-+}(q,\omega) = \sum_k \frac{f(\varepsilon_k - \Delta) - f(\varepsilon_{k+q} + \Delta)}{\varepsilon_{k+q} - \varepsilon_k + 2\Delta - \omega} , \qquad (2.2)$$

where $f(x)$ is the Fermi distribution function, and the intensity contours in the $\omega - q$ plane of these excitations are shown in Fig. 2.2b for an example of a parabolic energy dispersion. Here $\chi_0^{-+}(q,\omega)$ is the transverse dynamical susceptibility without the electron-electron interaction.

In the above picture an excited electron and an excited hole are assumed to move independently with each other in a common mean field. In actuality, however, they are scattering one another through the interaction between them. This problem can easily be treated within the random-phase approximation (RPA), where only scatterings conserving the momentum transfer are taken into account [2.6, 8]. We have the following familiar RPA result for the dynamical susceptibility

$$\chi_{RPA}^{-+}(q,\omega) = \chi_0^{-+}(q,\omega)/[1 - I\chi_0^{-+}(q,\omega)] , \qquad (2.3)$$

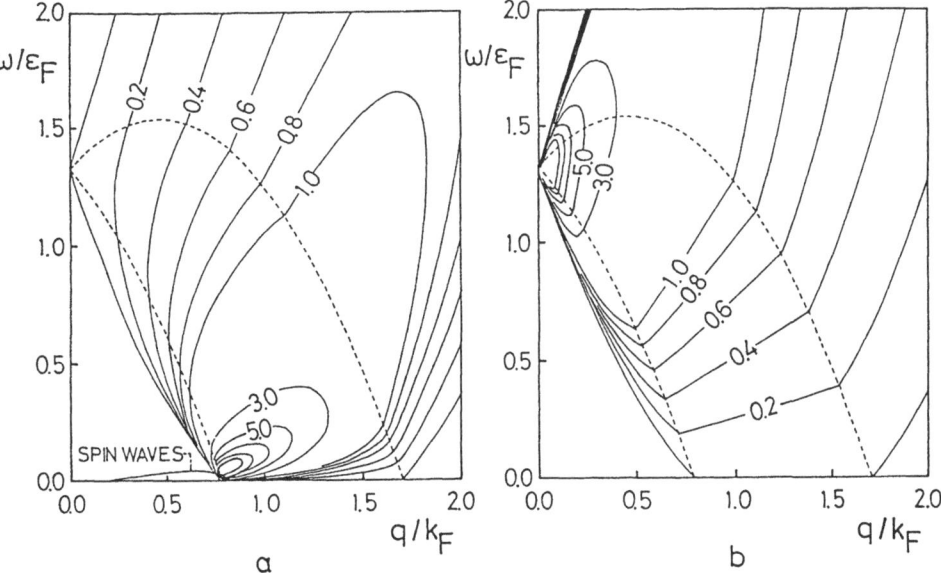

Fig. 2.2. (a) Intensity contours of the imaginary part of the RPA transverse dynamical susceptibility in a ferromagnetic electron gas. The spin wave dispersion is also shown. (b) Intensity contours for the Stoner excitations (no exchange enhancement)

where $I = U/N_0$, N_0 being the number of atoms, is the exchange interaction constant which is assumed to be independent of the wave vector. This corresponds to taking only the intra-atomic exchange interaction U in the tight-binding model (Hubbard model). We use this model throughout this text. As a consequence of the RPA theory we get the bound states corresponding to the spin-wave excitations in a region where the Stoner excitation is absent. This is shown in Fig. 2.2a. The intensity spectrum in the Stoner excitation continuum, given by Im $[\chi_{RPA}^{-+}(q, \omega)]$, is also changed from that of (2.2) by the exchange interaction, as shown by the contours in Fig. 2.2a. The intensity in the low-energy region is much exchange-enhanced from the Stoner excitation spectrum.

The RPA calculations of the spin-wave dispersions in real ferromagnetic metals such as Fe, Ni, Pd_2MnSn, etc. have been carried out by using calculated band structures. The results compare very well with the results of neutron-scattering experiments [2.9, 10]. Thus it is generally believed that the ground state and the elementary excitations from the ground state of ferromagnetic d-metals are fairly well described by the HF-RPA type theory. This belief is reinforced by a neutron scattering experiment observing a crossing of accoustic and optical modes of spin waves in the [100] direction in Ni at the energy predicted by an RPA calculation [2.11,12]. Such a crossing was not predicted in the [111] direction in accordance with the observation. The spin waves, in general, do not exist as a good mode of elementary excitations in the entire Brillouin zone; they are damped as the energy dispersion goes into the region of the Stoner excitation continuum. The broadening of the spin-wave scattering-intensity spectra in a region of relatively large calculated densities of Stoner excitations was observed by neutron-scattering experiments for Ni [2.12]. This is another support for the RPA theory.

We note here that the spin waves in an antiferromagnetic insulator compound can be described properly by the RPA theory, too. One can show that in the strong-correlation limit the spin-wave dispersion calculated by RPA agrees precisely with the conventional result based on the Heisenberg model with the Anderson kinetic superexchange interaction [2.1].

2.2.3 Finite-Temperature Properties

Physical properties at finite temperatures of various magnets are goverened by the magnetic excitations, as discussed in Sect. 2.2.2. At low temperatures those elementary excitations may be used to calculate the thermodynamical properties.

We first note that the Stoner theory is not reasonable since it corresponds to employing unrealistic magnetic excitations (Fig. 2.2b). The correct magnetic excitations are those, as shown in Fig. 2.2a, and the spin waves and exchange-enhanced spin fluctuations must be used in calculating the finite-temperature properties of a metallic magnet. This procedure is expected to give correct results but only in the low-temperature limit insofar as we use the RPA magnetic excitations which are correct only at $T = 0$. We already noted the failure of the

Stoner theory based on HFA at finite temperatures. The RPA theory fails at finite temperatures since it gives the spin fluctuations around the Stoner equilibrium state. In other words, the RPA theory reduces to the Stoner theory in the long-wavelength limit.

The correct theory must be a self-consistent one where the thermodynamical functions are calculated by using a renormalized expression for the exchange-enhanced spin fluctuations around the renormalized equilibrium state. In other words, the thermal equilibrium state and the spin density fluctuations must be calculated in a self-consistent fashion.

At present we have correct overall theories in the two limiting cases, the local-moment limit and the weakly and nearly ferromagnetic limit. The former is realized in the strong-correlation limit with an integral number of electrons per atom while the latter is realized when the substance is in the vicinity of ferromagnetic instability.

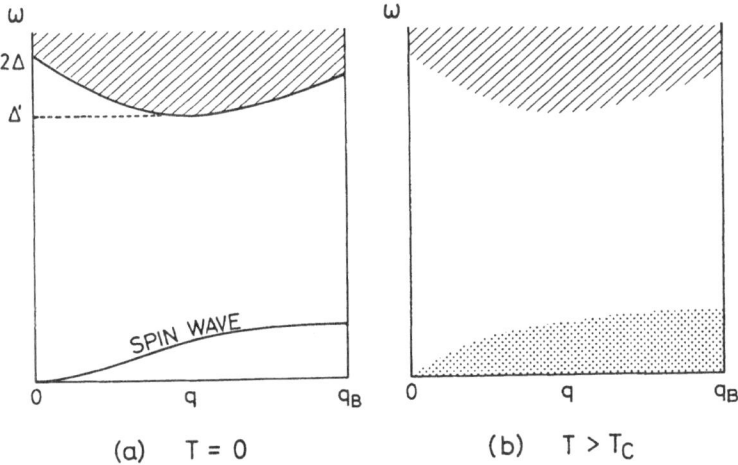

Fig. 2.3a,b. Magnetic excitations in a strongly correlated electron system. (a) ferromagnetic ground state; (b) paramagnetic state. The excitation has finite intensity in the shaded area

The magnetic excitations in the strong-correlation case are sketched in Fig. 2.3. At $T = 0$ the continuous spectrum of electron-hole pair excitations with opposite spins has a large energy gap 2Δ proportional to the electron-electron interaction energy. There are bound states corresponding to the spin waves, in the entire Brillouin zone and the energy dispersion of the spin waves in the strong correlation case is much smaller than the threshold energy Δ' of the continuum excitations. This situation is shown in Fig. 2.3a. In the bound state an electron and a hole are so tightly bound that they are mostly on the same atom. In other words, this electron-hole pair excitation is nothing but a spin flip in an atom. The transfer of a spin-flip excitation from atom to atom gives the spin-wave energy dispersion. The high-energy continuum of excitations, on the

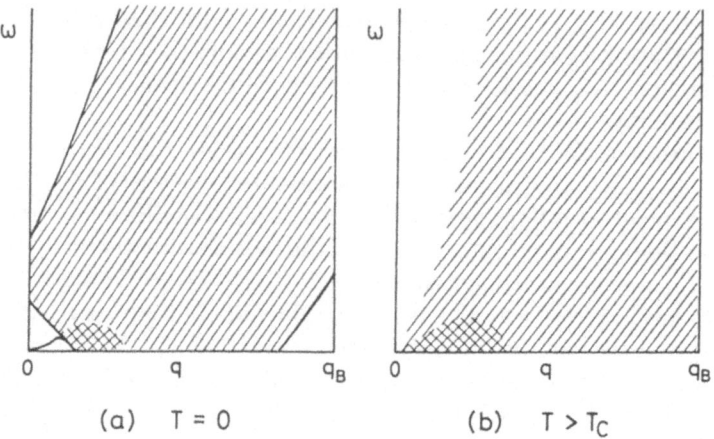

Fig. 2.4a,b. Magnetic excitations in a weak itinerant ferromagnet. (a) ferromagnetic ground state; (b) paramagnetic state. Intensity of excitations is strong in cross-hatched area

other hand, consists of polar excitations with separated electron-hole pairs; an electron and a hole are on different atoms. When the over-all energy dispersion of spin waves is much smaller than Δ' the system is regarded as consisting of local magnetic moments in the temperature range smaller than Δ'. The Heisenberg exchange constants are given by Fourier transforming the spin-wave energy dispersion. Above the Curie or Néel temperature we have a continuous energy spectrum of spin excitations with the widths of the order of $k_B T_C$ or $k_B T_N$, which is much smaller than Δ'. This situation is shown in Fig. 2.3b. Because of the good separation of low-energy magnetic excitations from the high-energy ones we have well-defined local moments in this case. This situation is realized in magnetic insulator compounds which are the Mott insulators, for 4f electrons in most of the rare-earth metals and for 3d electrons in some Heusler alloys. The magnetism and magnetic excitations in these systems are well described by using the local moment models of Heisenberg and sd types.

The magnetic excitations in a weakly ferromagnetic metal, on the other hand, are dominated by the continuous spectrum of exchange-enhanced spin-density fluctuations. The spin waves at $T = 0$ exist only in a small region of the q space. The situation is shown in Fig. 2.4a and b for $T = 0$ and $T > T_C$, respectively. In this case the magnetic excitations in the continuum are the overdamped modes of spin-density fluctuations and the energy spectrum generally extends continuously to a high-energy region. Since the local amplitude of the thermal spin-density fluctuation is generally small and the long-wavelength components with relatively small energy spreads dominate the thermal magnetic excitations, we can develop a correct theory in this case by making use of a long-wavelength approximation and an expansion in the amplitude of the spin density. Since this theory is rather new and may not be so familiar as the theories for the local-moment systems, we discuss it in Sect. 2.3.

The intermediate regime between the above-mentioned two limiting cases will be discussed in Sect. 2.4.

2.3 Self-Consistent Renormalization Theory for the Coupled Extended Modes of Spin Fluctuations

2.3.1 General Discussion

The spin-fluctuation theory of weak itinerant ferromagnetism was started in the early 1970s with a phenomenological mode-mode coupling theory of spin fluctuations [2.13] and a quantum-statistical-mechanical theory of self-consistent renormalization (SCR) of spin fluctuations [2.14]. The results of the latter theory have been shown to be derived by various other formalisms. Improvements of the original description were made by treating both transvers and longitudinal components of spin fluctuations on equal footing. At present the SCR theory is believed to be correct quantitatively when applied to weak itinerant ferromagnetism. The phenomenological theory was shown to correspond to the static approximation with a wave vector cut-off in the SCR theory. Here we will not go into details of the theory, referring to [2.1] for interested readers, but give a simple argument leading to the correct results. Throughout this text we use energy units for temperature and frequency, setting $\hbar = 1$ and $k_B = 1$. Magnetization and susceptibility are given in units of $2\mu_B$ and $4\mu_B^2$, respectively.

Let us start from the Stoner theory with the Landau expansion for the free energy [2.15]:

$$F[M] = \left(\frac{1}{2\chi_0} - I\right)M^2 + \frac{1}{4}gM^4 - Mh, \quad \text{with} \tag{2.4}$$

$$M = \frac{1}{2}(N_\downarrow - N_\uparrow) , \quad h = 2\mu_B H_0 ,$$

$$2\chi_0 = \varrho\left(1 - \frac{\pi^2}{6}RT^2 + \dots\right) ,$$

$$g = F_1\varrho^{-3}\left(1 + \frac{\pi^2}{6}R_1T^2 + \dots\right) , \tag{2.5}$$

where H_0 is the external magnetic field, ϱ the density of states at the Fermi level, and F_1, R, R_1 the constants expressed in terms of $\varrho(\varepsilon)$ and its derivatives [2.1]. From (2.4) we get

$$-2(\alpha - 1)/\varrho + gM^2 = h/M ,$$

$$-2(\alpha - 1)/\varrho + 3gM^2 = 1/\chi_\parallel , \quad \text{with} \tag{2.6}$$

$$\alpha = 2I\chi_0 = \alpha_0\left(1 - \frac{\pi^2}{6}RT^2 + \dots\right) \tag{2.7}$$

where $\chi_{\|}$ is the differential magnetic susceptibility parallel to the external field. Thus we get

$$M(0) = \varrho[2(\alpha_0 - 1)/F_1]^{1/2} ,$$

$$T_C^{Stoner} = [6(\alpha_0 - 1)/\alpha_0\pi^2 R]^{1/2} ,$$

$$1/\chi_{Stoner} = (\alpha_0\pi^2 R/3\varrho)(T^2 - T_C^2) . \tag{2.8}$$

The magnetic susceptibility above T_C does not obey the Curie-Weiss law.

As already mentioned, this theory must be improved by taking account of the effects of exchange-enhanced spin fluctuations. Although they vanish on direct average their mean-square amplitudes always remain finite. Let us use the following functional average expression for $F[M]$

$$e^{-F[M]/T} \sim \int \prod_q{}' dS_q \exp\left(-\Psi[\{S_q\}]/T\right) , \tag{2.9}$$

where the prime in $\prod'$ means to omit $q = 0$, and

$$\Psi[\{S_q\}] = \sum_q \left(\frac{1}{2\chi_0(q)} - I\right) |S_q|^2 + \frac{1}{4}gV^3 \int dr[S(r)]^4 - \boldsymbol{M}\cdot\boldsymbol{h}$$

$$= \sum_q \left(\frac{1}{2\chi_0(q)} - I\right) |S_q|^2$$

$$+ \frac{1}{4}g \sum_q \sum_{q'} \sum_{q''} (S_q\cdot S_{-q'})(S_{q''}\cdot S_{q'-q''-q}) - \boldsymbol{M}\cdot\boldsymbol{h} . \tag{2.10}$$

Here $S(r)$ is the spin density, S_q its Fourier transform, we have $\boldsymbol{M} = -S_0$ and we take terms up to the fourth order in the spin density. $\chi_0(q)$ is the wave-vector-dependent susceptibility of the system without electron-electron interaction. The fourth-order term represents the coupling between the spin-fluctuation modes and the coupling constant is approximated by the value in the long-wavelength limit.

This type of description is allowed insofar as we can treat $\{S_q\}$ as macroscopic variables. In weak itinerant ferromagnetism where only the spin-density components with small q take part, we may safely use this formalism. Taking the derivative with respect to M of (2.9) we get the following thermal equilibrium condition

$$\frac{\partial F}{\partial M} = \left[-\frac{\alpha - 1}{\chi_0} + g \sum_q{}' \left(3\langle|S_{q\|}|^2\rangle + 2\langle|S_{q\perp}|^2\rangle\right)\right]M$$

$$+ gM^3 - h = 0 , \tag{2.11}$$

where $\|$ and $\perp$ indicate the components parallel and perpendicular to the external field direction, respectively. Introducing the mean-square local amplitudes of the spin density S_L^2 and that of the spin fluctuation m^2 by

$$m_\nu^2 = S_{L\nu}^2 - (M_\nu/N_0)^2 = N_0^{-2} \sum_q{}' \langle |S_{q\nu}|\rangle^2 \;,$$

$$m^2 = \sum_\nu m_\nu^2 = S_L^2 - (M/N_0)^2 \;,$$

(2.12)

where ν takes $\parallel$ and $\perp$, we rewrite (2.9) as follows

$$\left(-\frac{(\alpha-1)}{\chi_0} + \frac{5}{3} g N_0^2 \tilde{m}^2 \right) M + g M^3 = h, \quad \text{with}$$

$$\tilde{m}^2 = \frac{3}{5}(3m_\parallel^2 + 2m_\perp^2) \;.$$

(2.13)

The inverse longitudinal magnetic susceptibility is obtained by differentiating (2.13) with respect to M. We get

$$-\frac{(\alpha-1)}{\chi_0} + \frac{5}{3} g N_0^2 \left(\tilde{m}^2 + M \frac{\partial}{\partial M} \tilde{m}^2 \right) + 3 g M^2 = \frac{1}{\chi_\parallel} \;.$$

(2.14)

Equations (2.13, 14) have terms with spin fluctuations in contrast with (2.4). The derivative of $\tilde{m}^2$ with respect to M is important below T_C [2.16]. Now the mean-square local amplitude of the spin density is expressed in terms of the dynamical susceptibilities by using the fluctuation-dissipation theorem as follows

$$m_\nu^2 = \frac{1}{2\pi} \int_{-\infty}^{\infty} d\omega \coth\left(\frac{\omega}{2T}\right) N_0^{-2} \sum_q{}' \mathrm{Im}\,\{\chi_\nu(q,\omega)\}$$

$$= \frac{1}{2\pi} \int_{-\infty}^{\infty} d\omega \, \mathrm{sgn}\{\omega\} \left(1 + \frac{2}{e^{|\omega|/T} - 1} \right) N_0^{-2} \sum_q{}' \mathrm{Im}\,\{\chi_\nu(q,\omega)\}$$

$$= (m_\nu^2)_0 + \frac{2}{\pi} \int_0^{\infty} d\omega \frac{1}{e^{\omega/T} - 1} N_0^{-2} \sum_q{}' \mathrm{Im}\,\{\chi_\nu(q,\omega)\} \;,$$

(2.15)

where the first and the second terms on the right-hand side represent the contributions of zero-point and thermal-spin fluctuations, respectively. In the weakly ferromagnetic limit the temperature dependence of $\chi(q,\omega)$ is expected to be very weak in the dominating part of the Brillouin zone, except for a small region around the origin, and thus the zero-point contribution $(m_\nu^2)_0$ is practically temperature independent. In what follows we regard the terms with $(m_\nu^2)_0$ to be included effectively in the term with α_0 and therefore m_ν^2 is redefined by the thermal-spin fluctuation or the second term of (2.15) only. The magnetization M(0) at T = 0 is given by the first equation of (2.8) with α_0 including the above-mentioned effect of zero-point spin fluctuations. Let us discuss the problems above and below T_C separately.

2.3.2 Theory Above T_C

Above the Curie temperature both χ and $S^2_{L\nu}$ are isotropic and thus we get the following equation for the susceptiblitiy

$$\frac{1}{\chi} = \frac{(1-\alpha)}{\chi_0} + \frac{5}{3}gN_0^2S_L^2(T) \ ,$$

$$S_L^2(T) = \frac{3}{\pi}\int\limits_0^\infty d\omega \frac{1}{e^{\omega/T}-1}N_0^{-2}\sum_q \mathrm{Im}\{\chi^{-+}(q,\omega)\} \ . \tag{2.16}$$

We now need a proper expression for $\chi(q,\omega)$. The use of an RPA expression (2.3) in (2.16) leads to an inconsistency, since (2.3) in the static long-wavelength limit ($q\rightarrow 0$, $\omega\rightarrow 0$) does not agree with χ in the first expression of (2.16). It is essential to remove this inconsistency in order to get any meaningful theory of renormalization (for further details, see [2.1]).

We now express the dynamical susceptibility as follows

$$\chi^{-+}(q,\omega) = \frac{\chi_0^{-+}(q,\omega)}{1 - I\chi_0^{-+}(q,\omega) + \lambda(q,\omega)}$$

$$= \frac{\chi_0^{-+}(q,\omega)}{(1-\alpha) + \alpha[1 - \chi_0^{-+}(q,\omega)/\chi_0^{-+}(0,0)] + \lambda(q,\omega)} \ , \tag{2.17}$$

where $\lambda(q,\omega)$ or $\bar{\chi}(q,\omega) = \chi_0(q,\omega)/[1 + \lambda(q,\omega)]$ is properly defined by a set of the Feynman diagrams.

In the weakly and nealy ferromagnetic limits, where the denominator of (2.17) for $q\rightarrow 0$ and $\omega\rightarrow 0$ is vanishingly small or we have $|\alpha - 1|\ll 1$, the long-wavelength (small q) components of spin fluctuations with small frequencies play a predominant role and we can use the following expansion form:

$$\chi_0(q,\omega)/\chi_0(0,0) = 1 - Aq^2 - B(\omega/q)^2 + iC\omega/q + \dots \ , \tag{2.18}$$

where the expansion coefficients are given from the band structure near the Fermi surface. We can also make a great simplification by approximating $\lambda(q,\omega)$ by $\lambda(0,0)$. This is allowed since $\lambda(0,0)$ though small is quite important compared with $|1 - \alpha|$ while the q- and ω-dependent part of λ is small compared with the second term in the denominator of (2.17). Comparing the uniform static limit of (2.17) with the first expression of (2.16) we get

$$\lambda(q,\omega)\rightarrow\lambda(0,0) = \frac{5}{3}\chi_0gN_0^2S_L^2(T) = \frac{5}{3}\frac{\chi_0}{N_0}\bar{F}_1S_L^2(T) \ ,$$

with

$$\bar{F}_1 = F_1(N_0/\varrho)^3 = N_0^3g \ . \tag{2.19}$$

Now (2.16–19) should be solved for $S_L^2(T)$ or for χ. We rewrite them in the following convenient form

$$S_L^2 = N_0^{-2} \sum_q \langle |S_q|^2 \rangle_T = \frac{2}{\pi} \int_0^\infty d\omega \frac{1}{e^{\omega/T} - 1} 3N_0^{-2} \sum_q \mathrm{Im}\{\chi(q,\omega)\} \ ,$$

$$\chi(q,\omega) = \frac{\chi_0(q,\omega)}{\alpha_0 - \alpha + \frac{5}{3}\frac{\chi_0}{N_0}\overline{F}_1\{S_L^2 - \frac{3}{5}[M(0)/N_0]^2\} + \alpha[1 - \chi_0(q,\omega)/\chi_0(0,0)]} \ .$$

$$(2.20)$$

Note that we have $\chi(q,\omega) = \chi^{-+}(q,\omega)/2$. By using (2.18) in (2.20) and leaving only the leading terms in the expansions, we obtain

$$\frac{N_0}{\chi} = \frac{5}{3}\overline{F}_1 \left[S_L^2(T) - \frac{3}{5}\left(\frac{M(0)}{N_0}\right)^2 \right] + (\alpha_0 - \alpha)N_0/\chi_0 \ ,$$

$$S_L^2(T) = \frac{3v_0\Gamma_0}{2\pi^3\overline{A}\alpha_0} \int_0^{q_B} dq\, q^3$$

$$\times \int d\omega \frac{\omega}{e^{\omega/T} - 1} \cdot \frac{(1 + \delta)}{\Gamma_0^2 q^2[q^2 + (N_0/2\alpha_0\chi\overline{A})]^2 + \omega^2} \ , \qquad (2.21)$$

with

$$\overline{A} = N_0 A/\varrho \ , \quad \Gamma_0 = A/C \ , \quad q_B = (6\pi^2/v_0)^{1/3} \ , \qquad (2.22)$$

where v_0 is the volume per magnetic atom. Equation (2.21) contains only five parameters, which can be measured experimentally, aside from $\alpha_0 \simeq 1$ and $(\alpha_0 - \alpha) \sim (T/T_F)^2$.

The Curie temperature is given by setting $1/\chi = 0$ in (2.21). Neglecting the small Stoner contribution $(\alpha_0 - \alpha)$ yields

$$S_L^2(T_C) = \frac{3}{5}[M(0)/N_0]^2 \ , \qquad (2.23)$$

$$T_C = 1.419\, p_s^{3/2}(\overline{A}/\alpha_0 v_0)^{3/4}\Gamma_0^{1/4} \ , \quad \text{with} \qquad (2.24)$$

$$p_s = 2M(0)/N_0 \ . \qquad (2.25)$$

Note that the mean-square local amplitude of the spin density at $T = T_C$ is just 3/5 of the ferromagnetic moment per magnetic atom squared at $T = 0$. This is a general result for isotropic weak itinerant ferromagnets. The magnetic susceptibility above T_C is obtained by solving (2.20 or 21). There are various approximate analytical expressions for $1/\chi$ and numerical solutions for some particular values for the parameters. We will not go into details of these expressions here but note that $S_L^2(T)$ increases almost linearly with temperature and give rise to the Curie-Weiss susceptibility through (2.21). This mechanism is

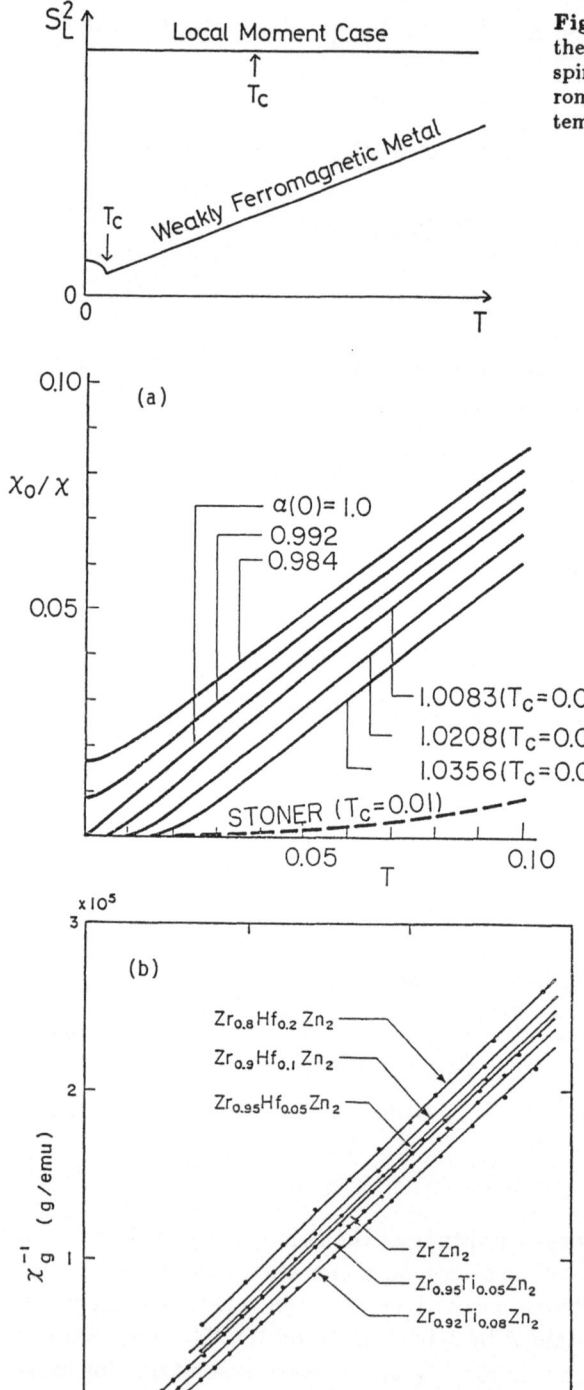

Fig. 2.5. Temperature dependence of the mean-square local amplitude of the spin density S_L^2 in a weak itinerant ferromagnet and in a local moment system

Fig. 2.6a,b. Inverse magnetic susceptibility of weak itinerant ferromagnets above T_C. (a) results of calculations by using the SCR theory for an electron-gas like band; (b) results of measurements [2.17]

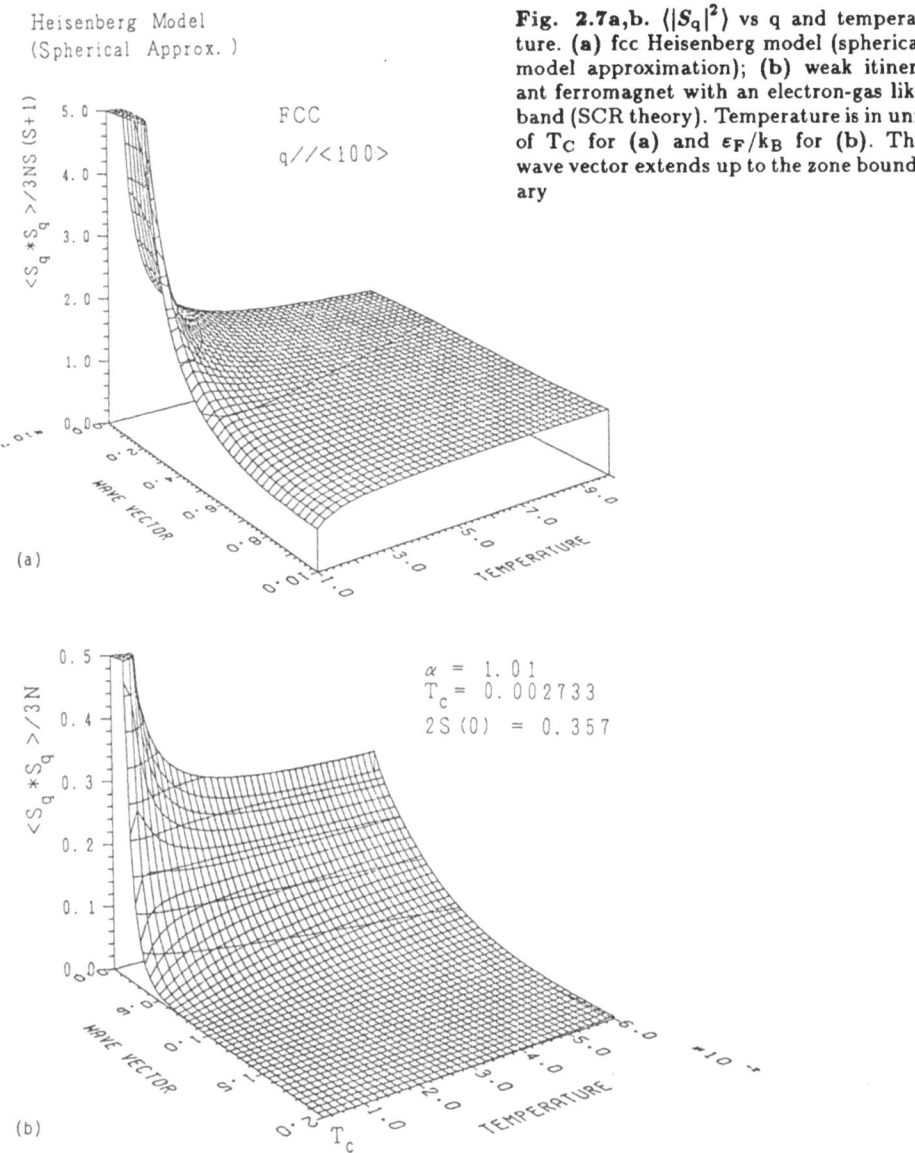

Heisenberg Model
(Spherical Approx.)

FCC

$q // <100>$

<S_q * S_q> / 3NS (S+1)

5.0
4.0
3.0
2.0
1.0
0.0

WAVE VECTOR

0.01
0.2
0.4
0.6
0.8
0.01

TEMPERATURE

1.0
3.0
5.0
7.0
9.0

(a)

<S_q * S_q> / 3N

0.5
0.4
0.3
0.2
0.1
0.0

$\alpha = 1.01$
$T_c = 0.002733$
$2S(0) = 0.357$

WAVE VECTOR

T_c

TEMPERATURE

1.0
2.0
3.0
4.0
5.0
6.0

(b)

Fig. 2.7a,b. $(|S_q|^2)$ vs q and temperature. (a) fcc Heisenberg model (spherical model approximation); (b) weak itinerant ferromagnet with an electron-gas like band (SCR theory). Temperature is in units of T_C for (a) and ϵ_F/k_B for (b). The wave vector extends up to the zone boundary

entirely different from the conventional one based on the local moment model where we have $S_L^2 = $ constant. We sketch in Fig. 2.5 the temperature dependence of S_L^2 in the two cases. Some results of numerical calculations [2.14] and of experiment [2.17] for $1/\chi(T)$ are shown in Fig. 2.6a and b, respectively. According to a detailed analysis the temperature dependence of $1/\chi$ is $(T - T_C)^2$ in a very small region in the direct vicinity of T_C, and then we have a region of $(T^{4/3} - T_C^{4/3})$ which is fairly large in the limit of small M(0). Finally we have a region of linear dependence on T. When M(0) is not too small we have

29

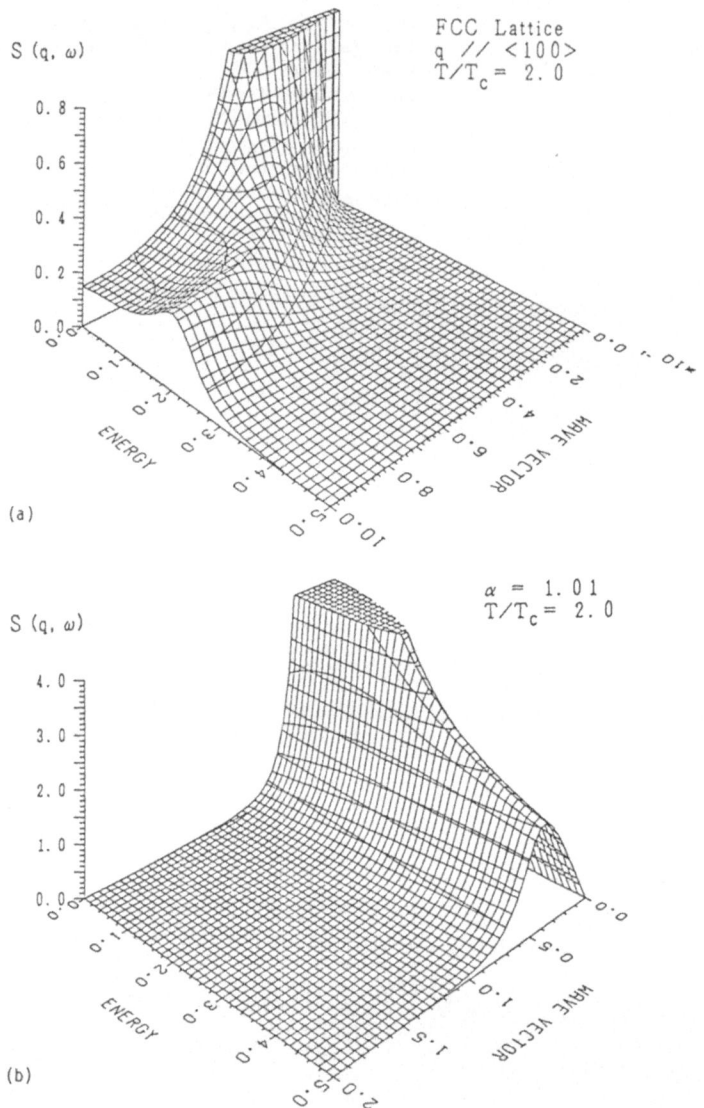

$S(q, \omega)$

FCC Lattice
q // $\langle 100 \rangle$
$T/T_C = 2.0$

0.8

0.6

0.4

0.2

0.0

ENERGY

WAVE VECTOR

(a)

$S(q, \omega)$

$\alpha = 1.01$
$T/T_C = 2.0$

4.0

3.0

2.0

1.0

0.0

ENERGY

WAVE VECTOR

(b)

Fig. 2.8a,b. Fourier transform of the space-time correlation functions, $S(q, \omega)$ above T_C. **(a)** fcc Heisenberg model (three-pole approximation); **(b)** weak itinerant ferromagnet (SCR theory). The energy is in units of $k_B T_C$ and the wave vector extends up to the zone boundary

a very good Curie-Weiss behaviour for $1/\chi$ in a wide temperature range except for the close vicinity of T_C.

Figures 2.7a and b show $\langle |S_q|^2 \rangle$ as a function of q and T for a Heisenberg ferromagnet and a weak itinerant ferromagnet above T_C, respectively. In a Heisenberg system all the spin-fluctuation components have almost equal amplitudes well above T_C, say at $2T_C$, corresponding to disordered local moments. In the weak itinerant ferromagnet, on the other hand, the long-wavelength components dominate the spin fluctuations even much above T_C; the spin fluctuations are localized in q-space or extended in real space.

Figures 2.8a and b show the Fourier transform of the correlation function

$$S_{\nu\nu}(q,\omega) = \frac{1}{V} \int\limits_{-\infty}^{\infty} dt \int d\mathbf{r}\, e^{i(\omega t - q \cdot r)} \langle \delta S_\nu(\mathbf{r} + \mathbf{r}', t - t') \delta S_\nu(\mathbf{r}', t') \rangle$$

$$= 2(1 - e^{-\omega/T})^{-1} \operatorname{Im} \{\chi(q, \omega + is)\}\ , \tag{2.26}$$

as a function of q and ω at $T = 2T_C$ for the both systems, respectively. For a Heisenberg magnet, the energy distribution of spin fluctuations is limited within $\sim k_B T_C$ while it is extended to much higher energy than $k_B T_C$ in a weak itinerant ferromagnet. We have seen that the nature of spin fluctuations in weak itinerant ferromagnets makes a strong contrast with that in local-moment systems. We may regard these systems to belong to mutually opposite limiting cases.

These specific properties of spin fluctuations in weak itinerant ferromagnets should reflect themselves in various physical properties. This problem has been studied theoretically and a number of qualitatively new predictions were made as to the physical properties of weak itinerant ferro- and antiferromagnets and all of them so far have been borne out by subsequent experimental investigations. Furthermore even quantitative agreement has been obtained between theory and experiment whenever such a comparison was possible. We will come back to this subject in Sect. 2.3.4, after a brief discussion of the theory below T_C in the following subsection.

2.3.3 Theory Below T_C

Below the Curie temperature we need to solve (2.13–15) for M and m^2. Here we deal with both the transverse and longitudinal components of spin fluctuations on equal footing. Such a treatment below T_C, improving over the previous ones, was presented recently [2.16].

The expression for the transverse dynamical susceptibility is given by

$$\chi^{-+}(q,\omega) = \frac{\chi_{M0}^{-+}(q,\omega)}{h\chi_{M0}^{-+}(0,0)/2M + I[\chi_{M0}^{-+}(0,0) - \chi_{M0}^{-+}(q,\omega)]}\ , \tag{2.27}$$

where χ_{M0}^{-+} is the transverse susceptibility of non-interacting electrons with a fixed value M of magnetization, while the longitudinal component is more complicated. If one fixes the chemical potential an external magnetic field parallel to the magnetization induces a charge density together with a magnetization. For a system of non-interacting electrons we have the following well-known results for the dynamical suceptibility

$$\chi_{0\varrho}(q,\omega) = 2\chi_{0z}(q,\omega) = \frac{1}{2}[\chi_{0\downarrow}(q,\omega) + \chi_{0\uparrow}(q,\omega)]\ ,$$

$$\chi_{0z\varrho}(q,\omega) = 2\chi_{0z\varrho}(q,\omega) = \frac{1}{2}[\chi_{0\downarrow}(q,\omega) - \chi_{0\uparrow}(q,\omega)]\ ,$$

with

$$\chi_{0\sigma}(q,\omega) = \sum_k \frac{f(\varepsilon_{k\sigma}) - f(\varepsilon_{k+q\sigma})}{\varepsilon_{k+q\sigma} - \varepsilon_{k\sigma} - \omega - is} \ ,$$

$$\varepsilon_{k\sigma} = \varepsilon_k - \frac{1}{2}\sigma h \ , \tag{2.28}$$

where χ_ϱ and χ_z are the charge and longitudinal spin susceptibilities, respectively and $\chi_{\varrho z}$ and $\chi_{z\varrho}$ are the cross terms.

This expression in the limit of $q \to 0$ and $\omega \to 0$ leads to a change in the total number of electrons due to the external magnetic field, and is not reasonable [2.18]; the co-existence of an overlapping conduction band does not help since its density of states is usually much smaller than that of the magnetic d bands. Thus in the static long-wavelength limit we have to fix the total number of electrons rather than the chemical potential. We then get the following result

$$[\chi_{0z}(0,0)]^{-1} = \sum_\sigma [\chi_{0\sigma}(0,0)]^{-1} \ . \tag{2.29}$$

For $q \neq 0$ we need to consider the effect of charge screening explicitly and the Hubbard model is not appropriate for this purpose; the RPA expression for the dynamical susceptibility in the static long-wavelength limit has a non-vanishing cross term, in particular it reduces to (2.28) rather than (2.29) for vanishing interaction I. A simple approximation to remedy this discrepancy is to assume complete screening of the local induced charge. Let us consider the following dynamical mean-field equations for the charge and spin densities

$$\langle \delta \varrho(q,t) \rangle + \langle S_z(q,t) \rangle$$
$$= \chi_{0\uparrow}(q,\omega) \left[-\frac{1}{2}h(q,t) - U\langle \delta \varrho(q,t) \rangle + J\langle S_z(q,t) \rangle \right]$$

$$\langle \delta \varrho(q,t) \rangle - \langle S_z(q,t) \rangle$$
$$= \chi_{0\downarrow}(q,\omega) \left[\frac{1}{2}h(q,t) - U\langle \delta \varrho(q,t) \rangle - J\langle S_z(q,t) \rangle \right] \ , \tag{2.30}$$

where $h(q,t)$ is the wave-vector-dependent external magnetic field oscillating with the angular frequency ω, and U and J are the Coulomb and exchange interaction constant, respectively. These equations are solved as follows

$$\chi_{\varrho z}(q,\omega) = \frac{\frac{1}{2}\lambda'}{(\lambda + U)(\lambda - J) - \lambda'^2} \ ,$$

$$\chi_z(q,\omega) = \frac{\frac{1}{2}(\lambda + U)}{(\lambda + U)(\lambda - J) + \lambda'^2} \ ,$$

with

$$\lambda = \tfrac{1}{2} \sum_\sigma \chi_{0\sigma}^{-1} \ ,$$

$$\lambda' = \tfrac{1}{2} \sum_\sigma \sigma \chi_{0\sigma}^{-1} \ . \tag{2.31}$$

Insofar as we assume U and J to be wave-vector independent, the correct result in the limit of $q \to 0$, $\omega \to 0$ is ensured only by taking $U \to \infty$. Thus we have

$$\chi_z^{RPA}(q,\omega) = \frac{\chi_{0z}(q,\omega)}{1 - 2J\chi_{0z}(q,\omega)} \ . \tag{2.32}$$

This approximation of complete charge screening is not quite satisfactory except for the limit of small q. Now taking account of the renormalization effect in a similar way to the case of transverse components we may write

$$\chi_z(q,\omega) = \frac{\chi_{0z}(q,\omega)}{\chi_{0z}(0,0)/\chi_{\parallel} + 2J[\chi_{0z}(0,0) - \chi_{0z}(q,\omega)]} \ , \tag{2.33}$$

where $\chi_{\parallel}$ is the uniform static longitudinal susceptibility. Now the problem below T_C can be treated by using (2.27) and (2.33) in (2.13–15). We have the following expansion forms for $\chi_0(q,\omega)$

$$\chi_{M0}^{-+}(q,\omega)/\chi_{00}(0,0) \simeq 1 - Aq^2 + iC\frac{\omega}{q} - \frac{1}{2}\left(\frac{M}{\varrho}\right)^2 (F_1 + \ldots) \ ,$$

$$\chi_{0z}(q,\omega)/\chi_{00}(0,0) \simeq 1 - Aq^2 + iC\frac{\omega}{q} - \frac{1}{2}\left(\frac{M}{\varrho}\right)^2 (3F_1 + \ldots) \ . \tag{2.34}$$

The problem now becomes an integro-differential equation for $\tilde{m}^2$. According to a numerical solution with certain approximations, the $M^2(T)$ curve decreases mostly as T^2 and a $(T_C^{4/3} - T^{4/3})$ behaviour is expected only in the close vicinity of T_C [2.16]. This seems to improve a previous result [2.14], obtained by using the transverse fluctuation components only, predicting a much larger range of $(T_C^{4/3} - T^{4/3})$ behaviour. For a convincing result, however, a more rigorous solution of (2.13–15) is required.

In addition, because of the above-discussed approximation for the charge screening effect, the theory below T_C has still room for improvements. Note, however, that the values of M(0) and T_C are not influenced by this approximation.

2.3.4 Comparison with Experiment

A number of experimental investigations have been carried out in a past decade to test the SCR theory of weak itinerant ferro- and antiferromagnetism. These studies include magnetic measurements, specific heat, thermal expansion, electrical and thermal resistivities, magnetoresistance, nuclear spin relaxation, and so on. All of these measurements almost unanimously supported the SCR theory at least qualitatively. Comparison between the theory and experiment is sometimes quantitative or parameter-free in the sense that a certain parameter value is determined consistently by more than two different measurements. Here we will not go into an extended survey of these investigations, referring to [2.1] for interested readers, but show only a few examples of convincing quantitative comparisons.

As was discussed in Sect. 2.3.2, at least the following parameters are necessary for the SCR theoretical description of weak itinerant ferromagnets

$$M(0), \overline{F}, \overline{A}, \Gamma_0 \ ,$$ (2.35)

in addition to $\alpha_0 \simeq 1$ and v_0. $M(0)$ and $\overline{F}_1$ are determined from the Arrott plots of the magnetization measurements and $\overline{A}$ and Γ_0 from neutron experiment which measures

$$\mathrm{Im}\,\{\chi(q,\omega)\} = \frac{\chi(0,0)}{1+q^2/\kappa^2}\frac{\omega\Gamma_q}{\omega^2+\Gamma_q^2} \ , \quad \text{with}$$

$$\kappa^2 = N_0/2\alpha\chi\overline{A} \quad \text{and} \quad \Gamma_q = \Gamma_0 q(\kappa^2+q^2) \ .$$ (2.36)

Γ_0 is also determined from the measurement of NMR-T_1 through the following theoretical expression

$$1/T_1 = KT(\varsigma/h)/(1+P\varsigma^3/h) \ , \quad \text{with}$$

$$K = (\gamma_N A_{hf})^2 v_0/4\pi^2\Gamma_0 \ ,$$

$$\varsigma = 2M(T)/N_0 \ ,$$

$$P = \alpha\overline{A}(8\pi^3/v_0 A)^2 \ ,$$ (2.37)

where γ_N is the nuclear gyromagnetic ratio, A_{hf} the hyperfine coupling constant and A the area of the Fermi surface of magnetic electrons. As a matter of fact an early test of the SCR theory was provided by T_1 measurements on $ZrZn_2$ and Sc_3In for various values of T and h [2.19]. The results were fitted almost perfectly by using (2.37) with only two adjustable parameters, aside from the one for the orbital current contribution to $1/T_1$ which is proportional to T. These results were one of the first convincing tests of the SCR theory.

An early parameter-free test of the SCR theory was made in the magneto-resistance study of Sc_3In [2.20]. The magnetic-field dependence of the coefficient of the T^2 term in the low-tempeature resistivity is expressed with only one parameter determined by NMR-T_1 measurements [2.21]. The quantitative agreement between the theory and experiment was satisfactory.

The thermal expansion of magnetic origin provides us with another parameter-free test of the theory [2.22], since it is given by

$$\omega_m = (\Delta V/V)_m = (D/B)[S_L^2(T) - S_L^2(0)] \, , \qquad (2.38)$$

or in more detail by

$$\omega_m(T) \simeq - (2D/5B)\{[M(0)]^2 - [M(T)]^2\} \, , \quad T < T_C \, ,$$

$$\omega_m(T) - \omega_m(T_C) = (3D/5Bg)1/\chi(T) \, , \quad T > T_C \, , \qquad (2.39)$$

where D and B are the magneto-volume coupling constant and the bulk modulus, respectively, and D/B is determined from the forced magneto-striction measurement. $M(T)$, g and $\chi(T)$ are, of course, given from magnetic measurements. Although there is a serious problem of separating ω_m from the nonmagnetic contribution ω_{nm} in the measurements, various methods of analyses were presented. The results reported so far on $ZrZn_2$ [2.23], MnSi [2.24], Ni_3Al [2.25], and $TiBe_{1.8}Cu_{0.2}$ [2.26] seem to be in good quantitative accord with the theoretical results deduced from (2.39).

Table 2.1a,b. Quantiative comparisons between the SCR theory and experimental results on weak itinerant ferromagnets. (a) Curie temperature and the effective moment value deduced from the magnetic susceptibility above T_C; (b) nuclear spin relaxation time

Table 2.1a

	p_s	v_0	$\bar{F}_1[k_B]$	$\bar{A}[k_B Å^2]$	$\Gamma_0[k_B Å^3]$	T_C^{obs} [K]	T_C^{calc} [K]	p_{eff}^{calc}	p_{eff}^{obs}
MnSi	0.4	23.67	3.52×10^3 [a]	1.14×10^3 [9)]	$580^{27)}$	30	$32^{29)}$	$2.0^{29)}$	2.12^f
Ni_3Al	0.075^b	15.06	1.33×10^5 [b]	1.24×10^4 [10)]	$5740^{28)}$	41.5^b	$40^{16)}$	$1.7^{16)}$	1.3^b
$ZrZn_2$	0.12^c	50.57	$1\ 54 \times 10^4$ [d]	1.06×10^4	$1720^{19)}$	21.3^c	–	$2.0^{29)}$	1.44^g
Sc_3In	0.045^e	30.84	1.63×10^5 [e]	7.64×10^3	$1850^{19)}$	5.5^e	–	$0.97^{29)}$	$0.66^{19)}$

[a] The value at T = 50 K; D. Bloch et al.: Phys. Lett. **51A**, 259 (1975).
[b] F.R. de Boer: Dissertation, University of Amsterdam (1969) unpublished. The value for $\bar{F}_1$ is at T = 4.2 K.
[c] S. Ogawa: J. Phys. Soc. Jpn. **40**, 1007 (1976).
[d] The value at T = 30.0 K[17)].
[e] J. Tekeuchi, Y. Masuda: J. Phys. Soc. Jpn. **46**, 468 (1979). The value for $\bar{F}_1$ is at T = 10.87 K.
[f] H. Yasuoka et al.: J. Phys. Soc. Jpn. **44**, 842 (1978).
[g] M. Kontani et al.: J. Phys. Soc. Jpn. **39**, 665 (1975).

Table 2.1b

	$\gamma_N[10^3$ /Gs]	$A_{hf}[kOe]$	$K^{calc}[k_B/s]$	$K^{obs}[k_B/s]$	$T_1^{calc}[\mu s]$ (T = 200~300 K)	$T_1^{obs}[\mu s]$ (T = 200~300 K)
MnSi	6.644	$- 275^a$	3.5×10^{15}	–	$41\sim44^{29)}$	35 ± 5^a
Ni_3Al	6.979	$- 15.2^b$	$7.5 \times 10^{11\ 29)}$	7.0×10^{11} [b]	–	–

[a] H. Yasuoka et al.: J. Phys. Soc. Jpn. **44**, 842 (1978)
[b] T Umemura, Y. Masuda: J. Phys. Soc. Jpn. **52**, 1439 (1983)

Recent neutron-scattering measurements on MnSi [2.27] and Ni_3Al [2.28] have proved the form of (2.36) for the wave-vector dependence of the damping constant and determined the parameter values for $\bar{A}$ and Γ_0, enabling us to make quantitative test of the SCR expression (2.24) for the Curie temperature, and (2.37) for the NMR-T_1, etc. We show the comparisons in Table 2.1a and b, respectively, where the agreements are extremely good, convincing us of quantitative correctness of the theory. The neutron measurements on MnSi also give a quantitative result on $S(q, \omega)$ in the entire Brillouin zone with ω/k_B and T up to $10\,T_C$, showing a characteristic behaviour presented in Fig. 2.8b.

As for the quantitative test of the magnetic susceptiblity [2.16, 29], we show in Table 2.1a the calculated and observed values for p_{eff}. The agreement is fair but not as good as for T_C or T_1, except for MnSi. This may probably be due to the fact that the measured values for $\bar{F}_1$ are not very accurate depending on the stoichiometry, etc., of the samples. It may be worth while to note that the $(T^{4/3} - T_C^{4/3})$ dependence of $1/\chi$ is expected to be observed in a rather wide temperature range when the spin fluctuations are strongly concentrated in the region of small q. This seems to be realized in Ni_3Al as was calculated from the theory and was observed experimentally [2.25]. As the distribution of the spin fluctuations in q-space extends, this temperature range becomes narrower and the temperature range of true Curie-Weiss behaviour tends to spread. When the contributions of the components near the Brillouin zone boundary, though small, become no longer negligible at very high temperatures, the $1/\chi - T$ curve tends to bend downward. This trend as expected from theoretical analyses [2.29] was actually observed in MnSi and $ZrZn_2$.

The temperature dependence of magnetization M(T) was calculated for Ni_3Al by using the above mentioned parameter values and was compared with the observed result [2.16]. The agreement is generally satisfactory except for a fine detail; the calculated curve has a little wider temperature range of T^2 dependence of $M^2(0) - M^2(T)$ than the observed result. Note that the theory and calculation below T_C include more approximations than those above T_C.

To sum, we may now safely conclude that the SCR theory or the self-consistent renormalization theory for the coupled modes of spin fluctuations is valid quantitatively at least when applied to weak itinerant ferromagnets.

2.3.5 Supplementary Discussions on the SCR Theory

In the preceding subsections we have discussed the SCR theory in its simplest form in the weakly ferromagnetic limit. This theory was first given by using the method of coupling constant integral and then the temperature Green's function formalisms, etc. If one makes a static approximation with a wave vector cut-off in the SCR theory, it reduces to the phenomenological mode-mode coupling theory of spin fluctuations. Although this is a crude approximation one can recover some of the SCR results in the form of the preceding subsections by giving a proper temperature dependence to the cut-off wave vector. This means that the mode-mode coupling constant for the long-wavelength spin fluctuations is given correctly by the phenomenological approach, the result we anticipated in Sect. 2.3.1. We also note here that the relevance of the reaction field concept to the SCR theory has been pointed out recently [2.30].

It should be noted here that the original formalism mentioned above contains an additional contribution of the leading order to the above-discussed results. This arises from the coupling among three modes of spin fluctuations, as shown by a diagram in Fig. 2.9. This contribution arises specifically in the vicinity of ferromagnetic instability and is absent in antiferromagnetic cases. This term, however, is shown to be small numerically at least for the band with a parabolic energy dispersion.

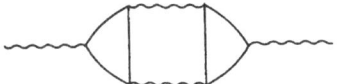

Fig. 2.9. A diagram showing the coupling among three modes of spin fluctuations; contributions of the leading order to the self-energy of spin fluctuation

The essential importance of the rotationally invariant treatment of the spin fluctuations was emphasized in connection with the theory of magneto-volume effect [2.22, 31] and is now apparent as was seen in the quantitative discussions given in this section.

The SCR theory was extended to antiferromagnetism [2.32] including helimagnetism. The general qualitative conclusions of the theory are the same as for the ferromagnetic case, if one shifts the Γ-point in q-space to the point of the antiferromagnetic ordering vector $\boldsymbol{Q}$. However, detailed results are different, since the staggered components of the spin density is not a constant of motion in contrast with the uniform component and the dynamical structure of the relevant components $(q \to Q \neq 0,\ \omega \to 0)$ of spin fluctuations is different from that in the ferromagnetic case $(q \to 0,\ \omega \to 0)$. Various characteristic features of weak itinerant antiferromagnetism as predicted by the SCR theory were confirmed by a number of experimental investigations [2.1].

For a nesting model of an antiferromagnet, the effect of spin fluctuations as a correction to the HFA-RPA results is relatively small [2.33]. In this case the spin fluctuation components in an extremely small part of the q-space is significantly enhanced, since $\chi_0(q)$ diverges logarithmically for $q \to Q$. The situation is somewhat similar to the BCS superconductors.

We finally note that the SCR theory in its general form can be applied to any itinerant-electron magnets, although the approximation is justified only in the weakly ferromagnetic case. Even for the intermediate regime this theory should make evident improvements over the HF-RPA theories at least. As will be discussed in the next section, this theory corresponds to using a mean mode-mode coupling approximation with a long-wavelength approximation for the coupling constant. In this context we may go on to extend this approximation so that it applies to more general cases.

2.4 A Unified Description of Magnetism – Adiabatic Approximation

2.4.1 General Considerations

We have seen in the preceding sections that there are two mutually opposite extremes in the picture of magnetism, the well-known local-moment case and the extended-moment case, as realized in weak itinerant ferro- and antiferromagnets. It is important to recognize that these mutually opposite extremes are connected continuously by varying the relative strength of the electron-electron interaction U/W, U and W being the intra-atomic exchange energy and the band width, respectively. Let us consider, for example, a system with a half-filled band. In the strong correlation limit (U/W≫1) we have a Mott insulator with local moments as is well known. In the weak correlation limit (U/W≪1) the system is a Pauli paramagnet. Beyond a certain critical value of U/W we generally have a phase of itinerant antiferromagnetism. Weak itinerant antiferromagnetism is realized slightly beyond this critical boundary. On increasing U/W the antiferromagnetic moment increases together with the band-energy gap and the area of Fermi surface decreases until it vanishes at another critical value of U/W (∼1), beyond which the system becomes an insulator. The spatial extension of the local moment decreases with still increasing U/W and we have a system with well-defined local moments. Similar situation is expected also in the case of metals with overlapping d or f and conduction bands, where there are ferro- as well as antiferromagnetism.

Thus we expect that many metallic magnets with itinerant d (or 5f) electrons belong to the intermediate regime between the two mutually opposite extremes. We show in Fig. 2.10 a chart of classification of various magnets according to the nature of spin fluctuations. An important piece of experimental evidence supporting this statement is the Rhodes-Wohlfarth plots [2.34] of the ratio p_C/p_s against T_C, where p_s is the ferromagnetic moment at $T = 0$ per magnetic atom and p_C is related to the Curie constant C through the relation

$$C = \frac{1}{3}N_0\mu_B^2 p_C(p_C + 2) = \frac{1}{3}N_0\mu_B^2 p_{eff}^2 \ . \tag{2.40}$$

We show an example of p_C/p_s vs. T_C plots in Fig. 2.11. Such a plot is meaningful since all the ferromagnetic metals above T_C show the Curie-Weiss behaviour of the susceptibility at least approximately. For a local moment system the p_C/p_s ratio should naturally be 1 while the ratio is very large for a weak itinerant ferromagnet where the CW law arises from a new mechanism without local moment, as discussed in Sect. 2.3; the ratio diverges as $p_s \to 0$ and for nearly ferromagnetic metals that show the CW susceptibility. An important point in Fig. 2.11 is that the ratio distributes almost continuously between these opposite limits, clearly indicating the importance of the intermediate regime.

SPIN FLUCTUATIONS

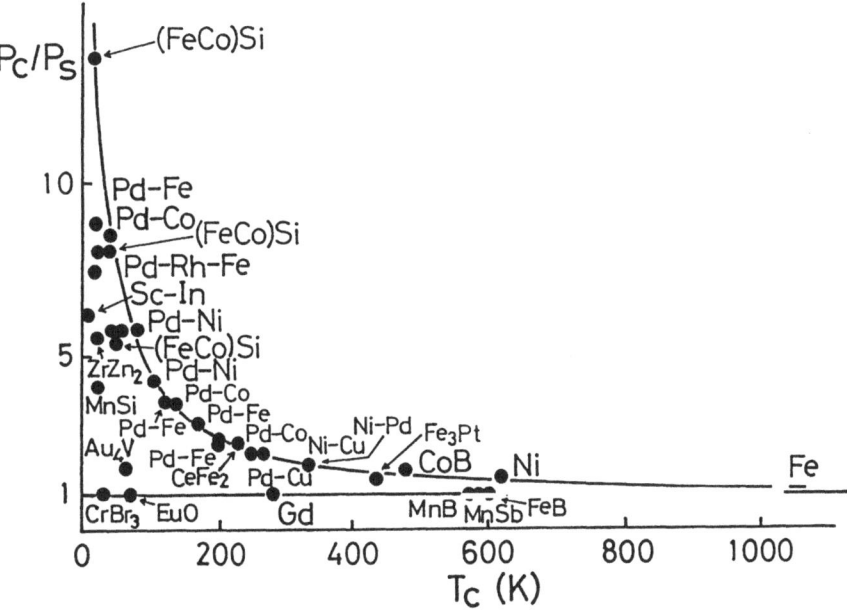

<table>
<tr><td></td><td>LOCAL IN
q-SPACE</td><td colspan="3"></td><td>LOCAL IN
REAL SPACE</td></tr>
</table>

SMALL

Nearly Ferromagnetic Metals(a)
Weakly Ferromagnetic(b) &
Antiferromagnetic(c) Metals

AMPLITUDE

Cr	MnSi		
CrB_2	α-Mn	$CeFe_2$	
γ-Mn	MnP	CoS_2	
Ni	Co	Fe	Fe_3Pt	$FePt_3, MnPt_3$ $FePd_3$

SATU-RATED

Local Moment
Systems(d)

(a) Pd, $HfZn_2$, $TiBe_2$, YCo_2, YRh_6B_4, Ni-Pt alloys, etc...

(b) Sc_3In, $ZrZn_2$, YNi_3, Ni_3Al, $Fe_{.5}Co_{.5}Si$, $LaRh_6B_4$, $CeRh_3B_2$,

$TiBe_{2-x}Cu_x$, Ni-Pt alloys, etc...

(c) β-Mn, V_3Se_4, V_3S_4, V_5Se_8, Cr-Mo alloys, etc...

(d) Magnetic Insulator Compounds, 4f-Metals, Heusler Alloys

(Pd_2MnSn, etc...).

Fig. 2.10. Classification of magnetic substances according to the nature of spin
fluctuations

Fig. 2.11. p_C/p_s vs T_C, the Rhodes-Wohlfarth plot

As was seen already, we now have well-established theories in the two extreme cases. Thus it is quite natural to think that a proper interpolation between these extremes should lead to a unified theory of magnetism. Theoretical investigations for a unified theory have been initiated several years ago. The problem is naturally a hard one and theories developed so far are mostly limited within an adiabatic approximation for the spin fluctuations. Before discussing these theories in the following subsections let us summarize the vitally important points in developing a general theory.

1. *Variable nature of* S_L^2, the mean square amplitude of the local spin density, is clearly one of the most important characteristics of itinerant electron magnetism in general, distinct from the localized-electron magnetism. The vital importance of this effect in weakly ferromagnetic metals has already been seen in Sect. 2.3. In a general theory, however, we cannot use an expansion form (2.10) since S_L^2 is not generally small.

2. *Spatial extension of the spin correlation* or *the short range order* (SRO) is another essential property to be put in the theory, since we need to interpolate between the extended and local moment limits. Persistence of strong SRO much above T_C in weak itinerant ferromagnets was seen already.

The first theory of interpolation was developed by using a static approximation in the Stratonovich-Hubbard functional integral method and the results include a unified explanation for the Curie-Weiss susceptibility covering both the two different mechanisms in the opposite extremes [2.35]. Some of the results were shown later to be derived by a simple method without having recourse to the functional integral formalism.

2.4.2 Functional Integral Theory

Let us discuss very briefly the functional integral theory for a unified picture, as mentioned in Sect. 2.4.1. The spirit of this approach is to consider the following form of the free energy functional as an extension of (2.10) for weak itinerant ferromagnets

$$\psi = - \sum_q V_q(S_L^2)\, |S_q|^2 + N_0 L(S_L^2)$$

$$= - \sum_{j,\ell} V_{j\ell}(S_L^2)(S_j \cdot S_\ell) + N_0 L(S_L^2) \ , \quad \text{with}$$

$$S_L^2 = N_0^{-2} \sum_q |S_q|^2$$

$$\sum_q V_q(S_L^2) = 0 \ , \tag{2.41}$$

where the functional forms for $V_q(S_L^2)$ and $L(S_L^2)$ are to be calculated from a given band structure. We calculate the partition function by taking a functional average of $e^{-\psi/T}$ with respect to both direction and length of the spin variables. Thus this model may also be regarded as a generalization of the Heisenberg model to include variation in size of the local moments.

In actuality we use the Stratonovich-Hubbard method to deal with the quantum systems. We confine ourselves here to the static approximation. For the Hubbard model we calculate $\psi[\xi]$ as a free energy of the non-interacting system (U = 0) under the influence of the magnetic field $(\pi UT)^{1/2}\xi_j$ at the site j, etc., and then take an average of $\exp\left(-\pi \sum_j \xi_j^2 - \psi[\xi]/T\right)$ over the field variables $\{\xi_j\}$, leading to the partition function $\exp\left(-F/T\right)$. Various methods of calculating $\psi[\xi]$ and a few methods of evaluating the functional integrals have been presented.

As a matter of fact, the functional integral method was applied many years ago to metallic magnetism [2.36–38] in order to give better grounds to the long-advocated local moment picture of magnetism [2.5]. We refer to [2.1] for these historical developments but just note that these theories make use of a local saddle point approximation combined with the coherent potential approximation (CPA) above T_C [2.37] or the Hartree-Fock approximation for the excited states [2.38], and thus the size of the local moment is fixed except for a small effect of $\sim(T/T_F)^2$. Clearly we have to go beyond the local saddle point approximation for a unified theory where it is essential to take into account the variation of S_L^2 with temperature.

One method with such an improvement of evaluation for the functional integral with ψ in the form (2.41) is to use the Lagrange multipliers in performing integrations over $\xi_q's$ first with a fixed value for $x = \langle |\xi_q|^2 \rangle$ and then to use a saddle point approximation for an integral over x [2.35]. The general consequences of this approach are the same as to be discussed in Sect. 2.5 by using an alternative approach to the functional integral method.

As for the evaluation of ψ from a given band structure, there are two types of approaches from the opposite sides. One is the long-wavelength approximation and the other is the local approximation based on CPA. Both lines of approaches have been developed in recent years to a significant extent. In particular, we have a closed form expression for ψ in the form of (2.41) based on CPA and applicable to any value of S_L^2 [2.39,40]. Note that this expression leads to the known correct results in the two extremes; the local moment limit and the weakly ferromagnetic limit.

Significant simplification is possible in the functional integral calculation if one contents oneself with neglect of the non-local character of spin fluctuations at the expense of a part of the interpolating nature of the theory. This leads to a molecular field or single site approximation [2.41,42]. It is essential here, too, to evaluate the functional integral beyond the saddle-point approximation.

Numerical studies of various metallic magnets by using calculated band structures have been performed by many investigators within the above-mentioned approximation methods. For the calculation of the local term of ψ in (2.41) or, in general, we need only the density of states, while for the non-local terms more detailed information for the band structure is required. Naturally the calculation is more laborious for the latter. A calculation by using both terms in full has been performed only for a simple cubic tight-binding band [2.40]. In actual calculations with realistic band structures that have been reported so far the non-local term is treated with simple assumptions [2.43] including its neglect [2.41,42], or they are calculated in detail at the expense of the neglect of the amplitude variation [2.44–47].

We will not go into details of these approaches, referring to [2.1] for interested readers, but discuss only the main consequences of recent investigations.

2.4.3 Mean Mode-Mode Coupling Theory of Spin Fluctuations – An Interpolation Theory

For brevity, let us confine ourselves here to $T > T_C$. We consider the dynamical susceptibility $\bar{\chi}(q, \omega; \{S(q', \omega')\})$ of a non-interacting electron system under the influence of the fluctuating exchange field arising from the spin fluctuations specified by a set of correlation functions $\{S(q, \omega)\}$. Then a simple mean field consideration leads to the following expression for the dynamical susceptibility [2.48]

$$\chi(q, \omega) = \bar{\chi}(q, \omega; \{S(q', \omega')\})/[1 - 2I\bar{\chi}(q, \omega; \{S(q', \omega')\})] \ . \tag{2.42}$$

Explicit forms for $\bar{\chi}(q, \omega; \{S(q', \omega')\})$ can be obtained with certain approximations in the Green's function formalism, etc. This equation generally includes integrals over q' and ω' and $S(q', \omega')$ is related with $\chi(q', \omega')$ through the fluctuation-dissipation theorem. Thus (2.42) is an integral equation for $\chi(q, \omega)$. Although such an equation was actually derived [2.49], it is not so easy to solve it without further simplifications.

An extreme simplification is made if the effect of the mode-mode coupling, or the coupling between different modes of spin fluctuations is taken into account only in average. We then have

$$1/\chi(q, \omega) = 1/\bar{\chi}(q, \omega; S_L^2) - 2I \ , \tag{2.43}$$

$$S_L^2 = 3\,T\,N_0^{-2} \sum_m \sum_q \chi(q, i\omega_m) \ , \tag{2.44}$$

where $\omega_m = 2\pi mT$, m being an integer. This leads to the SCR theory if we further assume $\bar{\chi}(q, \omega; S_L^2) = \chi_0(q, \omega)/[1 + \lambda(S_L^2)]$. Thus (2.43 and 44) can make a generalization of the SCR theory to allow larger amplitudes of spin fluctuations. The point of the problem is how to calculate $\bar{\chi}(q, \omega; S_L^2)$. For this purpose, we can think of various approximation methods including long-wavelength approximations and local approximations such as the coherent potential approximation (CPA).

Let us here pursue the formal aspects and general accounts of the theory a little further. The static uniform susceptibility given by

$$1/\chi = 1/\overline{\chi}(S_L^2) - 2I \tag{2.45}$$

may be expanded in the following form insofar as the variation of S_L^2 with temperature is small

$$1/\chi = (4N_0 I^2/3T_0)[S_L^2(T) - S_L^2(T_C)] \ , \quad \text{with}$$

$$N_0/3T_0 = -[\partial\overline{\chi}(S_L^2)/\partial S_L^2]_{T=T_C} \ , \tag{2.46}$$

where the relation $2I\overline{\chi}[S_L^2(T_C)] = 1$ is used. $1/T_0$ may be called the longitudinal stiffness constant, the stiffness against the change in the mean-square local amplitude of the spin density. It may be noteworthy that with this simplification we can still develop a dynamical theory of spin fluctuations, although such an attempt does not seem to be extended so far beyond the SCR theory.

At this point we make a still further simplification of using a static approximation; we take only m = 0 term in the summation over m. Thus we have

$$S_L^2 = 3TN_0^{-2} \sum_q \chi_q \ ,$$

$$1/\chi_q = 1/\chi + 1/\overline{\chi}_q(S_L^2) - 1/\overline{\chi}(S_L^2) \ . \tag{2.47}$$

We note here the following relation to hold:

$$V_q(S_L^2) = \langle 1/2\overline{\chi}_q(S_L^2)\rangle - 1/2\overline{\chi}_q(S_L^2) \ . \tag{2.48}$$

As was defined in (2.41), $V_q(S_L^2)$ is the Fourier transform of the effective Heisenberg exchange interaction constant between local spin densities.

At the Curie temperature we have

$$S_L^2(T_C) = 3T_C N_0^{-2} \sum_q [2V_0(S_L^2) - 2V_q(S_L^2)]^{-1}$$

$$= \frac{3T_C}{2N_0} \langle [V_0(S_L^2) - V_q(S_L^2)]^{-1}\rangle \ . \tag{2.49}$$

This expression in the case of a fixed value of S_L^2 reduces to the spherical-model approximation for the local-moment system. In the weakly ferromagnetic limit, where $\overline{\chi}_q(S_L^2)$ is replaced by χ_{0q}, the expression reduces to the result of a static approximation to the SCR theory.

The susceptibility above T_C is given from (2.46–48) as follows

$$\frac{1}{\chi} = \frac{4I^2}{N_0 T_0} \sum_q \left(\frac{T}{1/\chi + 2V_0 - 2V_q} - \frac{T_C}{2V_0 - 2V_q} \right) \ . \tag{2.50}$$

This expression, as a matter of fact, includes the Curie-Weiss susceptibilities of both the two origins.

When the longitudinal stiffness constant is infinite or $T_0 \to 0$, we have

$$S_L^2 = \frac{3T}{N_0^2} \sum_q \frac{1}{1/\chi + 2V_0 - 2V_q} = \frac{3T_C}{N_0^2} \sum_q \frac{1}{2V_0 - 2V_q} = \text{const} . \quad (2.51)$$

This is nothing but the spherical model equation for the magnetic susceptibility. If we further approximate $V_0 - V_q$ by $\langle V_0 - V_q \rangle = V_0$ we get the following molecular field results

$$1/\chi = (2V_0/T_C)(T - T_C) = 3(T - T_C)/N_0 S_L^2 ,$$

$$T_C = 2N_0 V_0 S_L^2 / 3 . \quad (2.52)$$

When the longitudinal stiffness is small or T_0 is large, on the other hand, we get the following result at least near T_C, neglecting $1/\chi$ in the first term of the right-hand side of (2.50):

$$1/\chi = [4N_0 I^2 S_L^2(T_C)/3T_C T_0](T - T_C) . \quad (2.53)$$

This is the same kind of the Curie-Weiss law as in weak itinerant ferromagnetism. The Curie constant is proportional to T_0, i.e., the longitudinal softness or the inversed longitudinal stiffness constant.

Thus we see that (2.50), or more generally (2.45–47), is a unified expression for the susceptibility interpolating between the two opposite extremes within a static approximation for the spin fluctuations. By varying the longitudinal stiffness constant (and the cut-off wave vector) we can connect the two extremes continuously.

For convenience let us rewrite (2.50) as follows in terms of the reduced variables and parameters

$$\tau_0 y = t \langle (y + \eta_q)^{-1} \rangle - 1 , \quad \text{with} \quad (2.54)$$

$$y = N_0 S_L^2(T_C)/3T_C \chi , \quad t = T/T_C$$

$$\tau_0 = 9T_C T_0 / 4[N_0 I S_L^2(T_C)]^2 = T_0/T_C \langle I/(V_0 - V_q) \rangle_{T=T_C}^2 ,$$

$$\eta_q = (V_0 - V_q) \langle (V_0 - V_q)^{-1} \rangle_{T=T_C} . \quad (2.55)$$

With these notations we also have

$$y_q = N_0 S_L^2(T_C)/3T_C \chi_q = y + \eta_q ,$$

$$S_L^2(T)/S_L^2(T_C) = 1 + \tau_0 y ,$$

$$\langle |S_q(T)|^2 \rangle / S_L^2(T_C) = t/y_q . \quad (2.56)$$

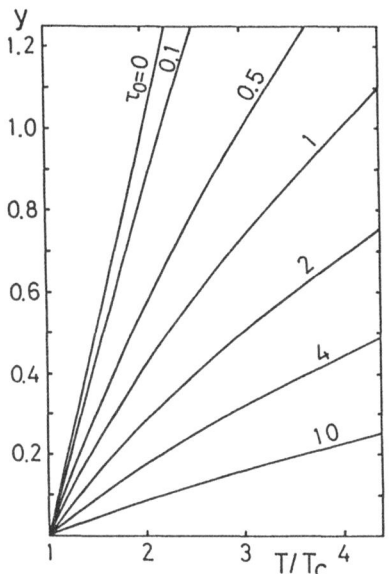

Fig. 2.12. Reduced inverse susceptibility due to a molecular field approximation vs reduced temperature for various values of the longitudinal stiffness constant

Thus when $\eta_q(S_L^2)$ is given we can calculate the temperature dependences of the susceptiblities y and y_q, $\langle |S_q|^2 \rangle$ and S_L^2 for various values for the reduced longitudinal stiffness constant $1/\tau_0$.

Leaving discussions for a parametrization of the theory and evaluation of the parameters to [2.1] or to the original paper, we show some results of an example calculation. Figure 2.12 displays the $y-t$ curves for the molecular-field approximation, corresponding to setting $\eta_q = 1$. Here the Curie-Weiss law for the local-moment system with $p_C/p_s = 1$ (corresponding to $\tau_0 = 0$) is represented simply by $y = t - 1$. With increasing τ_0 the p_C/p_s ratio increases or the slope of the curve decreases and the curve tends to become somewhat convex upward at high temperatures. For large τ_0 the linearity of the $y-t$ curve becomes better again, although the static approximation with a constant cut-off wave vector (q_B in this example) is not quite satisfactory for weakly ferromagnetic metals.

In order to illustrate the relation between the short-range order and the longitudinal stiffness, we take the following example

$$\eta_q = 3(q/q_B)^2 \ , \tag{2.57}$$

neglecting for brevity the temperature or S_L^2 dependence of η_q. Fig. 2.13 shows $\langle |S_q|^2 \rangle / S_L^2(T_C)$ as a function of q and T for various values for the parameter τ_0. We see how the two limiting cases with local moments (no SRO) and with extended moments (strong SRO), as illustrated in Fig. 2.7a and b, respectively, are interpolated smoothly by varying the longitudinal stiffness constant.

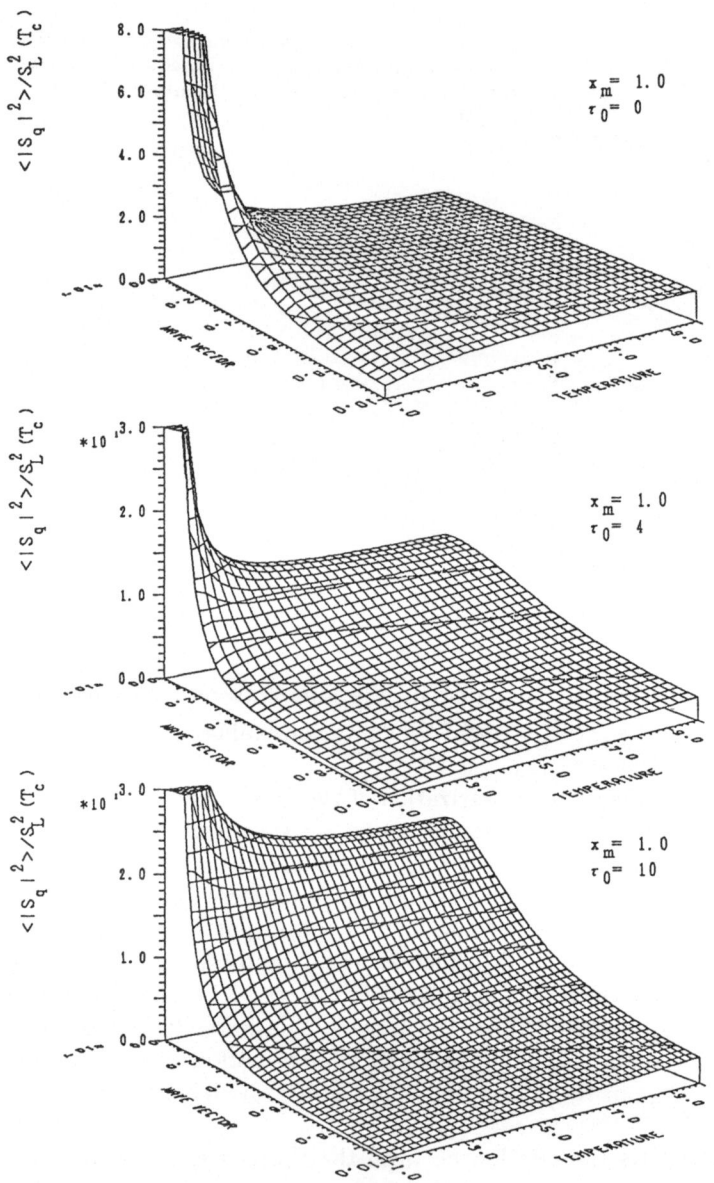

Fig. 2.13. $\langle |S_q|^2 \rangle$ as calculated by a unified spin-fluctuation theory with the longitudinal and transverse stiffness constants as parameters

It is worth while to note that strong short-range order and the CW law are compatible when S_L^2 increases with temperature while they are incompatible when S_L^2 is constant as is well-known for the Heisenberg systems. In other words, when the CW law is observed with the p_C/p_s ratio significantly larger than 1, we may conclude that S_L^2 is increasing with temperature.

2.4.4 Physical Properties in the Intermediate Regime

In addition to the above-discussed unified explanation for the Curie-Weiss susceptibility, the interpolation theory has been applied with at least qualitative success to various substances showing various characteristic behaviors. The most useful concept here is the variation of S_L^2 with temperature. Some typical curves for the $S_L^2 - T$ relation are sketched in Fig. 2.14. Examples for all of these curves in Fig. 2.14 have actually been obtained from calculations by using various band structures and various degrees of relative strength of the exchange interaction. The curves a and d are for a local moment system and a weak itinerant ferromagnet, respectively. The curve c shows a significant decrease of S_L^2 with increasing temperature below T_C, corresponding probably to the invar metals. The curves e and f show the rapid increase with temperature and saturation at high temperatures of S_L^2, leading to the concept of temperature-induced local moments. This concept seems to explain [2.35,50] the observed breaks in the plots of $1/\chi$ vs. T for CoS_2 and $CoSe_2$ [2.51]. The curves g and h are for nearly ferromagnetic metals and Pauli paramagnets, respectively. Unusual magnetic and thermal properties of a nearly ferromagnetic semiconductor FeSi [2.52] seem to be explained in terms of this theory [2.53] by an extremely rapid increase of S_L^2 at around a certain temperature followed by a saturation of it. We sketch in Fig. 2.15a expected temperatur variations of S_L^2 in CoS_2, $CoSe_2$ and FeSi and in b and c the observed and expected (dashed lines) behaviors of $1/\chi$ and $1/T_1$, respectively.

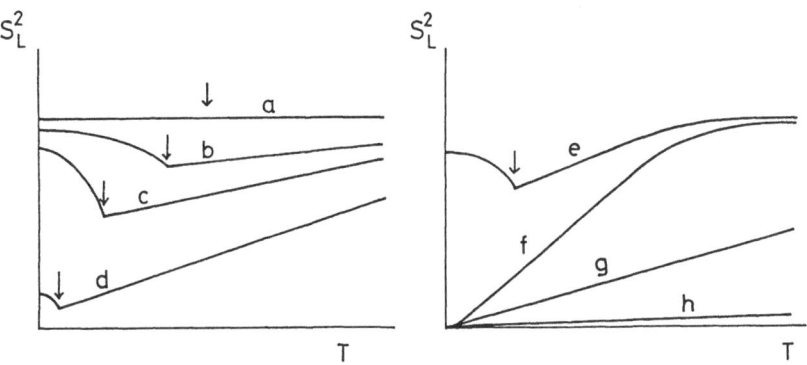

Fig. 2.14. Typical examples of temperature variation of the mean-square local amplitude of spin fluctuation S_L^2. See text for details

Since this interpolation theory continuously covers between small and large amplitudes of the spin density, it is suitable for a description of the Mott transition. A full phase diagram in the $T - U$ plane was calculated for the first time for a system with a half-filled tight-binding band [2.40]. We show the result in Fig. 2.16.

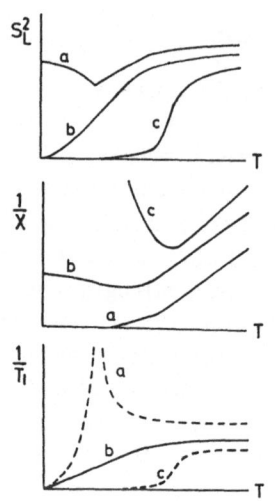

Fig. 2.15. Temperature variations of S_L^2, $1/\chi$ and $1/T_1$ for (a) CoS_2, (b) $CoSe_2$ and (c) FeSi

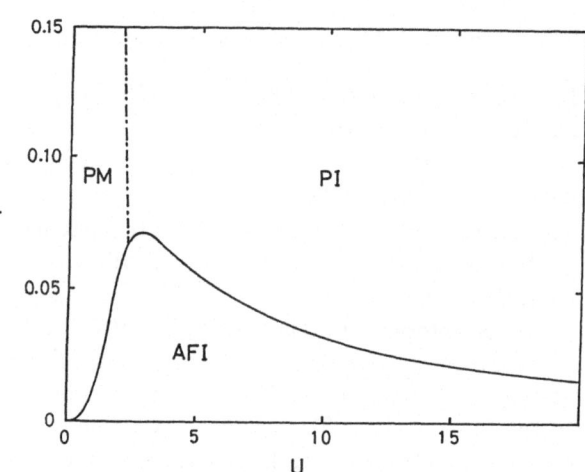

Fig. 2.16. Calculated phase diagram for the half-filled simple cubic tight-binding band. (AFI) antiferromagnetic phase; (PM) paramagnetic metallic phase and (PI) paramagnetic insulator phase; U and T are in units of the transfer integral

Still another recent example showing the importance of the temperature variation of S_L^2 is an explanation of unusual temperature dependence of the magnetic susceptibility of MnAs which has been an enigma for a long time [2.54]. The transition between α and γ phases of iron seems to be related with the rapid increase of S_L^2 with temperature in the latter [2.55].

As for the quantitative aspects of the functional integral calculations, the values for T_C in Fe, Co and Ni have been explained within a factor of 2 to 3. The other calculated quantities also seem to be of nearly the same accuracy. However, the qualitative aspects of all the above-mentioned phenomena seem to be well explained in terms of given band structures.

There is an attempt to describe the theory with the use of a relatively small number of parameters including the longitudinal and transverse (exchange) stiffness constants for the spin density [2.48], as was partly discussed in Sect. 2.4.3. Such a simple theory, though having rather limited validity, seems to be useful for systematic understanding of itinerant-electron magnetism. Here we will not go into details of these and other subjects of interest, which have been investigated intensively in recent years. For interested readers we refer to [2.1] and to original papers.

The above-mentioned recent investigations have clarified a number of physical properties at finite temperatures of various metallic magnets on the basis of their band structures. In view of this success we may conclude that the idea of a spin fluctuation theory for the intermediate regime, interpolating between the two extremes, gives a valid direction of investigations toward establishment of a unified theory of magnetism. In particular, the variable nature of the local amplitude of spin density and the short-range magnetic order strongly associated with it are new important concepts essential to the theory of itinerant-electron magnetism in general. The existing theories based on the static approximation, though successful from a qualitative point of view, are not quite satisfactory not only from a quantitative point of view but also in describing many physical properties such as the cross section of neutron inelastic scattering, electrical resistivity, NMR relaxation, etc. In order to establish a real unified theory of magnetism it is essential to give a dynamical theory of spin fluctuations in the intermediate regime interpolating between the well-established two extremes.

2.4.5 Supplementary Discussions

A model with extremely strong short-range order in Fe and Ni have been advocated for some years as a local band theory [2.56]. This theory starts with an assumption of strong short-range order on the grounds of earlier reports on neutron observations of persistence of spin waves above T_C in Ni and Fe. Assuming a fixed amplitude of the local spin density, as is evident from the arguments by using the Landau-Lifshitz equation, this theory describes spin waves above T_C and deduce the degrees of short-range order in Fe and Ni so as to explain the experimental results of the above reports.

From the thermodynamical point of view this model is equivalent to the Heisenberg model with strong short-range order which is known to be incompatible with the Curie-Weiss susceptibility. Also, the earlier reports on neutron scattering experiments have been revised recently as will be discussed in the following section. For a detailed account of this theory including its recent extensions we refer to Chap. 4. Possible applicability of a simple Heisenberg model without strong short-range order to Fe has been discussed from time to time [2.57,58]. It may be worth while to note that both of these models belong to the limit of infinite longitudinal stiffness constant in the above-discussed unified picture. They differ with each other only in the degrees of short-range order. Although the concept of variable S_L^2 (or a finite value for the longitudinal stiffness constant) has been ignored in some of recent discussions [2.59], its consideration is essential in the theory of itinerant electron magnetism [2.60]. In particular, the degree of short-range order must be discussed with variable S_L^2, as was noted in Sect. 2.4.3. Needless to say the limitations of the static approximation must always be kept in mind.

There have been many investigations on the underlying electronic structure or the one-electron states in magnetic transition metals at finite temperatures [2.7]. The problem is to deal with the self-energy correction due to the spin fluctuations and has so far been treated within the adiabatic approximation. This problem is most directly associated with the photoemission spectra which are the subject of Chap. 3. We only note that one has to be very careful in drawing conclusions as to the nature of spin fluctuations from the photoemission spectra, since the information is less direct than from neutron experiments, as far as the spin fluctuations are concerned, and various approximate theoretical calculations are involved in its analysis.

2.5 Dynamical Properties of Spin Fluctuations

2.5.1 General Remarks

As was noted in the preceding section, it is now essential to clarify the dynamical properties of spin fluctuations in the intermediate regime in order to establish a unified theory of magnetism. We have already seen in Sect. 2.2 and 3, in particular Figs. 2.8a and b, the contrasting features of the spin correlation functions $S(q, \omega)$ in the well-established opposite extremes. In view of these results we clearly need to deal with $S(q, \omega)$ in the entire Brillouin zone for the study of intermediate regime. From experimental point of view neutron inelastic scattering measurements provide us with the most useful information [2.61]. We will discuss here on the properties of spin fluctuations in various metallic magnets by using mainly the results of neutron scattering experiment and then discuss the limitations of the static approximation and possible approaches toward a unified theory of dynamical spin fluctuations.

2.5.2 Spin Waves at Low Temperatures

As was noted in Sect. 2.2, the local-moment system shows well-defined spin wave excitations in the entire Brillouin zone with the average energy of the order of T_C, as was observed for example in EuO [2.62] and Heusler alloys such as Pd_2MnSn [2.63], etc.

In a weak itinerant ferromagnet MnSi, in contrast, spin waves were observed in an extremely small part around $q = 0$ of the Brillouin zone as was expected theoretically [2.27]. Spin waves have not yet been observed in any other weak itinerant ferromagnets. In Fe and Ni the spin waves were observed in a small but significant part of the Brillouin zone, i.e., $q < q_B/2$ for both Fe [2.64] and Ni [2.12]. This corresponds, however, to only a small fraction in volume of the entire zone. Their energy dispersions are very steep and increase monotonically up to the maximum observed values, $\sim 2T_C$ and $\sim 4T_C$ for Fe and Ni, respectively. For Fe the scattering intensity decreases gradually beyond 40 meV with increasing energy. In Ni broadening of the spin wave spectra was observed in [111] direction beyond half the zone boundary and was interpreted as due to the overlap with the Stoner-excitation continuum. We may expect from these behaviors of spin waves that both below and above T_C the low-energy magnetic collective excitations are not separated from the high-energy part arising from the independent particle-hole type excitations. This is considered to be one of the characteristic features of itinerant magnets including the intermediate regime. In the same sense Fe_3Pt may also be classified in the intermediate regime. On the other hand, $MnPt_3$, $FePt_3$, and $FePd_3$, where damped spin waves were observed in the entire zone and their energies are limited below $\sim T_C$, are expected to be much closer to or almost in the local moment regime.

2.5.3 Spin Correlations Above T_C

There are not many reported neutron measurements for $S(q, \omega)$ above T_C. Almost complete information on $S(q, \omega)$ is obtained only for EuO [2.65], Pd_2MnSn [2.66] as local moment systems and MnSi [2.27] as a weak itinerant ferromagnet (helimagnet to be exact but with an extremely small ordering wave vector $Q \simeq 0.03 \text{ Å}^{-1}$, becoming ferromagnetic under an external magnetic field of more than 6 kOe). As was mentioned already, these results are in very good accord with the theories in the two extremes.

Neutron measurements of $S(q, \omega)$ above T_C for substances in the intermediate regime, Fe and Ni in particular, have been the subject of intensive investigations in recent years. The controversy over the possible persistence of spin waves (in a region of relatively small q) in Ni and Fe above T_C seems to be converging, i.e., the spin waves in Ni and Fe observed below T_C in the limited regions of the Brillouin zone do not persist, at least as dominating magnetic excitations, above T_C [2.67,68]. Although some small structures were reported [2.68], $S(q, \omega)$ seems to be given mainly by a broad peak around $\omega = 0$ in a fairly large region of relatively small q, say $q < q_B/3$.

$$S(q, \omega) = \frac{2T\chi}{1 + q^2/\kappa^2} \frac{\Gamma_q}{\omega^2 + \Gamma_q^2} \frac{\omega/T}{1 - e^{-\omega/T}} , \quad \text{with} \tag{2.58}$$

$$\kappa^2 \propto 1/\chi , \quad \text{and} \quad \Gamma_q \propto q^2(\kappa^2 + q^2) . \tag{2.59}$$

From a microscopic point of view this expression has been discussed mainly by using the Heisenberg model, although the expression should also hold for itinerant magnets because of its phenomenological nature. As a matter of fact, the SCR theory gives just this form (2.58). Only difference is the linear q dependence of Γ_q against q^2 dependence in (2.59). However, a q^2 dependence is easily obtained for small q when the effect of finite lifetime of itinerant electrons is taken into account [2.1]. Thus it is hard to make distinction between localized and extended nature of spin fluctuations only from neutron measurements in a region of small q. In order to see the properties of spin fluctuations, in particular to what extent they are localized or extended, it is essential to investigate $S(q, \omega)$ in a region of large q or hopefully in the entire Brillouin zone.

In this context it is interesting to see recent low-resolution absolute scale measurements of polarized neutron scattering intensity in the entire Brillouin zone. For Ni at $T \simeq 2T_C$ [2.69] the scattering is reported to be rather heavily weighted in the side of small q but the intensity near the zone boundary is also significant up to the upper bound $\sim 3T_C$ of energy in the measurements. The integrated total intensity corresponds to the local moment of $0.8 \mu_B^2$, larger than the value $0.36 \mu_B^2$ expected from the classical ferromagnetic moment of $0.6 \mu_B$ per atom but smaller than $1.8 \mu_B^2$ expected from a quantum spin 1/2 on 60 % of the atoms. For Fe at $T = 1.25 T_C$ [2.70], the essential contribution of scattering intensity is reported to be restricted to energies below 50 meV. The

51

integrated intensity up to ~2T_C corresponds to ~1.7 μ_B^2 per atom at the zone boundary and ~2.4 μ_B^2 on average over the entire zone. Half of this contribution is contained within the low-q region (q<0.7 Å^{-1}), indicating a significant degree of short-range order. These squared moment values are much smaller than 9 μ_B^2 expected from a local-moment value corresponding to the ferromagnetic moment at T = 0. Although more detailed measurements are highly desired, it seems clear that both Fe and Ni belong to the intermediate regime.

2.5.4 Limitations of the Static Approximation

Let us now consider to what extent the static approximation for the spin fluctuations is applicable for the calculation of thermodynamical properties of metallic magnets.

We first note that this approximation corresponds to the following evaluation of the mean square amplitude of the spin fluctuation

$$\langle |S_{q\nu}|^2 \rangle = \frac{1}{2\pi} \int\limits_{-\infty}^{\infty} d\omega\, S(q,\omega) = \frac{1}{\pi} \int\limits_{-\infty}^{\infty} d\omega \frac{1}{1 - e^{-\omega/T}}\, \text{Im}\{\chi_\nu(q,\omega)\}$$

$$\to \frac{T}{\pi} \int\limits_{-\infty}^{\infty} d\omega \frac{\text{Im}\{\chi_\nu(q,\omega)\}}{\omega} = T\chi_{q\nu}\,.$$

The last equality is due to the Kramers-Kronig relation. The approximation is thus valid when the energy spectra of the spin fluctuations are weighted predominantly in a range of energy smaller than T. This is true for T>T_C in the local-moment system but not in itinerant electron systems in general.

Let us for example consider a weak itinerant ferromagnet, where the above condition is satisfied for small q only. The energy spectra for the components with relatively large q extend to quite high energy of the order of the band width. The thermal spin fluctuations, however, are limited within the region of small q and ω because of the natural cut-off due to the Bose factor. In a static approximation where the Bose factor is replaced by T/ω we need a wave vector cut-off in order to get any meaningful results. A temperature-independent cut-off wave vector leads to only approximately correct results. The correct results are obtained within the static approximation only by introducing a temperature-dependent cut-off wave vector ($q_c \propto T^{1/3}$).

Similar situations should naturally be expected in the intermediate regime where the energy spectra of some components of spin fluctuations with relatively large q are considered to extend to quite high energy, since the low-frequency part and high-frequency part of spin fluctuations generally overlap in the intermediate regime, as may easily be inferred from Figs. 2.3 and 4. One might think that a cut-off wave vector may help in this case, too. This should be true qualitatively but it is hard to find a proper temperature dependence of the cut-off wave vector without studying the dynamical properties of spin fluctuations. Thus we see that the results of the static approximation are mainly of qualitative significance or semiquantitative at best.

2.5.5 Present Status and Future Prospect of the Theory

We summarize the present status of the dynamical theory of spin fluctuations in Fig. 2.17. We have fundamentally correct theories in the shaded area of the chart. In the low-temperature limit RPA covers both the weak- and strong-correlation regimes. At finite temperatures the SCR theory has been established for the weakly ferro- and antiferromagnetic regime. This theory is connected with RPA in the low-temperature limit (except for the renormalization effect due to zero-point spin fluctuations) and also in the weak correlation limit where the renormalization effect becomes insignificant.

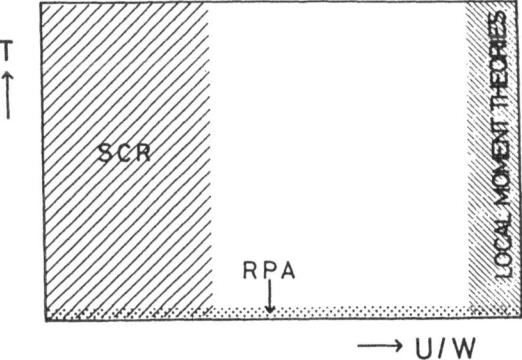

Fig. **2.17.** Ranges of validity for various approaches to the dynamical spin fluctuations

In the local moment case, on the other hand, we have a long history of developments of statistical mechanical theories for the Heisenberg spin Hamiltonians. Here we deal with well-defined local moments neglecting the high-energy inter-atomic excitations of electrons and thus the itinerant nature of electrons are not contained in the model, even though each moment may be interpreted as having a spatial extension toward neighboring atoms. The average magnetic excitation energy Γ in this model is comparable with the magnetic ordering energy T_C.

In itinerant electron systems of weakly magnetic and intermediate regimes, the purely magnetic low energy excitations cannot be well separated from the high energy polar excitations in contrast with the local moment regime. Thus all the magnetic excitations must be considered explicitly; Γ is much larger than the ordering energy T_C.

In weak itinerant ferromagnets with $T_C \ll \Gamma$ the quantum fluctuations with large energy spreads make temperature independent contributions to the renormalization of the thermal spin fluctuations dominated by small q components. The theory is thus well parametrized, as seen in Sect. 2.3. Although such a simplification is hardly expected, it seems reasonable to approach the intermediate regime by extending the RPA-SCR line of theories to include descriptions for the spin fluctuations with larger amplitude and shorter wave vectors.

Aside from the systems discussed in this chapter there are valance fluctuating systems and the dense Kondo systems, found in certain 4f and 5f compounds, where the relation $T_C \ll \Gamma$ holds. In the dense Kondo systems, with a small number of f electrons or holes per atom, the spin fluctuation is localized in real space owing to strong correlations between the f electrons and the effective mass enhancement of electrons is much larger than in weak itinerant magnets where the spin fluctuation is localized in q space. This problem of $s - f$ admixture combined with the strong $f - f$ correlations is another challenge in the field of magnetism.

2.6 Conclusion

After a long history of controversies over the local-moment model and the itinerant electron model for metallic magnetism, we now believe to have a unified picture of magnetism, as has been discussed in the preceding sections. Such a situation has been realized by recent theoretical and experimental developments in this area, in particular the establishment of the theory of mutually interacting modes of extended spin fluctuations in itinerant electron systems which is correct quantiatively at least in the weakly ferromagnetic regime. This is a newly established limiting case opposite to the local-moment regime and important specific characteristics of itinerant electron magnetism are revealed here in a striking way.

It seems apparent that we need to establish a unified theory of magnetism by interpolating between the two well-established mutually opposite extremes. Theoretical efforts toward this goal have so far been limited within the adiabatic approximation for the spin fluctuations. In view of a number of qualitatively successful explanations of various characteristic properties of certain metallic magnets on the basis of their band structures, we may conclude that the physical picture and the approach are essentially correct. We now need to establish a dynamical theory of spin fluctuations in the intermediate regime in order to attain the final goal of this long-standing problem.

Acknowledgements. The author would like to thank Prof. H. Fukuyama, Dr. H. Hasegawa and Dr. Y. Takahashi for their useful comments on the manuscript.

References

2.1   For a review see: T. Moriya: *Spin Fluctuations in Itinerant-Electron Magnetism*, Springer Ser. Solid-State Sci., Vol. 56 (Springer, Berlin, Heidelberg 1985)
2.2   For a review see: P.W. Anderson: In *Solid State Physics*, **14**, 99 (Academic, New York 1963)

2.3 See the following reviews: K. Yosida: In *Progress in Low Temperature Physics* **4**, 265 (North-Holland, Amsterdam 1964);
R.J. Elliott: In *Magnetism*, **29**, 385 (Academic, New York 1965)

2.4 H. Suhl (ed.): *Magnetism*, Vol. 5 (Academic, New York 1973)

2.5 W. Marshall (ed.): *Theory of Magnetism in Transition Metals* (Academic, New York 1967)

2.6 T. Izuyama, D.J. Kim, R. Kubo: J. Phys. Soc. Jpn. **18**, 1025 (1963)

2.7 B.L. Gyorffy: In Metallic Magnetism II, ed. by H.Capellmann, Topics Cur. Phys. (Springer, Berlin, Heidelberg);
V.L. Morizzi, J.F. Janak, A.R. Williams: *Calculated Properties of Metals* (Pergamon, New York 1978)

2.8 C. Herring: In *Magnetism*, **4**, 362 (Academic, New York 1966)

2.9 J.F. Cooke, J.W. Lynn, H.L. Davis: Phys. Rev. B**21**, 4118 (1980) and references therein;
J.F. Cooke, J.A. Blackman, T. Morgan: Phys. Rev. Lett. **54**, 718 (1985)

2.10 Y. Kubo, S. Ishida, J. Ishida: J. Phys. Soc. Jpn. **50**, 47 (1981)

2.11 H.A. Mook, D. Tocchetti: Phys. Rev. Lett. **43**, 2029 (1979)

2.12 H.A. Mook, D. McK. Paul: Phys. Rev. Lett. **54**, 227 (1985)

2.13 K.K. Murata, S. Doniach: Phys. Rev. Lett. **29**, 285 (1972)

2.14 T. Moriya, A. Kawabata: J. Phys. Soc. Jpn. **34**, 639 (1973); **35**, 669 (1973); Proc. ICM'73 **4**, 5 (Moscow 1973)

2.15 D.M. Edwards, E.P. Wohlfarth: Proc. Roy. Soc. A**303**, 127 (1968)

2.16 G.G. Lonzarich: J. Magn. Magn. Mat. **45**, 43 (1984);
G.G. Lonzarich, L. Taillefer: J. Phys. C **18**, 4339 (1985)
See also an earlier work: S.G. Mishra: Dr. disseration (ITT, Kampur 1977), unpublished;
S.G. Mishra, T.V. Ramakrishnan: Inst. Phys. Cont. Ser. **39**, 528 (1978)

2.17 S. Ogawa: Researches of Electrotechnical Laboratory No. 735 (1972)

2.18 D.J. Kim. H.C. Praddaude, B. Schwartz: Phys. Rev. Lett. **23**, 419, 1270(E) (1969)

2.19 M. Kontani, T. Hioki, Y. Masuda: Solid State Commun. **18**, 1251 (1975); Physica **86B**, 399 (1977);
Y. Masuda, T. Hioki, A. Oota: Physica **91B**, 291 (1977)

2.20 T. Hioki, Y. Masuda: J. Phys. Soc. Jpn. **43**, 1200 (1977)

2.21 K. Ueda: Solid State Commun. **19**, 965 (1976)

2.22 T. Moriya, K. Usami: Solid State Commun. **34**, 95 (1980)

2.23 S. Ogawa: Physica **119B**, 68 (1983) and references therein

2.24 M. Matsunaga, Y. Ishikawa, T. Nakajima: J. Phys. Soc. Jpn. **51**, 1153 (1982)

2.25 K. Suzuki, Y. Masuda: J. Phys. Soc. Jpn. **54**, 326, 630 (1985)

2.26 G. Creuzet, I.A. Campbell, J.L. Smith: J. Physique **44**, L547 (1983)

2.27 Y. Ishikawa, G. Shirane, J.A. Tarvin, M. Kohgi: Phys. Rev. B**16**, 4956 (1977);
Y. Ishikawa, Y. Noda, C. Fincher, G. Shirane: Phys. Rev. B**25**, 254 (1982);
Y. Ishikawa, Y. Uemura, C.F. Majzak, G. Shirane, Y. Noda: Phys. Rev. B**31**, 5884 (1985)

2.28 N.R. Bernhoeft, G.G. Lonzarich, P.W. Mitchell, D. McK. Paul: Phys. Rev. B**28**, 422 (1983)

2.29 Y. Takahashi, T. Moriya: J. Phys. Soc. Jpn. **54**, 1592 (1985)

2.30 M. Cyrot: In *Electron Correlation and Magnetism in Narrow Band Systems*, ed. T. Moriya, Springer Ser. Solid-State Sci., Vol. 29 (Springer, Berlin, Heidelberg 1981) p. 51; J. Magn. Magn. Mat. **45**, 9 (1984)

2.31 T. Moriya: Physica **119B**, 330 (1983)

2.32 H. Hasegawa, T. Moriya: J. Phys. Soc. Jpn. **36**, 1542 (1974)

2.33 H. Hasegawa: J. Low Temp. Phys. **31**, 475, 845 (1978)

2.34 R.R. Rhodes, E.P. Wohlfarth: Proc. Roy. Soc. **273**, 247 (1963);
E.P. Wohlfarth: J. Magn. Magn. Mat. **7**, 113 (1978)

2.35 T. Moriya, Y. Takahashi: J. Phys. Soc. Jpn. **45**, 397 (1978);
J. Physique **39**, C6-1466 (1978)

2.36 J.R. Schrieffer, W.E. Evenson, S.Q. Wang: J. Physique **32**, C1 (1971) and references therein

2.37   M. Cyrot: J. Physique **33**, 125 (1972);
        Phil. Mag. **25**, 1031 (1972)
2.38   H. Capellmann: J. Phys. F**4**, 1466 (1974); Z. Phys. B**34**, 29 (1979)
2.39   T. Moriya, J. Magn. Magn. Mat. **14**, 1 (1979)
2.40   T. Moriya, H. Hasegawa: J. Phys. Soc. Jpn. **48**, 1490 (1980)
2.41   J. Hubbard: Phys. Rev. B**20**, 4584 (1979);
        B**23**, 5974 (1981)
2.42   H. Hasegawa: J. Phys. Soc. Jpn. **49**, 178, 963 (1980)
2.43   K. Usami, T. Moriya: J. Magn. Magn. Mat. **20**, 171 (1980)
2.44   P.J. Lin-Chung, A.J. Holden: Phys. Rev. B**23**, 3414 (1981)
2.45   C.S. Wang, R.E. Prange, V. Korenman: Phys. Rev. B**25**, 5766 (1982)
2.46   T. Oguchi, K. Terakura, N. Hamada: J. Phys. F**13**, 145 (1983)
2.47   J. Staunton, B.L. Gyorffy, A.J. Pindor, G.M. Stocks, H. Winter: J. Magn. Magn.
        Mat. **45**, 15 (1984) and references therein
2.48   T. Moriya: Prog. Theor. Phys. Suppl. **69**, 323 (1980); J. Phys. Soc. Jpn. **51**, 420,
        2806 (1982)
2.49   T. Moriya: J. Phys. Soc. Jpn. **40**, 933 (1976)
2.50   Y. Takahashi, M. Tano: J. Phys. Soc. Jpn. **51**, 1792 (1982)
2.51   K. Adachi, K. Sato, M. Takeda: J. Phys. Soc. Jpn. **26**, 631 (1969)
2.52   V. Jaccarino, G.K. Wertheim, J.H. Wernick, L.R. Walker, S. Arajs: Phys. Rev. **160**,
        476 (1967)
2.53   Y. Takahashi, T. Moriya: J. Phys. Soc. Jpn. **46**, 1451 (1979)
2.54   K. Motizuki, K. Katoh: J. Phys. Soc. Jpn. **53**, 735 (1984)
2.55   H. Hasegawa, G.D. Pettifor: Phys. Rev. Lett. **50**, 130 (1983)
2.56   V. Korenman, J.L.Muray, R.E. Prange: Phys. Rev. B**16**, 4032, 4048, 4058 (1977);
        R.E. Prange, V. Korenman: Phys. Rev. B**19**, 4691, 4698 (1979)
2.57   D.M. Edwards: J. Magn. Magn. Mat. **15–18**, 262 (1980)
2.58   B.S. Shastry, D.M. Edwards, A.P. Young: J. Physique C**14**, L665 (1981)
2.59   D.M. Edwards: Proc. Workshop on 3d Metallic Magnetism, Grenoble (1983) 1;
        J. Magn. Magn. Mat. **36**, 213 (1983)
2.60   T. Moriya: J. Magn. Magn. Mat. **45**, 79 (1984)
2.61   K.R.A. Ziebeck: In *Metallic Magnetism II*, ed. by H. Capellmann, Topics Cur. Phys.
        (Springer, Berlin, Heidelberg)
2.62   J. Als-Nielsen, O.W. Dietrich, L. Passell: Phys. Rev. B**14**, 4897 (1976)
2.63   Y. Noda, Y. Ishikawa: J. Phys. Soc. Jpn. **40**, 690, 699 (1976)
2.64   C.K. Loong. J.M. Carpenter, J.W. Lynn, R.A. Robinson, H.A. Mook: J. Appl. Phys.
        55, 1895 (1984)
2.65   H.A. Mook: Phys. Rev. Lett. **46**, 508 (1981)
2.66   G. Shirane, Y.J. Uemura, J.P. Wicksted, Y. Endoh, Y. Ishikawa: Phys. Rev. B**31**,
        1227 (1985)
2.67   G. Shirane, O. Steinsvoll, Y.J. Uemura, J. Wicksted: J. Appl. Phys. **55**, 1887 (1984),
        and references therein
2.68   H.A. Mook, J.W. Lynn: J. Appl. Phys. **57**, 3006 (1985)
2.69   P.J. Brown, H. Capellmann, J. Déportes, D. Givord, K.R.A. Ziebeck: J. Magn. Magn.
        Mat. **31–34**, 295 (1983) and private communications
2.70   P.J. Brown, H. Capellmann, J. Déportes, D. Givord, S.M. Johnson, J.W. Lynn,
        K.R.A. Ziebeck: J. Physique **46**, 827 (1985)

3. 3d-Metallic Magnetism and Spin-Resolved Photoemission

E. Kisker
With 44 Figures

The principles of the novel experimental technique of spin- and angle-resolved photoemission are outlined. Recent results on fundamental problems in 3d-transition metal magnetism are discussed: It will be demonstrated that a detailed picture of the occupied electronic structure of ferromagnets can be obtained. The basic quantity, the exchange splitting, has been resolved explicitly for Fe and Ni, and band dispersions have been observed. In the vicinity of core-electron excitation thresholds, resonant photoemission and Auger electron emission are observed. It will be shown that new insight into the physics of these many-electron processes is obtained. Inelastic electron excitation processes are also observed in the photoelectron spectra. An example is the low-energy secondary electron cascade. It is shown that the secondary electrons are highly spin-polarized, which can be understood in terms of interaction of the excited electrons with electrons from the valence bands (Stoner excitations). The comparatively high intensity and high spin polarization of the secondary electrons, also observed in excitation with a primary electron beam, have allowed magnetic domains to be imaged with very high resolution. The high-temperature properties of ferromagnets are subject to controversy in the literature. It will be shown that spin-resolved photoemission data on Fe(100) provide evidence for short range order in excess of 4 Å. Results on the valence band electronic structure of the 3d transition metals obtained by other related methods such as spin-polarized inverse photoemission and spin-resolved energy loss spectroscopy are also discussed. Finally the prospects for future work are given.

3.1 Background

A considerable amount of experimental and theoretical effort has been spent in the past 50 years to understand the physics of ferromagnetism. One of the reasons for this interest certainly is the extensive application of magnetism in motors, transformers, and data storage devices, for example. Magnetism appears also to be of importance in many catalytic processes, as in the Fischer-Tropsch gasoline synthesis. A theoretical understanding might lead to the possibility to "tailor" materials for specific purposes.

Most technical applications make use of the fact that a certain material is permanently magnetic. But sometimes advantage is taken of the ferromagnetic to paramagnetic phase transition at the Curie temperature (T_C). Most important certainly is the recent effort in developing the so-called perpendicular magnetic recording technology for data storage. It is believed that when the magnetic domains in a data storage device are oriented perpendicular to the plane of the recording medium, a larger bit density can be obtained than with magnetic domains oriented parallel to the film plane (as in common magnetic tapes and disks). In order to reduce the magnetic field necessary for reversing the magnetization direction, the film is locally heated close to its Curie temperature ("Curie temperature writing") by focused light from a laser diode.

An obvious application of the ferromagnetic to paramagnetic phase transition is in temperature regulation. As an example, we mention certain kinds of soldering irons. A ferromagnetic material with a Curie temperature equal to the desired operating temperature is in thermal contact with the soldering tip. A permanent magnet is attracted when the temperature falls below T_C, and operates a switch which controlls the electric power to the heating element.

Ferromagnetism manifests itself as a macroscopic effect. It is known from the Einstein-de-Haas experiment that the electron spin – which is certainly a microscopic quantity – is the reason for ferromagnetism. While the low-temperature properties can be well described by modern computational methods based on the local-density approximation, the generalization to describe also high-temperature properties is a matter of controversy. These problems include understanding fundamental properties, such as the value of the Curie temperature. There is at present no theory on finite-temperature magnetism which is generally accepted. The novel method of spin- and angle-resolved photoemission has recently provided a guideline towards an understanding of these problems. We will show that a detailed picture on the electronic structure of ferromagnets at low *and* at elevated temperatures can be obtained by this method.

In this review we will concentrate on spin-resolved photoemission from the 3d ferromagnets Fe and Ni, rather than to give a complete overview on the large field of spin-polarized electron spectroscopy. Details on the large amount of work in spin-polarized photoemission also from magnetic insulators, magnetic semiconductors and non-magnetic materials can be found in a recent review by *Siegmann* et al [3.1]. It should be mentioned that in the case of non-magnetic solids, spin polarization of photoemitted electrons occurs as a consequence of spin-orbit interaction. The GaAs spin-polarized electron photocathode [3.2] is based on this effect. Spin-orbit interaction in heavy metals like Pt has recently been studied by *Eyers* et al. [3.3].

It had been pointed out in 1930 by *Fues* and *Hellman* [3.4] that electrons emitted from ferromagnets should be spin polarized as a consequence of ferromagnetism. The electronic properties of ferromagnets can then be studied by analyzing the emitted electrons by means of electron spectroscopy. Methods employed are field emission and spin- and angle-resolved photoemission. While field emission allows only to test states in the vicinity of the Fermi level, angle-resolved photoemission allows band dispersions to be determined to binding energies of several eV. Angle-resolved photoemission has been pioneered during the period from 1960 to the 1970's by *Kane* [3.5], *Berglund* and *Spicer* [3.6], *Gerhardt* and *Dietz* [3.7], *Smith* and *Traum* [3.8], *Feibelmann* and *Eastman* [3.9], and others. It has been shown in careful studies on the prototype material Cu that the bulk electronic structure can be tested quantitatively by this method [3.10,11]. For a recent review on angle-resolved photoemission, see [3.12].

Much effort in experimental work on spin-polarized electron physics has been spent in studies of the β decay [3.13] and in atomic physics [3.14]. Spin-polarized photoelectrons from solids were observed for the first time by *Busch* et al. [3.15]. It would not have been possible to measure the spin polarization energy-selectively along the photoelectron energy distribution curve (EDC) because a magnetic field had been applied to align the magnetization direction parallel to the electron optical axis. In 1976 a very moderate energy resolution of 500 eV has been obtained in a spin-polarized electron scattering experiment on EuO by *Chrobok* and *Hoffmann* [3.16], also exposing the sample to a strong longitudinal magnetic field. Shortly after, it was shown in spin-polarized field emission from W/EuS that highly polarized electrons are emitted even from samples in magnetic remanence (without using an aligning external magnetic field) [3.17]. An energy resolution of about 0.1 eV was comparatively easy to obtain. Working without an aligning longitudinal magnetic field also turned out to be possible in spin-polarized electron scattering [3.18]. The final step towards *spin-, energy-* and *angle-resolved* photoemission has been done in 1980 when it was shown that it is possible even in threshold photoemission to rely on the magnetic remanence of suitably shaped samples [3.19]. The key point was to use samples with a remanent magnetization oriented *perpendicular* to the electron-optical axis ("transverse magnetization geometry"). An energy resolution of 0.7 eV was thereby immediately achieved.

As a first test, the novel method of spin- and angle-resolved photoemission has been used in a study on Fe films evaporated on to a rocksalt single crystal [3.20]. The dependence of the spin polarization on binding energy of the primary photoelectrons resembled that of the calculated Fe density of states (DOS) since the films were (probably) polycrystalline. An unexpected increase of the spin polarization in the low-energy secondary electron distribution has also been observed for the first time [3.20], in parallel to a similar observation on electron-beam excited secondary electrons [3.21]. The spin-resolving photoelectron spectrometer was used subsequently in a study on the spin polarization of the so-called 6 eV satellite of Ni employing synchrotron radiation from the ACO storage ring at Orsay [3.22]. In 1983, the Ni exchange splitting was resolved explicitly for the first time using a laboratory resonance NeI resonance lamp [3.23]. Combined with synchrotron radiation from the BESSY storage ring in W. Berlin, the method of spin- and angle-resolved photoemission has been developed to a high degree of perfection in 1983-1984 [3.24].

3.2 The Electronic Structure of Ferromagnets

It will be demonstrated in this review that the ground state (T = 0 K) properties of the 3d metallic ferromagnets can be calculated quite accurately, as compared to experimental data, although there are still remaining problems regarding the lattice type, the exact value of the lattice constant, or the bulk modulus [3.25]. It is, however, not yet possible to calculate a fundamental quantity as the Curie temperature from first principles (T_C = 631 K for Ni, 1044 K for Fe, and 1390 K for Co). Even the qualitative picture of the microscopic mechanism leading to the ferromagnetic to paramagnetic phase transition is a matter of controversy.

3.2.1 Low-Temperature Properties (Band Model)

The itinerant electron band-picture was developed in 1936 by Stoner [3.26]. Due to its success to describe the low-temperature properties it is still the basis for the present understanding of 3d-transition-metal ferromagnetism at low temperatures. This model easily accounts for the non-integral magnetic moments of the 3d-transition metals. Stoner showed that the ferromagnetic state is lower in energy than the paramagnetic one if

$$IN(E_F) > 1 \tag{3.1}$$

I is the intraatomic Coulomb integral, and $N(E_F)$ is the paramagnetic density of states (DOS) at E_F. Equation (3.1) is the famous Stoner criterion. If it is fulfilled, the DOS becomes spontaneously spin-split. The result is an excess of electrons with their magnetic moments parallel to the spontaneous macroscopic magnetic field (majority-spin ($\uparrow$) electrons) as compared those with opposite magnetic moments (minority-spin ($\downarrow$) electrons). The two groups of electrons differ in energy by a characteristic quantity, the exchange splitting Δ ($\approx 1.8\,eV$ for Fe, $\approx 0.5\,eV$ for Ni). With the local-spin density functional (LSDF) theory it has become possible to calculate the ferromagnetic electronic structure at low temperatures (T = 0 K) in detail. The resulting DOS of Fe is shown in Fig. 3.1 [3.27]. The magnetic moment is largely determined by the spin-splitting of the d-electrons. The Fe valence band comprises 8 electrons, of which 7.4 have d-character. In the low-temperature ferromagnetic state, there are 4.8 3d$\uparrow$-electrons and 2.6 3d$\downarrow$-electrons. This yields a magnetic moment of about 2.2 μ_B/atom. The magnetic moment of the 0.6 4s-electrons per atom is very small. The spin-split DOS of Co and Ni differ from those of Fe by the fact that the magnetic moments and mean exchange splittings are smaller. A characteristic difference to Fe is that in Co and Ni the majority-spin bands are fully occupied and separated from the Fermi energy by the so-called Stoner gap. Co and Ni are therefore referred to as "strong" ferromagnets.

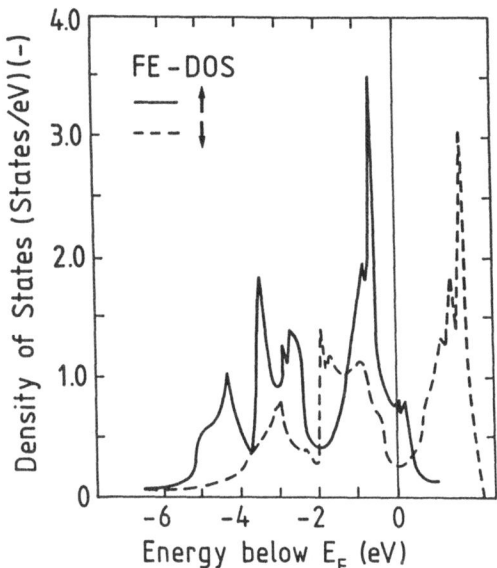

Fig. 3.1. The calculated spin-split density of states at $T = 0\,K$ of Fe (after *Moruzzi* et al. [3.27])

3.2.2 Models for 3d-Metallic Magnetism at Finite Temperatures

It is not yet possible to predict from first principles the magnetic and electronic structure of the 3d-transition metals at elevated temperatures. Several models have been proposed. The Stoner itinerant-electron model, describing successfully the low-temperature properties, has been generalized towards elevated temperatures. However, even when considering the 3d-electrons to be itinerant, local magnetic moments might exist at the lattice sites [3.28]. In the modern itinerant-electron theories on the electronic structure at elevated temperatures, the localization of the magnetic moments bridges the gap to the localized-electron Heisenberg model which describes well the magnetism of the rare earths and their compounds. Theories based on the existence of local magnetic moments above T_C are distinguished by the amount of correlation assumed to exist between neighbouring moments. These pictures are often justified by the use of the same experimental data [3.29].

a) Stoner Model

In the Stoner model, the exchange splitting is reduced proportional to the spontaneous magnetization when approaching T_C which is determined by the condition [3.30]

$$I \int N(E)|\partial f(T_C)/\partial E|dE = 1 \tag{3.2}$$

where $N(E)$ is the (paramagnetic) DOS function, and $f(T)$ is the Fermi function. It has been pointed out by *Gunnarson* [3.31] that a Curie temperature 5–10 times larger than that as observed would result from Eq. 3.2. The reason basically is that it requires too much energy (the exchange splitting) in the band picture to flip a spin, as compared to thermal energies kT_C which are only of the order of 100 meV. Ni might be a special case because it has been found that the exchange splitting is much less than that as calculated by state-of-the-art theories. Taking the smaller experimental value for the exchange splitting, a T_C of the right magnitude is obtained from Eq. (3.2) [3.32].

Recently, *Jarlborg* and *Peter* [3.33] suggested improving the Stoner model at elevated temperatures by introducing a broadening to the DOS to simulate electron-phonon coupling. It was shown that this reduces the Curie temperature considerably.

b) Fluctuating Band Model

It is generally believed [3.28] that for Fe, local magnetic moments exist also *above* T_C. For Ni, this appears to be more questionable (see *Edwards* [3.34] for an illustration of the difference between Fe and Ni). A controversy is also on the spatial extent of magnetically correlated regions above T_C in Fe. Are regions as large as 20–30 Å are magnetically coupled even above T_C as it has been suggested [3.35,36]? If this were the case, the magnetization would be constant over a sufficiently large number of atoms so that in each of these clusters an electronic structure similar to the ground state one is obtained. The direction of the local exchange fields is considered to fluctuate from cluster to cluster. This picture is referred to as the "fluctuating band model" [3.37,38]. It has been criticized by *Edwards* [3.39] on thermodynamical grounds. The fluctuating band theory with its large spin correlation length has been applied to interpret neutron scattering data above T_C, and has become highly controversial in context with the now questionable existence of spin waves above T_C [3.40,41]. Predictions have also been made in this model for expected changes in photoelectron energy distribution curves from Ni above T_C [3.42].

c) Disordered Local Moment Model

Within the disordered-local moment (DLM) theory, the local magnetization direction at the lattice sites is allowed to vary randomly [3.43] above T_C. No short-range order is anticipated. Restricting the magnetic moment directions to be "up" or "down", the system is similar to a random binary A–B alloy. This similarity allowed the use of calculation algorithms (KKR–CPA) which had originally been developed for calculating the electronic structure of random alloys. A Curie temperature close to the experimental values for Fe and Ni has been obtained. It turned out that the electronic structure is quite different from that in the low-temperature ordered state although the magnitude of the magnetic moments decreases only slightly up to T_C for Fe. This is seen in Fig. 3.2, where the theoretical $T = 0$ K DOS is compared with that as calculated in the DLM picture for Fe above T_C. An important prediction was that for bands with certain symmetry (e.g., Δ_5) the exchange splitting should vanish above T_C, while bands with different symmetries (e.g. Δ_2) are predicted to remain spin-split [3.44].

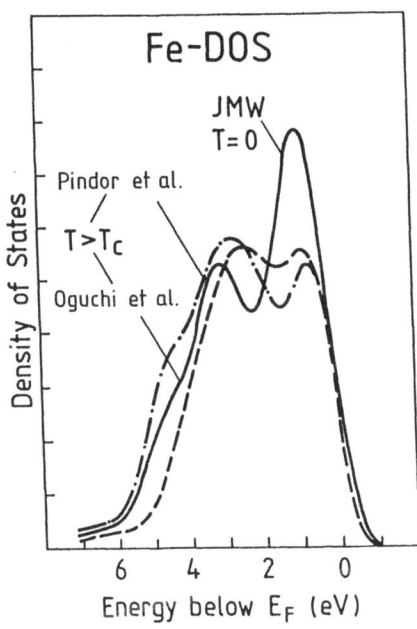

Fig. 3.2. Comparison of the spin-summed ferromagnetic T = 0 K DOS of Fe [3.27] with the calculated DOSs of Fe above T_C [3.43]. The DOSs are convoluted with a 0.8 eV resolution function

Recently, also photoelectron energy distribution curves from Fe(100) above T_C have been calculated based on the DLM theory [3.45]. This will be discussed further in Sect. 3.8.2a.

d) Cluster Model

The three models discussed so far might be limiting cases. A unified theory has been developed by *Moriya* and *Takahashi* [3.46] which takes into account short-range correlations *and* longitudinal fluctuations of the magnetic moment. In practice, the theory has only been employed parametrized to realistic materials [3.47].

Another approach was developed by *Haines* et al. [3.48]. The magnet was modeled by a 2000-atom cluster. Short-range order was introduced parametrized, represented by the order parameter λ. Long-range order also was implemented to simulate temperatures below T_C. The electronic structure of this cluster was then calculated within the tight-binding approximation. In order to compare with recent spin-resolved photoemission data, a formalism to calculate photoemission spectra was also included. Calculated photoemission spectra for a series of short-range order parameters then enabled to determine the short-range order parameter by comparison with theories. This will be discussed further in Sect. 3.8.

e) Supercell Model

A similar approach was used quite recently by *Jarlborg* and *Peter* [3.49] modelling the magnetic structure above T_C by a periodic arrangement of "supercells" containing a variable amount of spin-flipped Fe layers. They calculate the DOS by a Linear Muffin-Tin Orbital approximation for various degrees of disorder.

3.3 Concepts of Spin- and Angle-Resolved Photoemission from Ferromagnets

3.3.1 Models for the Photoemission Process

Photoemission might be approximated by a single-step model treating photon absorption, the transport of the photoelectrons to the surface, and the escape into vacuum as a single step. In this approach the photocurrent is given by [3.50]

$$I(h\nu, E_i) \propto (E_f - E_v)^{1/2} \sum_{i,f} |\langle \Phi_f | \mathbf{A}\mathbf{P} | \Phi_i \rangle|^2 \delta(E_f - h\nu - E_i) \qquad (3.3)$$

where $\Phi_{i,f}$ are the initial (final) states with energies $E_{i,f}$, E_v is the vacuum energy, A is the light vector potential and $h\nu$ is the light energy.

Peaks in the spectra which are assigned to certain initial states generally have a finite width even in an experiment with perfect energy- and angle resolution. This is caused by the lifetime broadening due to the relaxation of the photoemission hole. The effect can be modelled by introducing a complex self-energy Σ_i and replacing the δ-function in (3.3) by the expression [3.50]

$$A_i = 1/\pi \, \text{Im} \, \Sigma_i(E)/\{[E_f - E_i - \text{Re} \, \Sigma_i(E)]^2 + \text{Im} \, \Sigma_i(E)^2\} \qquad (3.4)$$

The self-energy Σ_i has the effect of broadening and shifting of the peaks in photoelectron energy distribution curves (EDCs) as compared to one-particle binding energies.

Within the one-step model, energy distribution curves can be calculated for specified conditions, and very good agreement with experimental data has been obtained. If, as the basis for the calculations, parametrized electronic structures are used, the parameters (as the exchange splitting) can then be determined by comparing with the experimental data.

Unlike the more approximative three-step model, which treats the processes as photoexcitation, transport and escape into vacuum separately, the one-step model shows, in agreement with experimental data, that even when no energetically allowed final state is available, photoemission occurs nevertheless (band gap emission).

3.3.2 Experimental Methods to Test the Band Structure

a) Threshold Photoemission

Because of the difficulties with magnetic fields at the sample, no angle and (electron) energy resolution have been achieved in the first spin-polarized photoemission experiments. Rather, the spin polarization of the total photoyield has been measured as function of photon energy. However, even in this case information on the initial state can be obtained when using light with energy close to the photothreshold value and employing a single-crystalline sample. The photocurrent sets in when $h\nu = \Phi$, Φ being the work function of the sample. At $h\nu = \Phi$ the electrons can leave the solid only perpendicular to the surface. The energy distribution is also very narrow. It is then known that the initial state is at E_F, and that the internal k vector is on a line in the Brillouin zone corresponding to the surface normal direction. However, with increasing photon energy, the emission angle opens up quickly ("escape cone effect"), and the initial state becomes gradually more difficult to identify.

b) Angle-Resolved Photoemission

In spin- and angle-resolved photoemission the photoelectrons are analyzed for their emission angle α, their kinetic energy E, and their spin σ. The photoelectrons are created by irradiating the solid with monochromatic light of energy $h\nu$. The width of the energy distribution curves is $h\nu - \Phi$, Φ being the work function of the sample. The energy distribution will then look schematically as shown in Fig. 3.3. The primary photoelectrons are observed on the high-energy side of the energy distribution. Somehow they reflect their initial quantum state inside the solid at the detector in vacuum. The most prominent feature is the peak of secondary electrons at low kinetic energy, which is due to cascade processes created by the primary photoelectrons. In the intermediate energy range, for certain conditions, structures due to Auger electrons are observed.

The interesting properties of the primary electrons as they appear at the detector are their kinetic energy E, their polar and azimuthal angles of emission (Θ and Φ), and their spin σ (see Fig. 3.4a). It will be seen that these quantities reflect the state of the electron within the solid, and even without peforming a full quantum mechanical calculation to interpret the photoemission data, knowledge on the electronic structure can be obtained just by "looking" at the experimental data.

From the kinetic energy E, the binding energy E_B is determined as $E_B = E_F - E$, E_F being is the Fermi energy. The Fermi energy is determined from the high-energy cut-off of the energy distribution curve, provided states at E_F are actually observed in the EDC. In the absence of inelastic scattering effects (which is largely true for the primary photoelectrons), there is a one-to-one correspondence between the initial state energy and peaks in the EDC.

The internal wave vector k^{int} of the electron has also to be determined for testing calculated E(k) dispersions. This is not always straightforward because of electron diffraction effects at the potential jump at the surface. It is only the

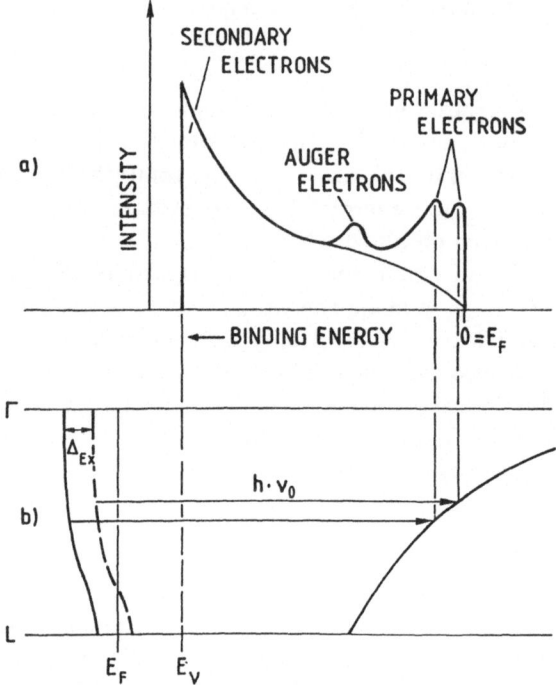

Fig. 3.3a,b. Model for angle-resolved photoemission from solids. (a) Typical photoelectron energy distribution curve as obtained with photon energies between roughly 10 and 100 eV energy from a transition metal (schematic). Contributions from primary photoelectrons and scattered (secondary) electrons are indicated. When the photon energy exceeds a core-level excitation threshold, Auger electrons might also be observed. (b) Explanation of the primary photoelectron peaks in angle-resolved photoemission in terms of direct transitions in a band structure picture. Emission from a single-crystal is assumed, resulting in selection of initial states along a specific direction in k-space. $h\nu$ is the monochromatic light energy, Δ_{ex} the exchange splitting of a metal as Fe, Co, or Ni. E_F is the Fermi energy, E_V is the vacuum level

component $k_\parallel$ parallel to the surface which is conserved (Fig. 3.4a) for smooth surfaces [3.51]. For normal emission, however, $k_\parallel$ is zero, and diffraction at the surface does not change the emission angle. For normal emission, the direction of k^{int} inside the solid therefore coincides with the surface normal direction. Knowing the final-state dispersions, the value of k^{int} then follows from the law of energy conservation (Fig. 3.3). Because of ease of interpretation, we will discuss in the following paragraphs mainly normal-emission data. It is evident from Fig. 3.3 that in the direct-transition model (conservation of k,) the initial state k-vector can be selected by chosing light of appropriate energy. This shows the advantage of using synchrotron radiation, which is tunable over a wide photon energy range if suitable optical monochromators are available.

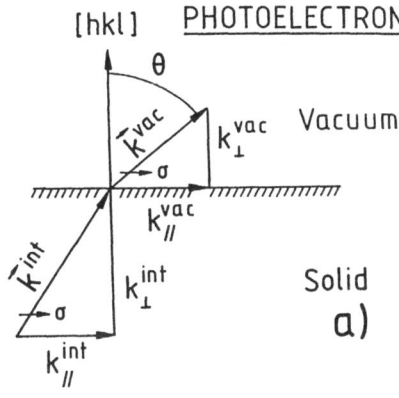

[hkl]   PHOTOELECTRON

Fig. 3.4a,b. (a) Diffraction of the photoelectron wave with wave vector k and spin
vector σ at the surface of a smooth single-
crystal. (b) Geometrical parameters of the
light beam with vector potential A

Frequently, a free-electron-like dispersion [3.52] is assumed for the final
state. It will be shown that some knowledge on the final state dispersions
exceeding the free-electron approximation can be obtained by measuring the
absorbed electron current to the sample as function of electron energy by employing a low-energy electron gun [3.53]. The absorbed current shows relative
minima when the energy coincides with band gap energies at the zone boundaries.

Methods have been worked out to determine k^{int} absolutely. We mention
here the "zero slope" method [3.54] which makes use of the fact that for group-
theoretical reasons the binding energy has an extremal value as function of
emission angle for internal k-vectors along high-symmetry directions.

Often bands of different symmetry are overlapping in energy and wave
vector. They might be distinguished by using linearly polarized light and ex-
ploiting dipole selection rules. Therefore, also the geometry of the incident light
has to be considered for the interpretation of the data (Fig. 3.4b). The dipole
selection rules allow only bands of certain symmetry to be detected [3.55].

3.3.3 Description of Spin Polarized Electrons

Electrons emitted from ferromagnets are generally spin-polarized due to the
ferromagnetic exchange splitting. It will be shown later that the spin of the

primary photoelectrons is actually conserved in the photoemission process for elements as Fe, Co, and Ni with only minor spin-orbit interaction.

The spin direction is quantized with respect to the direction of the magnetization. We describe the spin polarization of an ensemble of electrons as

$$\boldsymbol{P} = \mathrm{tr}\varrho\boldsymbol{\sigma} \qquad (3.5)$$

where σ is the Pauli spin matrix vector and ϱ is the spin density matrix. Often (and mostly throughout this article) it suffices to define the degree of polarization as the difference of the numbers $(N^\uparrow, N^\downarrow)$ of electrons with their magnetic moments parallel and antiparallel to the quantization axis, devided by the sum of the numbers, e.g.

$$P = (N^\uparrow - N^\downarrow)/(N^\uparrow + N^\downarrow) \qquad (3.6)$$

By definition, the spin polarization is positive when electrons with their magnetic moments parallel to the internal magnetization direction are dominating.

Instead of discussing the spin polarization as a function of binding energy E, it is often more instructive to consider the spin-resolved energy distribution curves (SREDCs) $N^\uparrow(E)$, $N^\downarrow(E)$. Knowing the spin-sensitivity $1/S$ of the spin analyzer (S is referred to as the "Sherman function"), the spin-resolved EDCs are obtained from the left and right count rates of the spin detector as

$$N^{\uparrow,\downarrow}(E) = 1/2\{N_0(E) \pm 1/S[N_1(E) - N_2(E)]\} \qquad (3.7)$$

where N_1, N_2 are the count rates in the spin-sensitive electron detectors, and $N_0 = N_1 + N_2$. The spin-sensitivity has to be determined by performing an extrapolation of the asymmetry $(N_1 - N_2)/(N_1 + N_2)$ towards zero thickness of the gold target foil.

In discussing the spin-resolved EDCs obtained at elevated temperatures, it is important to be aware that only the projection of the local spin polarization onto the spin-sensitive direction(s) of the spin detector is measured. The local magnetization might fluctuate in amplitude ("longitudinal fluctuations") and in angle ("transverse fluctuations") at elevated temperatures. This results in characteristic changes in the spin-resolved EDCs at high temperatures, to be discussed in Sect. 3.8.1.

3.3.4 Surface Sensitivity

The excited photoelectrons interact strongly by elastic and inelastic scattering processes with the electron system of the solid. Therefore, the primary photoelectrons which might have been created deep inside the solid due to the large penetration depth of the light, are attenuated drastically on their way towards the surface. Only those which have been created close to the surface will be detected in vacuum. The escape depths depends on the kinetic energy [3.56]. Its minimum value of about 5 Å or about 2 atomic layers is obtained at 30–80 eV electron energy. About 40 % of the photoelectrons would then originate from the first layer assuming exponential attenuation. This shows that generally bulk- and surface effects are intermixed in the photoelectron spectra. Calculations of the electronic structure indicate that the bulk electronic structure is approached very close below the surface (neglecting surface reconstruction). The break in crystal symmetry at the surface results mainly in a modification of the electronic structure of the topmost layer. Therefore, by photoemission, a quantity which is close to the bulk electronic structure is generally sampled. This has been worked out in detail by comparing angle-resolved photoemission data from Cu with calculated band structures [3.10,11]. However, for certain photoelectron wave vectors and binding energies, surface states have been found to contribute strongly to the photoemission intensity.

3.4 Apparatus for Spin- and Angle-Resolved Photoemission

An apparatus for spin- and angle-resolved photoemission consists basically of four major parts: The sample and its preparation stage, a light source, an electron spectrometer, and a spin analyzer, see Fig. 3.5.

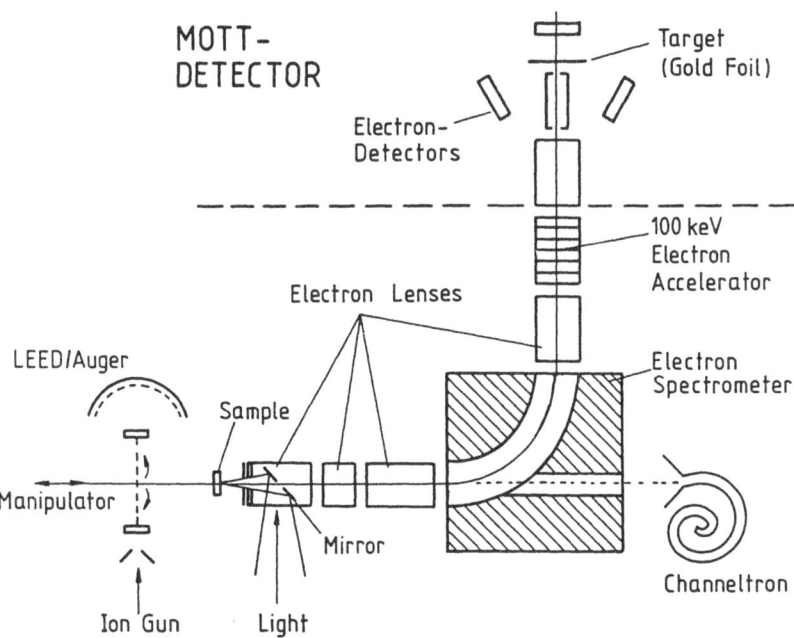

Fig. 3.5. Schematics of an apparatus for spin-resolved photoemission

3.4.1 Ferromagnetic Samples and Their Preparation

It is important that the sample is magnetically saturated within the probed area since otherwise spin-effects would be averaged out. In the first experiments, the samples had therefore been immersed in a strong magnetic field [3.15] parallel to the electron extraction direction ("longitudinal magnetization configuration"). This for electron optical reasons is not the desirable arrangement since it prevents resolution of the angle of emission and analysis of the kinetic energy [3.57]. (However, it has the advantage of providing magnetic saturation of almost any kind of magnetic material.) It turned out that samples magnetized perpendicular to the electron-optical axis (transverse magnetization geometry) actually provide the only practicable sample configuration for spin- and energy resolved photoemission experiments [3.19,20]. Since many materials can be magnetized *remanently*, magnetic stray fields which would deflect the photoelectron trajectories can be reduced to a neglible amount by shaping the samples suitably. Sample shapes with a high magnetic remanence are long, thin plates, picture-frame shaped closed magnetic loops (see Fig. 3.6), or evaporated thin films, which all have been employed in spin- and energy resolved photoemission from various materials. A remanence of 80–100 % of the saturation magnetization in the remanent state is easily obtained with this kind of samples. Epitaxially grown thin films can indeed be magnetized remanently as a single Weiss domain.

The sample preparation stage has to provide an ultrahigh vacuum environment and to allow to clean the sample surface by ion etching and annealing cycles. For obtaining a clean and most perfect surface, the samples might have to be heated to temperatures exceeding the Curie temperature. Even if the sample would have been magnetically saturated before heating, it would be demagnetized after cooling down. Provisions have therefore to be made to remagnetize the sample *in situ* after heating. This is performed either by bringing the sample temporarily close to a permanent magnet [3.19] or to a small coil

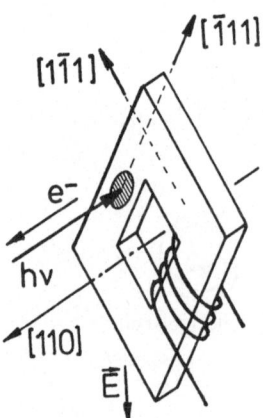

Fig. 3.6. The "picture-frame"-configuration (here Ni(110)) for obtaining remanently magnetically saturated samples with a minimized external magnetic stray field. The coil serves for magnetizing the sample by applying a current pulse

which is mounted inside the preparation chamber. A current pulse of length of about 1 ms through this coil suffices then to saturate the sample magnetically or to reverse the magnetization (which is generally done for verifying the observed effects). If the samples need not be heated, they might be magnetized prior to insertion into vacuum, as in studies on thin epitaxial films of Fe(110) and bcc Co(110) on GaAs [3.58,59].

3.4.2 Light Sources and Electron Spectrometers

Various types of light sources have been employed: High-pressure arc lamps, rare-gas resonance lamps, and synchrotron radiation. Arc lamps are of advantage in photothreshold experiments, whereas for light of higher energy ($>10\,\mathrm{eV}$) synchrotron radiation is often preferable due to its tunability, its polarization, and "cleanliness" (emerging from the storage ring under ultra-high vacuum). Synchrotron radiation offers a broad energy spectrum from the infrared to the order of 1000 eV (depending on the storage ring specifications). The radiation has therefore to be monochromatized. For this purpose, "beam lines" with optical monochromators are attached to the storage ring (for a description of a beam line used in spin-resolved experiments, see [3.60]). The beam lines are optimized for a certain optical band-pass, and differ in design considerably depending on the desired photon energy range.

The electron spectrometer serves for angle and energy analysis and is frequently based on a spherical deflector [3.20,62]. Energy- and angle resolutions of about 0.1 eV and of the order of a few degrees, respectively, have been achieved.

3.4.3 Spin Detection

In spin-resolved photoemission experiments on ferromagnets, a Mott detector is used for spin analysis. Electrons are accelerated to about 100 keV energy and scattered on a gold foil of thickness of about 100 nm (for details, see e.g. [3.61,63]). Due to spin-orbit interaction, a left-right asymmetry ($N_1 \neq N_2$) occurs if the beam is spin-polarized perpendicular to the scattering plane [3.14]. The two electron detectors shown in Fig. 3.5 will then display different count rates. Any spurious asymmetry present also for unpolarized electrons has to be eliminated either by comparing with the asymmetry as observed with unpolarized electrons or by a test measurement after exchanging the Au foil for an Al target in which spin-orbit interaction is negligible. The polarization sensitivity $1/S$ (cf. Eq. 3.7), is usually determined by extrapolating the measured asymmetry A to zero film thickness by measuring it for a set of Au targets with different thicknesses. With infinitesimal small film thickness, only single scattering events occur for which the polarization sensitivity has been calculated [3.14].

With two electron detectors, as shown in Fig. 3.5, only one component of the spin polarization vector can be measured (which is perpendicular to the scattering plane). This is often sufficient since the magnetization direction can generally be aligned parallel to the spin-sensitive direction of the Mott detector. However, one has to be aware that under certain circumstances, the direction of the magnetization vector might not stay constant during the experiment. A reason might be a dependence of the magnetic anisotropy constants on such parameters as temperature or thickness of thin film samples. A rotation of the spin polarization vector with temperature has been observed in spin-resolved field emission from EuS on W in the vicinity of the Curie temperature [3.64]. Another example has been found in thin film studies of Fe(110) and GaAs(110) [3.65] where the easy magnetization direction changes from {110} to {100} at about 100 Å film thickness [3.66]. How to measure the three components of the spin polarization vector has been discussed elsewhere [3.67].

Due to the fact that measuring the spin polarization is a scattering experiment in itself, the count rate in the electron detectors is only 10^{-4} to 10^{-3} of the current entering the spin detector. Other (no more efficient yet) methods of spin detection are the absorbed-current detector [3.68] or the spin-polarized low-energy diffraction detector ("SPLEED"-detector) [3.69]. The Mott detector, once it has been successfully implemented, is of advantage because of its long-term reliable operation. Its characteristics are largely insensitive to the photoelectron energy and angle because of the 100 keV acceleration. However, there is little doubt that the Mott detector will be replaced gradually in the future by low-energy detectors which are smaller, easier to construct, and do not require the high voltage.

3.5 The Electronic Structure of Ni and Fe at Low Temperatures

3.5.1 Photothreshold Experiments

A number of materials have been investigated by spin-polarized photothreshold photoemission. The one studied most is certainly Ni, but Fe(110), Cr(100), and the half-metallic ferromagnet NiMnSb have also been investigated.

a) Ni

A prediction of the Stoner model for the ferromagnetism of Ni was that it is a "strong" ferromagnet. This expresses the fact that the majority-spin 3d-bands are fully occupied, their top being separated from the Fermi energy by the so-called "Stoner gap" δ. The Stoner gap has been estimated to be of the order of 0.1 eV in Ni and it is determined by $X_5^{\uparrow}$ (see Fig. 3.7). Since only the minority spin bands cross through E_F, the spin polarization from any Ni surface should start at a negative value when $h\nu = \Phi$. It should become positive at higher photon energies due to the loss of angle resolution and the dominant character of the majority-spin electrons within the DOS. This had been predicted by *Wohlfarth* [3.70,71] and has been verified in a pioneering experiment by *Eib* and *Alvarado* [3.72] (see Fig. 3.8).

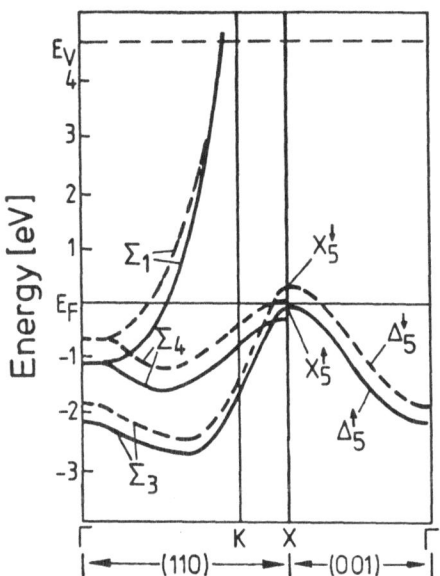

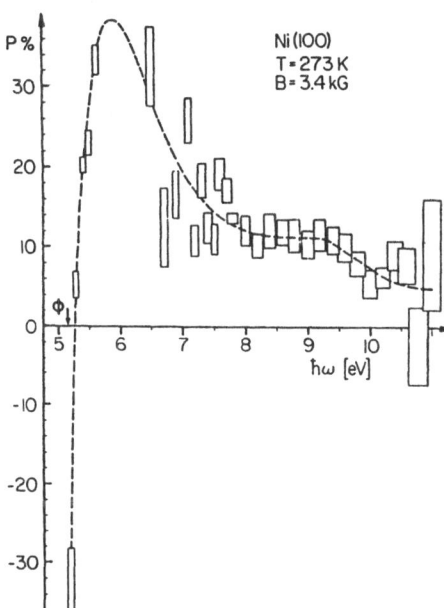

Fig. 3.7. Band structure of Ni along the Γ−K−X and X−Γ directions (after *Wang* and *Callaway* [3.74])

Fig. 3.8. The measured spin polarization vs. photon energy in a photothreshold experiment on Ni(100) (after *Eib* and *Alvarado* [3.72]). Φ indicates the photothreshold

A "rigorous" photoemission calculation has been performed by *Moore* and *Pendry* [3.73] treating the exchange splitting as a parameter. It turned out that the data actually *cannot* be reconciled with the value of the exchange splitting for Ni as calculated by LSDF theory, i.e. 0.6–0.8 eV [3.74]. Rather, a smaller value of only 0.3 eV had been inferred from the data by comparing with a photoemission calculation, treating the exchange splitting as parameter (see Fig. 3.9). The small value of the exchange splitting was consistent with subsequent spin-polarized photothreshold results on Ni(111) [3.75] and with data from angle-resolved and spin- *and* angle-resolved photoemission (see Sect. 3.5.2 for further discussion).

In both experiments mentioned on Ni(100) and Ni(111), the value of the spin polarization at photothreshold is about −50 %, although the Stoner model would demand for −100 %. The comparatively small observed value had been attributed to the limited bandwidth of the light which could not be reduced further because of intensity problems. Only in an experiment on Ni(110), the expected high spin polarization of close to −100 % has actually been observed [3.19], see Fig. 3.10. This was possible because of the larger separation of the top of the $\Sigma_4^\uparrow$-symmetry bands from the Fermi energy and because the Ni(110) work function is much smaller than that of the (100) and (111) surfaces, resulting in higher emission intensity which allowed reduction of the light bandwidth

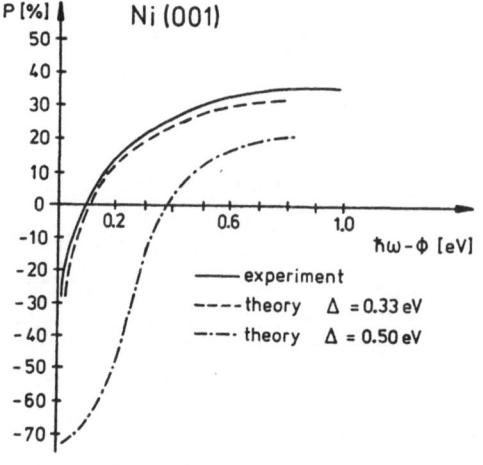

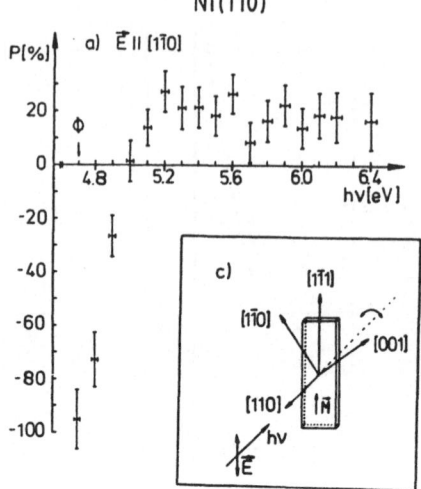

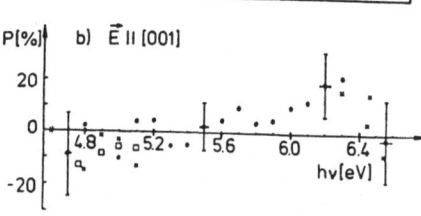

Fig. 3.9. Comparison of the experimental data on Ni(100) from Fig. 3.8 with the results obtained from a "rigorous" photoemission calculation with the ferromagnetic exchange splitting Δ as a parameter (after *Moore* and *Pendry* [3.73])

Fig. 3.10a–c. The measured spin polarization vs. photon energy in a photosthreshold experiment on Ni(110) with a transverse magnetized sample [3.19]. (a) Data with light electric vector E $\parallel$(110) (b) Data with electric vector E $\parallel$(100) (c) Geometric configuration of the sample with respect to the light polarization

and to completely resolve the minority-spin emission from initial states in the Stoner gap.

In the experiment on Ni(110), use also had been made of the geometry of the (110) face resulting in a dependence of the allowed initial state symmetry on the azimuth of the light electric vector. With E$\parallel$(110), initial states of Σ_4 symmetry are dipole allowed while with E$\parallel$(100), initial states of Σ_3 are allowed [3.55]. This manifests itself in the strong dependence of the observed spectra on the polarization direction of the light (see Fig. 3.10).

However, although the high negative spin polarization at the photothreshold for Ni(110) with E$\parallel$(110) is easily understood, it is less obvious to reconcile the shape of the P(E) curve with the calculated band structure. From the band structure (see Fig. 3.7) the crossover energy from negative to positive spin polarization should occur at energy exceeding 0.6 eV. On the other hand, if $X_2^{\downarrow}$ is very close to E_F (or slightly below) as inferred from angle-resolved and spin-resolved photoemission data [3.76, 23], a positive spin polarization is not expected at all at higher photon energies on the basis of emission from the two exchange-split Σ_4-symmetry bands alone. The unexpected small crossover energy of 0.33 eV and the observation of the comparatively large positive value of the spin polarization at higher photon energies has been interpreted by *Clauberg* [3.77] as caused by the opening of the escape cone with photon energy and the resulting pick-up of non-normal emission from states throughout the Brillouin zone. On the other hand, it remains a puzzle that the P(E) curve can be fitted perfectly within a two-level model with an exchange splitting of 0.3 eV and $X_2^{\downarrow}$ at 0.1 eV above E_F [3.78].

b) Fe and Ni-Fe Alloys

Spin-resolved photothreshold experiments on Fe(111) have been reported by *Eib* and *Reihl* [3.79] and were found to be in good agreement with the band picture of Fe. For a discussion of data from $Ni_{40}Fe_{60}$ and $Ni_{80}Fe_{20}$ alloys see [3.80,1].

c) Cr(100)

A very interesting study on the oxidation of Cr by spin-resolved photothreshold technique has been reported by *Meier* et al. [3.80]. Cr is antiferromagnetic, and no spin polarization should be observed for clean Cr if the escape depths of the photoelectrons is large. However, upon oxygen adsorption, spin polarization has been observed [3.80]. This had been interpreted as oxygen-induced ferromagnetism. Interestingly, the surface of Cr(100) has been predicted to be ferromagnetic [3.82,83]. Evidence for this prediction has been found by electron-capture experiments by *Rau* and *Eichner* [3.84]. Recent angle-resolved photoemission experiments on Cr(100) also have been interpreted in terms of surface ferromag-

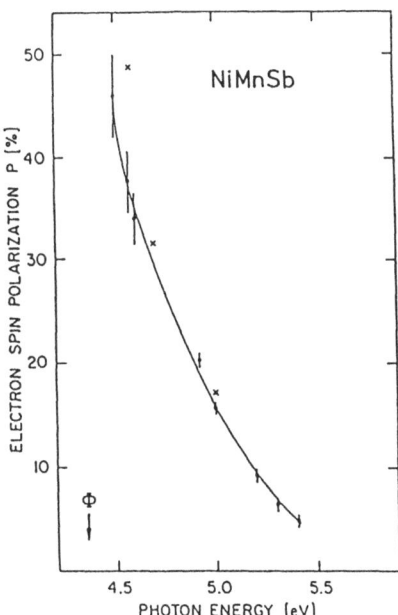

Fig. 3.11. Spin-resolved photoemission data in a photothreshold experiment from the Heusler alloy NiMnSb (after *Bona* et al [3.86]). ϕ indicates the photothreshold

netism [3.85]. The exchange split states might therefore be observable directly by spin- and angle-resolved photoemission.

d) NiMnSb

Recently, spin-resolved photothreshold spectroscopy has been employed to study the Heusler-alloy NiMnSb [3.86]. The data have been compared with predictions of the electronic structure of this half-metallic ferromagnet [3.87]. Instead of the expected +100 % spin polarization at the photothreshold, only 60 % has been observed, although the dependence on photon energy resembled closely the expected shape (see Fig. 3.11). It was concluded that the reduced spin polarization suggests a smaller gap in the minority-spin band structure than predicted.

3.5.2 Spin- and Angle-Resolved Photoemission at Low Temperatures

The photothreshold experiments have neither been explicitly angle- nor energy resolved, and the difficulty in interpreting the Ni(110) data demonstrates that it is actually necessary to improve the experimental capabilities towards explicit angle- and energy resolution employing light of higher energy, i.e. to extend the established technique of angle-resolved photoelectron spectroscopy by analyzing also the spin.

a) Spin-Split Electronic States of Ni

A spin-assignment to peaks in angle-resolved energy distribution curves (EDCs) from Fe, Co, and Ni has been made even without explicitly measuring the spin, guided by a comparison with bandstructure calculations. In N_i, discrepancies concerning the values of the exchange splitting (in agreement with the spin-resolved photothreshold work discussed in the previous section) and the d-bandwidth have been found (for a review see, e.g. [3.88]). It was shown further that the exchange splitting differs for states of e_g and t_{2g}-symmetry ($\Delta = 0.18$ and $0.3\,eV$, respectively), see Fig. 3.12 [3.76].

By spin-resolved photoemission, the exchange splitting of Ni at the X-point has been resolved for the first time explicitly by *Raue* et al. [3.23]. The data are shown in Fig. 3.13a. The exchange-split states ($X_2^\uparrow$, $X_2^\downarrow$, see Fig. 3.7) are clearly resolved, and the splitting is 0.18 eV in agreement with the earlier spin-averaged data. The data show furthermore that $X_2^\downarrow$ is very close to E_F.

The initial states relevant for Fig. 3.13a are of e_g character. With different azimuth of the light polarization, S_3-symmetry bands are tested which are mainly of t_{2g} character at X. Data obtained for this condition are shown in Fig. 3.13b. $X_5^\uparrow$ is found at 0.1 eV below E_F, and a peak from $X_5^\downarrow$ is not observed. Rather, the data suggest that it is located above E_F in agreement with the calculated band structure.

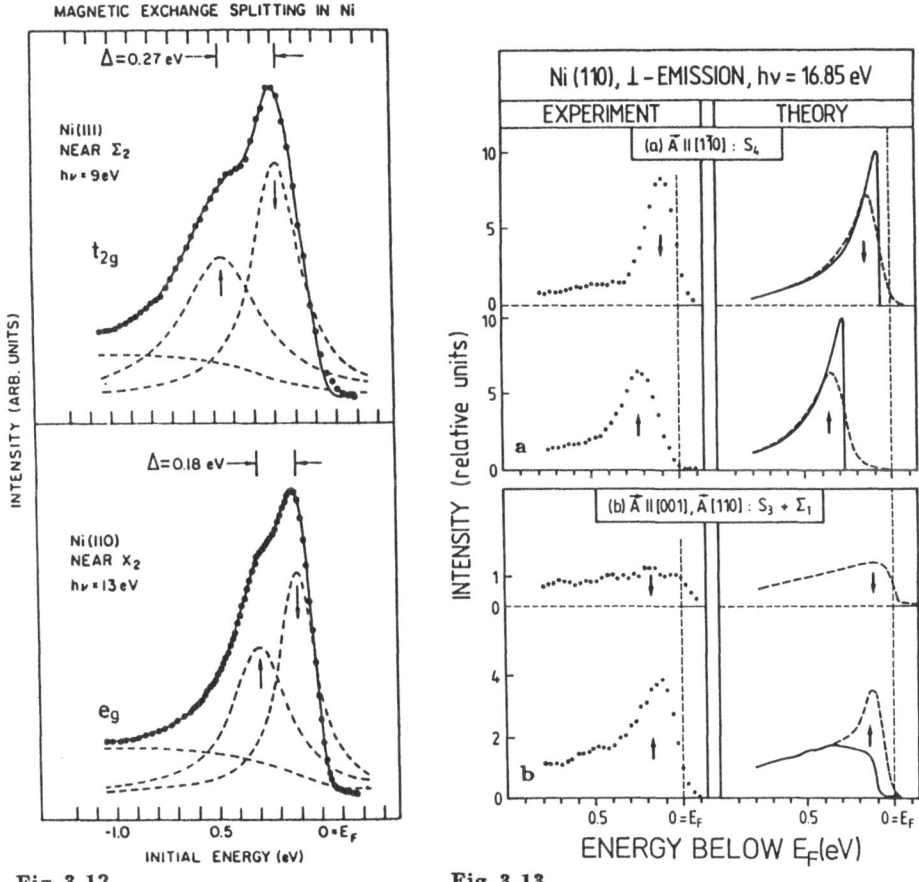

MAGNETIC EXCHANGE SPLITTING IN Ni

Fig. 3.12   Fig. 3.13

Fig. 3.12. Angle-resolved energy distribution curves for off-normal emission from Ni(111) (*top*) and for normal emission from Ni(110) (*bottom*). The dashed lines indicate a decomposition into Lorentzians representing the spin-split initial states. The initial states are of t_{2g} (*top*) and e_g (*bottom*) symmetry (after *Heimann* et al. [3.76])

Fig. 3.13a,b. Spin- and angle-resolved EDCs for normal emission from Ni(110) at room temperature for two orientations of the light electric vector. (a) $E \parallel (110)$, (b) $E \parallel (100)$ (after *Raue* et al. [3.23]). The experimental data in the left panels are compared with calculated curves in the right panels

A difference in exchange splitting between e_g and t_{2g} symmetry states is not inherent in the LSDF method. It has been shown by *Liebsch* [3.89] that the small value of the exchange splitting, its different values for states of e_g and t_{2g} symmetry, the experimentally observed reduced d-band width, and the occurrence of the 6 eV-satellite (to be discussed in Sect. 3.6) are a consequence of correlations among d electrons. *Oles* and *Stollhoff* [3.90] using a model Hamiltonian and non-spherical exchange and correlation terms also were able to show that the exchange splitting depends on the symmetry. The

77

values they obtained for the splitting are also considerably smaller than those obtained by standard LSDF theory.

It has been suspected that the *local* approximation might be a reason for the discrepancies between the calculated electronic structure and the experimental data. It has been shown by *Wang* [3.91] that, by introducing non-local corrections to the LSDF approximation, an exchange splitting of only 0.4 eV is obtained, and $X_2^{\downarrow}$ is found close to E_F. However, the d-band width is not reduced in this approach.

Recently, *Victora* and *Falicov* [3.92] have calculated an "exact" density of emitted states by taking into equal account bandstructure effects and electron-electron interactions. Very good agreement with the photoemission data is obtained in every respect.

Considerable progress has also been made in studying the unoccupied bands of Ni by spin-polarized and spin-averaged inverse photoelectron spectroscopy [3.93,94]. In Fig. 3.14, spin-resolved inverse photoemission intensities for different angles of incidence of the polarized electron beam on Ni(110) are displayed as obtained by *Unguris* et al. [3.93]. The large density of states of minority-spin holes above E_F (see Fig. 3.7) results in a strong yield of photons when the electron beam is ↓-spin-polarized. The peak occurs 0.3 eV above E_F for $\Theta = 0°$, in very good agreement with the band calculations.

Due to the discrepancies between peak positions in the photoemission data and the calculated band energies, two band structures have to be kept in mind

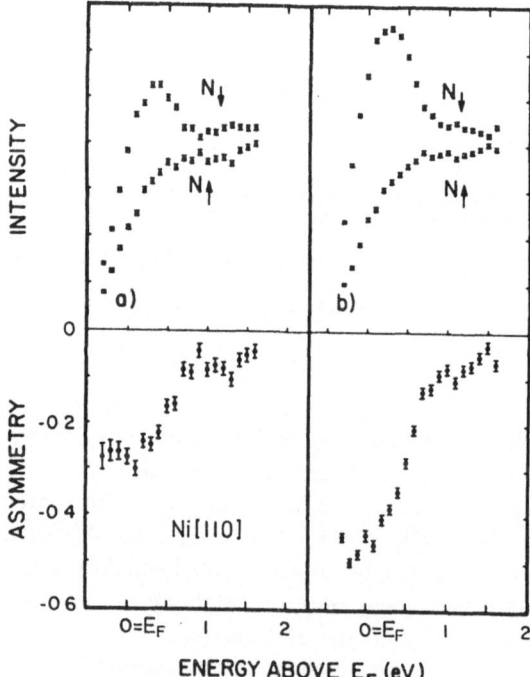

Fig. 3.14a,b. Spin-resolved inverse photoemission data on Ni(110). Shown in the top panels is the photon yield as function of electron energy (referred to the Fermi energy) of a spin-polarized electron beam for opposite spin orientations. (a: $\Theta = 0°$, b: $\theta = 20°$) (after *Unguris* et al. [3.93])

for Ni: The "experimental" and the calculated ones. For summaries on the observed discrepancies see *Callaway* [3.88] and *Martensson* and *Nielsson* [3.95].

b) Spin-Split Electronic States of Fe(100)

The bandstructure for high-symmetry directions of Fe as calculated recently by *Hathaway* et al. [3.25] is shown in Fig. 3.15. The Fe electronic structure has been tested along high-symmetry directions by angle-resolved photoemission [3.96–100]. The conclusion was that there is generally good agreement between experimental peak positions and calculated energy bands [3.98,100]. In a more global test as valence band XPS provides, data on Fe(100) by *Kirby* et al. [3.101] compared well with the calculated ferromagnetic Fe DOS.

In a first spin- and angle-resolved experiment on Fe(100), the states $\Gamma'^{\uparrow}_{25}$, $\Gamma'^{\downarrow}_{25}$, and $\Gamma^{\uparrow}_{12}$ have been partially resolved [3.102]. Data obtained at BESSY (W. Berlin) with improved energy- and angle resolution on the same crystal are shown in Fig. 3.16 [3.103]. The minority-spin EDC displays only one peak at 0.4 ± 0.2 eV binding energy, while the majority-spin EDC displays two broader

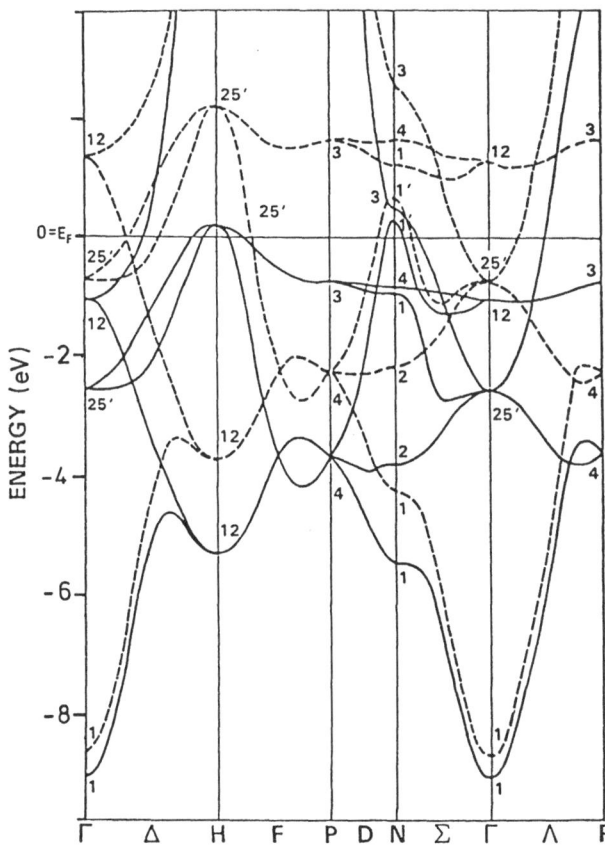

Fig. 3.15. Calculated band structure of Fe along high-symmetry directions (after *Hathaway* et al. [3.25])

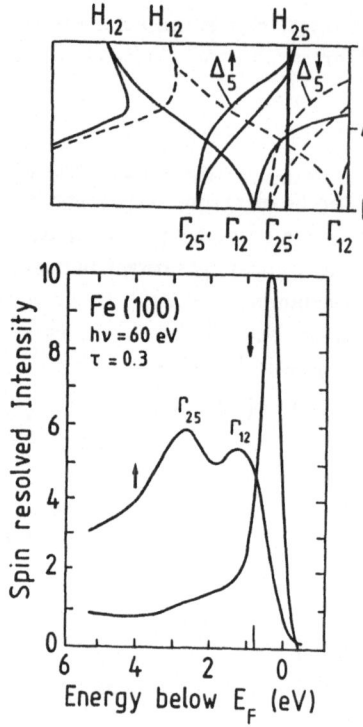

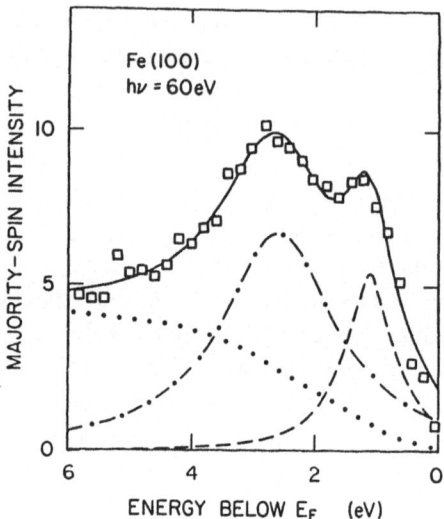

Fig. 3.17. Fit of the majority-spin photoelectron EDC from Fe(100) at 60 eV photon energy with two Lorentzians and an inelastic background. The solid line is the sum of the two Lorentzians and the background.

Fig. 3.16. Spin- and angle-resolved photoelectron EDCs from Fe(100) for emission centered around the surface-normal direction with predominantly s-polarized light at 60 eV photon energy. (Averaging spline-fits to experimental data similar to those shown in Fig. 3.32). Top: Comparison with a calculated band structure (after *Callaway* and *Wang* [3.104])

peaks, one at 1.2 eV, and a second one at 2.6±0.2 eV. A fit of the majority-spin EDC by two Lorentzians and an inelastic background as the sum over the Lorentzians (normalized to fit the intensity at the low-energy side) is shown in Fig. 3.17. The initial state k-vectors are along $\Gamma - H$ because of normal emission from the Fe(100) surface.

To identify the initial state k vector the final state dispersion has to be known. Often free-electron-like states are used to interpret the photoemission data. It has already been pointed out in Sect. 3.1.2 that a method to determine the final state dispersion is *absorbed electron current spectroscopy*. Such data, obtained on Fe(100), are shown in Fig. 3.18a [3.53]. The final state dispersion as determined from this data is shown in Fig. 3.18b. It is seen that at 60 eV photon energy, direct phototransitions occur near Γ. It is expected furthermore that at lower photon energies, direct transitions throughout the Brillouin zone will occur, allowing the dispersion of the Δ_5-symmetry bands across the Brillouin zone to be followed. Actually, due to lifetime effects, the lines should be interpreted as the centers of bands several eV wide. This diminishes the

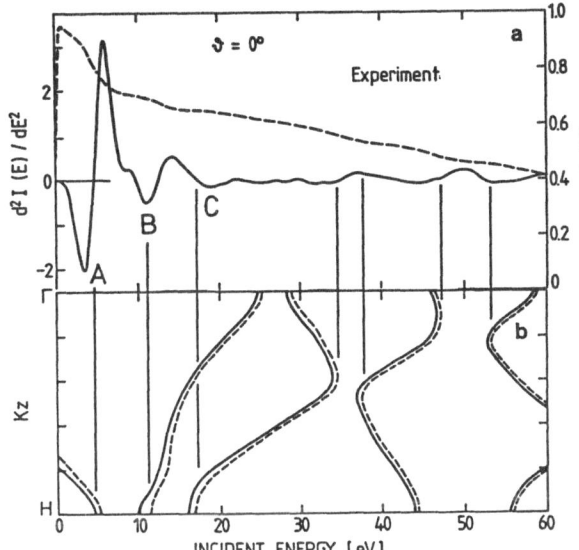

Fig. 3.18. *(a):* (- - -) The electron current absorbed by a Fe(100) single crystal at normal incidence as function of primary electron energy (after *Tamura* et al. [3.53]) (—) second derivative. Zero energy corresponds to the vacuum energy. *(b):* Band structure along $\Gamma - H$ above the vacuum energy as determined from the absorbed electron current data

difference in the initial state k-vectors of exchange-split states. The absorbed electron current measurements have also allowed to determine the imaginary part of the inner potential [3.53].

From the final state dispersion (Fig. 3.18), it is concluded that the initial states are close to $\Gamma'^{\downarrow}_{25}$, $\Gamma^{\uparrow}_{12}$, and $\Gamma'^{\uparrow}_{25}$ (cf. Fig. 3.16). The difference in binding energy between peaks labeled as $\Gamma'^{\uparrow}_{25}$ and $\Gamma'^{\downarrow}_{25}$ is the exchange splitting near Γ which appears to be 2.2 eV in the raw data. The comparison with calculated band structures [3.25,27,104] reveals agreement within 0.5 eV between the peak positions and critical point (Γ) energies. The data have also been analyzed within a full photoemission calculation based on a ferromagnetic potential by *Feder* et al. [3.105].

In view of the dipole selection rules, emission from $\Gamma^{\uparrow}_{12}$ is forbidden. That it appears strongly in the EDCs (see Fig. 3.18) might have one of three reasons: the limited angular resolution, the imperfection of the crystal surface, or relativistic effects [3.106]. Due to the high density of states of $\Gamma^{\uparrow}_{12}$, even small imperfections in the experimental conditions will give a large contribution from this state. For a similar reason as $\Gamma'^{\uparrow}_{12}$ is observed, an indication for $H^{\downarrow}_{12}$ is seen in the low-energy tail of the ↓-EDC at about 3.3 eV binding energy (see Fig. 3.16).

Several remarks are necessary at this point. To a good approximation, the data can be explained in terms of the bulk band structure. However, a quantitative interpretation imposes difficulties. This is because the minority-spin peak is very sharp and close to E_F. The experimental resolution (≈ 0.4 eV) and the effect of the Fermi energy cut-off have to be considered. Fitting the minority-spin EDC with a Lorentzian and considering the cut-off at E_F and the effect of the experimental resolution, the positions of the exchange split minority-spin and majority-spin peaks are determined to be at 0.07 and 2.5 eV, respectively. The splitting (2.4 eV) is considerably larger than that predicted by self-consistent calculations (1.8–2.1 eV at Γ [3.25,27,104]). In view of the surface sensitivity at 60 eV photon energy (which has been inferred directly from measuring the dependence of spin polarization on temperature [3.103], see Sect. 3.8.2), we might therefore conclude [3.107] that the data exhibit the predicted enhanced exchange splitting of surface resonances at $\bar{\Gamma}$, the center of the 2D Brillouin zone [3.108]. Actually, the positions of the exchange split states as determined when considering the finite energy resolution agree quantitatively

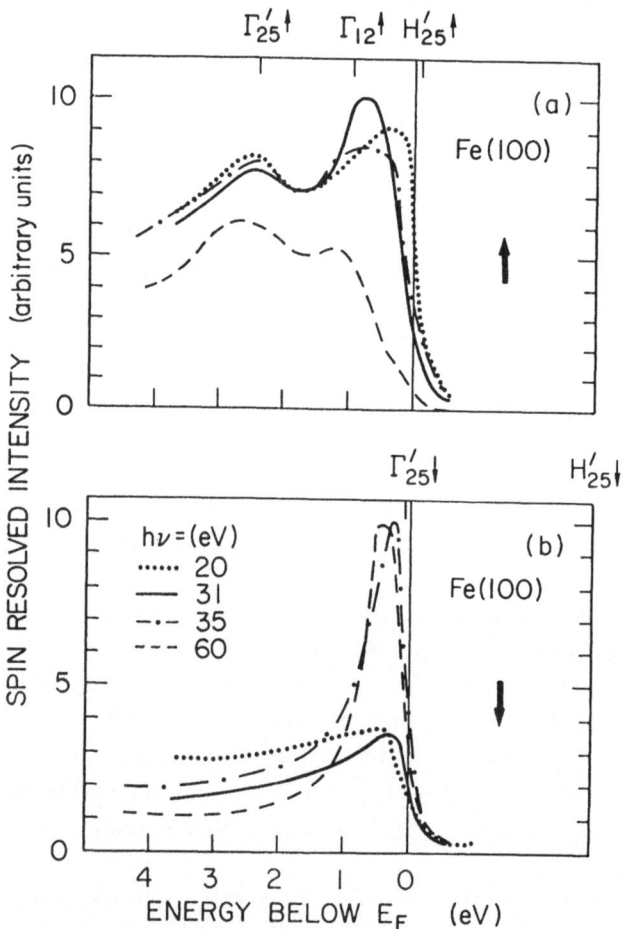

Fig. 3.19a,b. Spin- and angle-resolved photoelectron EDCs from Fe(100) with the photon energy varied between 20 and 60 eV. *Top:* majority-spin EDCs. *Bottom:* Minority-spin EDCs. In each pair of majority- and minority-spin EDCs, the height of the larger peak (majority or minority) is normalized to 10 units. Indicated are also critical point energies of the sampled Δ_5-symmetry bands (after *Kisker* et al. [3.103])

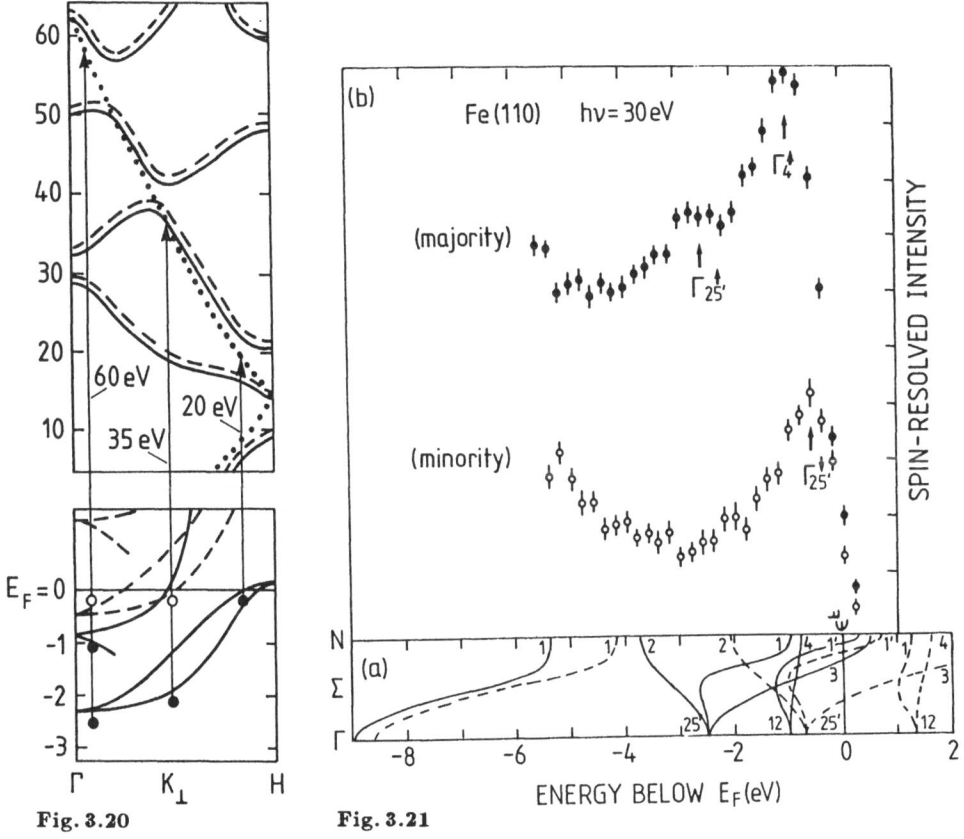

Fig. 3.20. Identification of the k-vectors along $\Gamma - H$ in a direct transition model for the spin-resolved EDCs shown in Fig. 3.19. The dots represent free-electron final states

Fig. 3.21. Spin-resolved photoelectron energy distribution curves from Fe(110) at 30 eV photon energy and comparison with calculated energy bands along $\Gamma - N$ [3.58]

with calculated band energies at $\overline{\Gamma}$ which are concentrated to 70 % in the first two layers [3.108]. It had also been suggested by *Durham* et al. [3.45] that the large splitting might be caused by an enhanced surface exchange splitting.

It has been pointed out by *Feder* et al. [3.102] that self-energy effects as discussed in Sect. 3.3.1 could result in a shift of the position of $\Gamma'^{\uparrow}_{25}$ downwards in energy in the photoemission spectra (which would be opposite to the shifts in Ni). Better understanding of self-energy effects in photoemission is certainly necessary.

Strong changes occur in spin-resolved EDCs taken at photon energies between 20 and 60 eV, see Fig. 3.19 [3.24, 103]. Below 33 eV photon energy, the minority-spin intensity drops suddenly, and the dominating peak in the SREDCs becomes a majority-spin peak. This indicates that at 33 eV photon energy, the initial state k-vector passes the first half of the Brillouin zone where the minority-spin Δ_5-band crosses through E_F. The initial and final states corresponding to Fig. 3.19 are indicated in Fig. 3.20. It appears that the strongest transitions occur to the free-electron-like final state.

We note that the spin-summed EDCs, in contrast to the spin-resolved ones, are rather independent on photon energy. They always display a leading peak slightly below E_F, and a broader peak near 2.6 eV binding energy. This has lead in the past to the assumption that band dispersions are not observed in Fe [3.96,97].

Recently, also the Fe(110) surface of epitaxially grown films on GaAs has been investigated by spin-resolved photoemission [3.58]. The films had not been grown *in situ* but rather been cleaned by mild Ne$^+$ ion sputtering prior to measuring the spin-resolved EDCs. The data (Fig. 3.21) were found to be in very good agreement with the calculated bulk band structure of *Hathaway* et al. [3.25] along the $\Gamma - N$ direction (cf. Fig. 3.15).

3.6 Resonant Photoemission and Auger Electrons

A well-known feature in energy distribution curves from Ni is the so-called 6 eV satellite, which has been observed by *Hüfner* and *Wertheim* in x-ray photoelectron (XPS) EDCs [3.110], and by *Guillot* et al. in ultraviolet photoemission (UPS) [3.111]. This structure had attracted a considerable amount of interest because of its potential relation to the Ni many-body ground state, and has already been mentioned in Sect. 5.2 in context with the observed discrepancies between photoemission data and the ground state electronic structure calculations. It might suffice here to note that the structure is considered to represent the 3d^8 final state configuration with two holes in the 3d shell of a single Ni atom (see, e.g. *Penn* [3.112]). This interpretation can be followed back to a suggestion by *Mott* [3.113]. The interest was revived when it was found that this structure in photoelectron EDCs increases strongly in intensity when the photon energy is tuned across the 3p $-$ 3d transition threshold around 65 eV (see Fig. 3.22).

Exciting a 3p-core-electron into an unoccupied d-state will result in a resonant Auger transition, with also a final 3d^8 configuration according to the reaction (see Fig. 3.23).

$$3p^6 3d^9 \rightarrow 3p^5 3d^{10} \rightarrow 3p^6 3d^8 + e^- \tag{3.8}$$

It has been assumed here that the 3d^9 configuration is dominating in the ground state of Ni. Because of the same final state (d^8), these Auger electrons should have similar properties as the off-resonance 6 eV satellite.

It has been suggested by *Feldkamp* and *Davis* [3.114] that the emitted electrons should be highly spin polarized in an atomic picture because of the dominating ^{1}G term in the 3d^8 multiplett and because only minority-spin elec-

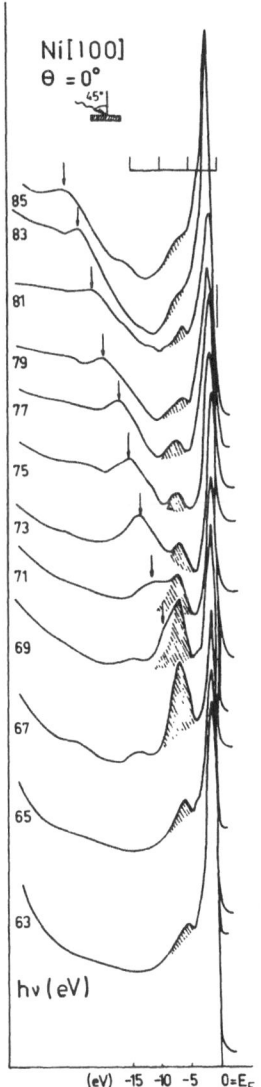

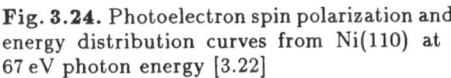

Ni[100]
Θ = 0°

85
83
81
79
77
75
73
71
69
67
65
63

hν (eV)

(eV) -15 -10 -5   0=E_F

Fig. 3.22. Photoelectron energy distribution curves from Ni(100) for photon energies between 63 and 85 eV (covering the 3p − 3d transition threshold) (after *Guillot* et al. [3.111])

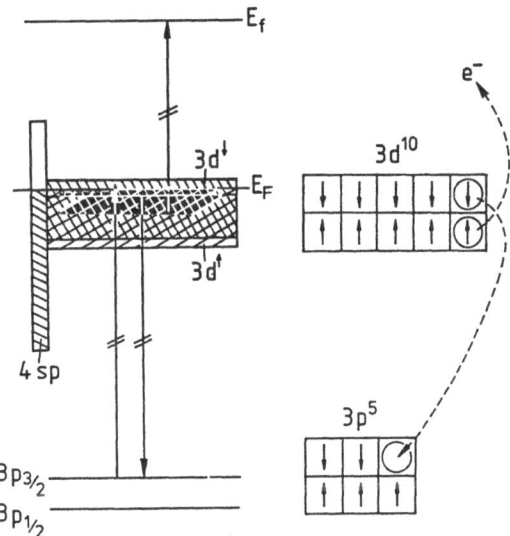

Fig. 3.23. Model for the resonant Auger transition in Ni (after *Clauberg* et al. [3.22])

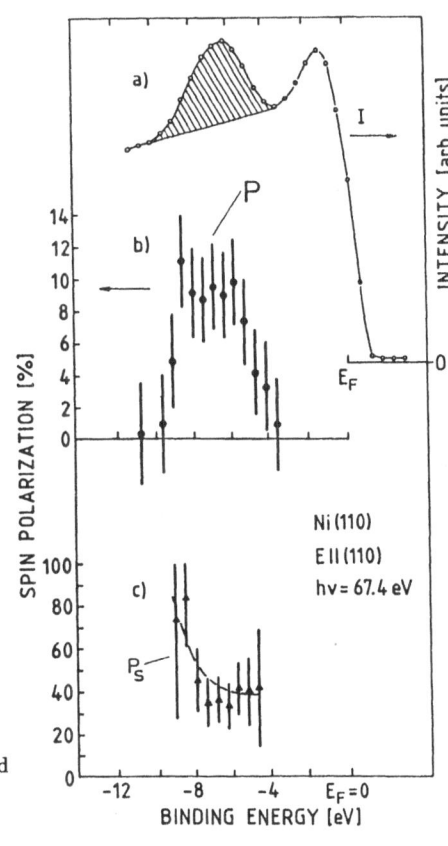

Fig. 3.24. Photoelectron spin polarization and energy distribution curves from Ni(110) at 67 eV photon energy [3.22]

trons can be excited to the empty d-electron DOS above E_F (see Fig. 3.23). This prediction has been verified in a spin-resolved photoemission experiment by *Clauberg* et al. [3.22] (see Fig. 3.24). It was found that the resonant structure is spin-polarized in excess of the average valence band polarization (6 %) in qualitative agreement with the predictions by *Feldkamp* and *Davis* [3.114].

The resonance of the 6 eV satellite and its spin polarization have been explained in an atomic picture. However, the close relation of the satellite to the MVV Auger decay is obvious from Eq. 9. If the state "after" photoexcitation is fully screened, the 3p-hole would recombine with a more or less undisturbed valence electron, and a similar valence electron would be ejected. The energy distribution would then resemble the self-convolution of the valence band DOS. To which extent the polarization of the Ni 6 eV satellite for photon energies exceeding the 3p-threshold can also be interpreted within the MVV Auger picture, is presently under investigation.

Similar as in Ni, a resonant satellite has been observed by *Chandesris* et al. in Fe(110) [3.115]. Here, the resonance occurs at about 53 eV which is the 3p-electron binding energy in Fe. It has been shown that the resonant feature actually can be explained well in the above mentioned MVV Auger picture [3.116,117]. The resonant photoemission intensity has been found to be centered at constant kinetic energy, suggesting that the structure is dominated by the Auger decay channel. A difference from Ni is also that a valence band satellite is not observed in valence band x-ray photoelectron spectra [3.101].

In Fig. 3.25, the spin-averaged EDC from Fe(100) taken at 56 eV photon energy is shown together with another one taken at 60 eV photon energy where

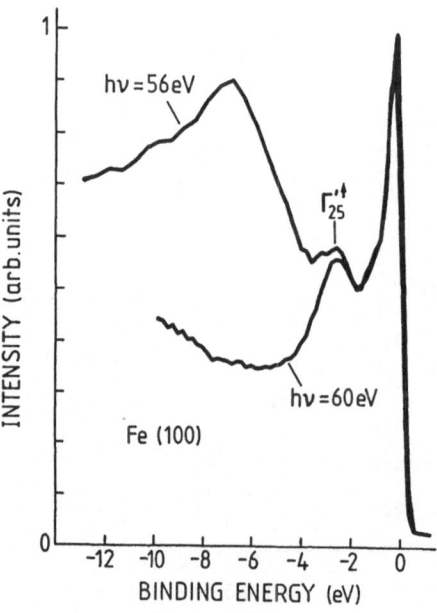

Fig. 3.25. Photoelectron energy distribution curves from Fe(100) at 56 eV and 60 eV photon energies [3.116]

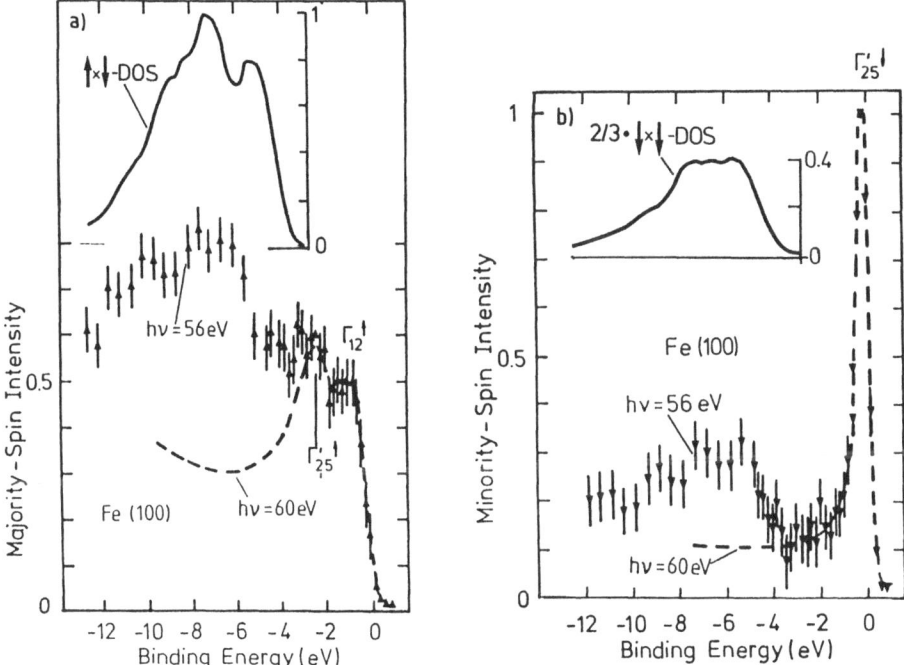

Fig. 3.26. (a) Majority-spin photoelectron EDCs from Fe(100) at 56 eV and 60 eV photon energies. (b) As above, but minority EDCs are shown. As insets, convolutions of the spin-split DOS are shown for comparison with the resonant intensity centered near 7 eV binding energy [3.116]

the Auger peak has moved to lower binding energy for estimating the resonant contribution. The spin-polarization analysis (see Fig. 3.26a,b for the spin-resolved EDCs) shows that the resonant photoemission intensity is related to convolutions of the spin-split DOS, the majority-spin intensity represented by the convolution of the majority-spin partial DOS with the minority-spin partial DOS, whereas the minority-spin distribution resembles closely the autoconvolution of the minority-spin DOS [3.116]. However, its intensity is smaller by a factor of about 2/3 than expected from the selfconvolution. This is explained by the following arguments: Auger recombination requires that the two electrons are close to the ion core. Because of the Pauli exclusion principle, this is unlikely if the two electrons are of equal spin [3.116]. According to the Fe DOS (Fig. 3.1), a predominantly minority-spin core hole was assumed in the theoretical analysis.

Also, electron-excited Auger electron spectroscopy has been shown recently to provide new information on electron correlation effects, on intrashell exchange interactions, and on local magnetization in composite systems. It is beyond the scope of this article to discuss the considerable amount of data already available. A recent review is given by *Landolt* et al. [3.118].

3.7 Secondary Electron Spin Polarization and Stoner Excitations

The effect of inelastic scattering of the primary photoelectrons is detectable as a background in the primary electron energy distributions. The background increases with increasing binding energy (see Fig. 3.17). A primary electron might transfer part of its energy to another valence band electron, resulting in two excited electrons, which then might undergo the same scattering process again. This leads to a cascade of low-energy electrons [3.119]. The energy distribution will be rather structureless due to the continuum of possible energy transfers, and will be peaked at the lowest possible energy, i.e. at E_F. Only electrons with energy exceeding the work function are able to escape into vacuum. The observed secondary electron energy distribution will therefore be peaked near zero kinetic energy.

Electron-electron scattering is governed by the Coulomb operator which conserves the spin. Due to the spin-asymmetry of the DOS, it is therefore expected to observe spin-effects in secondary electron emission. It has been shown that secondary electrons from metallic ferromagnets are spin polarized [3.20,120]. Data taken on a well-characterized Fe(100) single-crystal are shown in Fig. 3.27 [3.121]. It is observed that the spin polarization at low kinetic

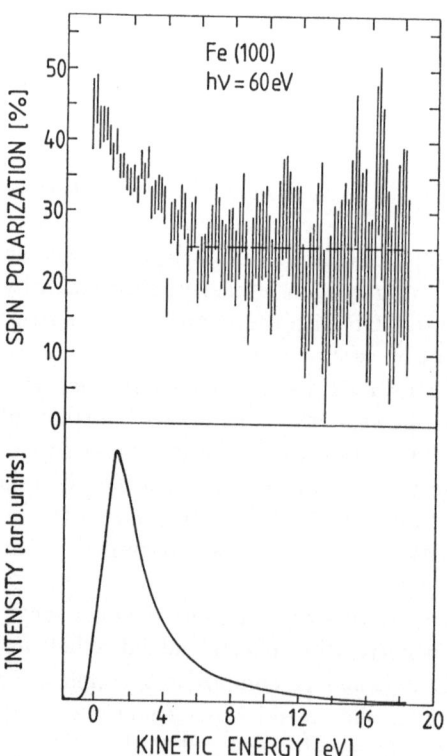

Fig. 3.27. (a) Secondary electron spin polarization (excited by photons of 60 eV energy) vs. kinetic energy. (b) Intensity distribution measured simultaneously with the spin polarization [3.121]

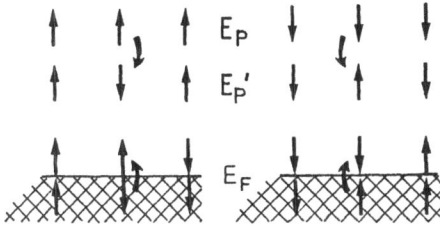

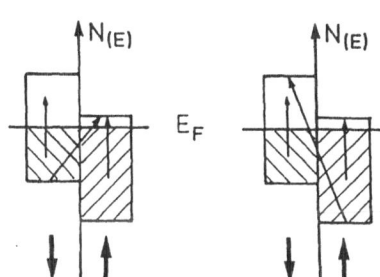

Fig. 3.28. Model to explain the enhanced secondary electron spin polarization at low energies. An electron with energy E_P is considered to be scattered inelastically by a valence band electron with energy loss $E_P - E_P'$. *Top:* Spin-flip and non-spin-flip channels for primary majority- and minority-spin electrons when interacting with a valence band electron. *Bottom:* Schematic of the DOS of Fe to demonstrate that the phase space for spin-flip scattering of a minority-spin primary electron is much larger than that for a majority-spin electron [3.121]

energy exceeds the average valence band spin polarization (27 %) by about a factor of two. The strong increase in spin polarization occurs below about 5 eV kinetic energy in Fe(100), Co(1010) and Ni(110) [3.121,122]. The effect has recently been confirmed for Fe(100) by *Allenspach* and *Landolt* [3.123], and a new structure in the dependence of polarization on energy has been observed at 10 eV kinetic energy.

The high spin polarization of the secondary electrons has been explained qualitatively in terms of preferential scattering of minority-spin electrons [3.121]. By spin-flip scattering events (the so-called Stoner excitations) they are converted to majority-spin electrons with lower energy. That this process leads to an enhanced positive spin polarization is a consequence of the dominance of minority-spin d-holes above E_F (Fig. 3.28a,b). Accordingly, the cross section for a majority-spin electron to excite via a spin-flip process a minority-spin valence band electron to the empty part of the majority-spin DOS is very small since there are only very few majority-spin holes in the 3d transition metals (Ni and Co are even considered to be strong ferromagnets with no majority-spin d-holes). The exchange scattering cross section increases with decreasing energy, enhancing the spin-asymmetry at low energies. The interpretation has recently been theoretically extended by *Penn* [3.124] and by *Glazer* et al. [3.125].

The fact that the highest spin polarization coincides with the highest intensity (unusual in polarized electron physics) has been used to construct a polarized secondary-electron scanning microscope [3.126,127]. A focused unpolarized electron beam is scanned across the sample, and the spin polarization of the secondary electrons is recorded as function of the position of the spot at the sample. By this method, the domain structure of the (100) surface of

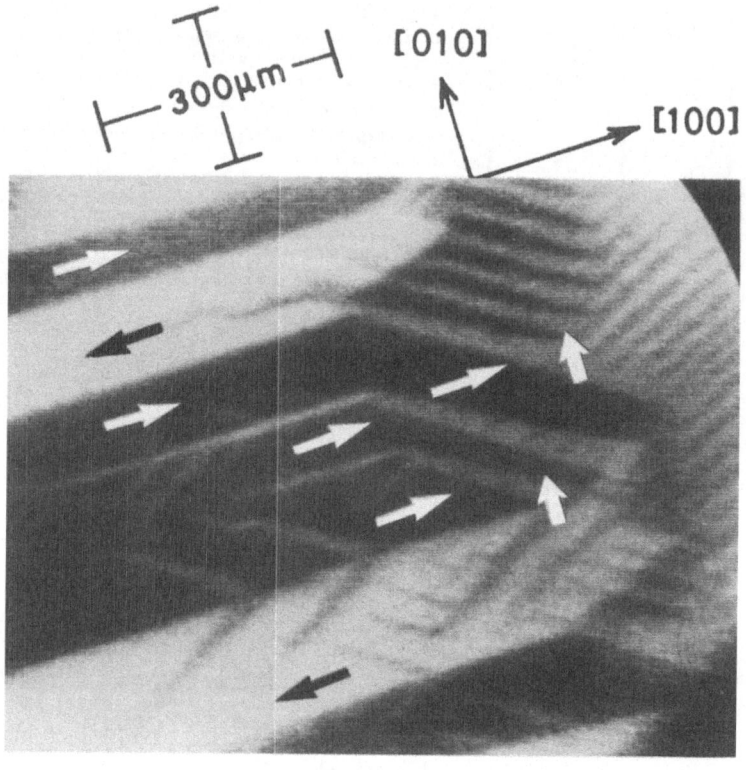

Fig. 3.29. Spin-polarized secondary-electron domain structure image from a silicon iron (001) surface (after *Koike* and *Hayakawa* [3.126])

a Fe-3% Si single crystal has been observed, see Fig. 3.29. Since the primary electron beam can be focused to about ≈ 50 Å, very small magnetic structures can be observed in principle.

Recently, Stoner excitations have been observed directly in spin-polarized electron energy loss experiments on Ni(110) by *Kirschner* et al. [3.128] and on a Fe-based metallic glass by *Hopster* et al. [3.129] (see Fig. 3.30). The experiment on Ni was a study of the spin asymmetry near 0.3 eV energy loss in excitation with a polarized primary electron beam. The experiment on Fe was performed in the opposite way using an unpolarized electron beam and analyzing the spin polarization of the inelastically scattered electrons. These studies have demonstrated that by electron energy loss spectroscopy, a region in the magnetic excitation spectrum can be studied which is at present not accessible by neutron scattering. Since by this kind of experiment excitations within the spin-split d-bands are induced (see Fig. 3.28), temperature-dependent studies are also of interest to obtain additional information on the changes occurring within the valence bands when approaching T_C. Recently, *Kirschner* [3.130] has

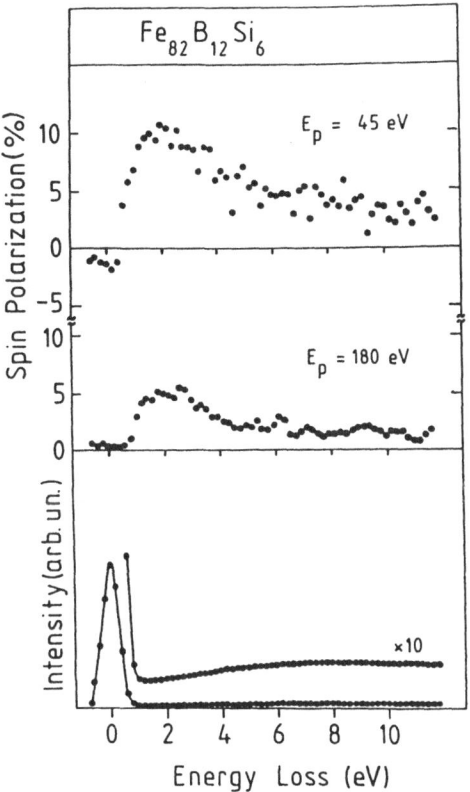

Fig. 3.30a–c. Results from spin-polarized energy loss spectroscopy on $Fe_{82}B_{12}Si_6$ (metallic glass). An unpolarized electron beam is used and the polarization of the inelastically scattered electrons is measured (after *Hopster* et al. [3.129]). **(a,b)** Spin polarization as function of energy loss for two primary beam energies E_P. **(c)** Electron energy loss spectrum

performed a "complete" electron-scattering experiment on Fe(110) by using a polarized primary beam and analyzing also the polarization of the inelastically scattered electrons.

3.8 Finite Temperature Ferromagnetism and Photoemission

The discussions have been centered so far on studies of the electronic structure at temperatures small compared to the Curie temperature T_C. It has been shown that, despite some quantitative discrepancies between experimental results and band structure calculations, especially for Ni, the overall understanding seems to be right. The situation is completely different at high temperatures, e.g. when approaching T_C. The question is by which microscopic mechanisms the magnetization is reduced with increasing temperature.

Different and controversial models on finite-temperature magnetism have been discussed in Sect. 3.2.2. Since spin- and angle-resolved photoemission has been shown to be capable of mapping the electronic structure at low temperatures, it should also be possible to study the changes in the electronic structure occuring at elevated temperatures. Prior to discussing recent experimental data, we might consider which temperature-induced changes in the spin-resolved energy distribution curves the different models would imply. There is one obvious result from any of these models: The spin separated EDCs become equal at the Curie temperature since no long-range quantization axis exists within the sample above this temperature. However, the different models predict different ways in which the spin-resolved EDCs would approach each other: In the Stoner model, the peak separation would be diminished proportional to the spontaneous magnetization. In the disordered-local moment picture the character of the changes should depend on the electron wave vector and on the wave function symmetry [3.44,45,131]. Certain states would show a spin-split behaviour above T_C, while others behave more Stoner-like. A similar prediction is made in the fluctuating band picture [3.132].

3.8.1 A Simple Model

Without specifying the size of spin-correlated regions, we might assume that the long-range magnetic order is reduced at elevated temperatures by transverse fluctuations of magnetic moments. At $T = 0\,K$ they are aligned either parallel or antiparallel to the magnetization direction. When the system becomes locally disordered by transverse fluctuations of magnetic moments (spin wave excitation), the *local* magnetization direction is instantaneously different from the *global* magnetization direction. Then the spins are no longer in pure states with respect to the quantization axis (z) defined by the Mott detector [3.133]. If we extract by means of a sufficiently fast probe an electron from a point in the crystal where the local magnetization direction is tilted by the angle Θ with respect to z, we observe spin mixing in the spin detector [3.134] (see *Capellmann* [3.35] and *Edwards* [3.135] for a discussion on timescales). This effect on the SREDCs is illustrated in Fig. 3.31, having assumed Lorentzian

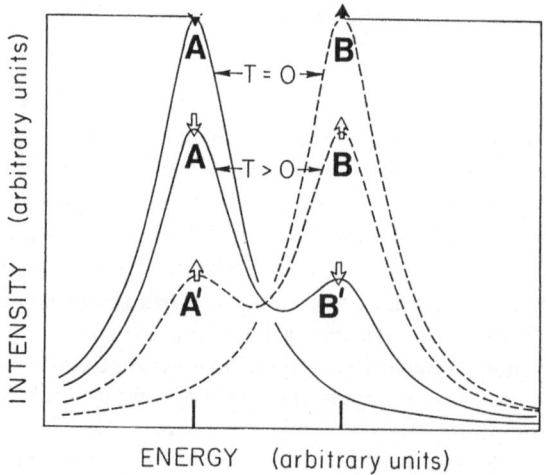

Fig. 3.31. Model spin-resolved photoelectron energy distribution curves for $T = 0\,K$ and at an elevated temperature (but below T_C). It is assumed that the electrons originate from regions with constant local magnetic moments the directions of which tend to fluctuate around the direction of the spontaneous magnetization at elevated temperatures. We refer to peaks A and B as "ordinary" peaks while A' and B' are referred to as "extraordinary" peaks [3.133]

lineshapes for the low-temperature SREDCs. The characteristic change is that in each of the SREDCs, a new peak emerges at the binding energy where the other one has its peak at low temperatures. The new peaks have been labeled as "extraordinary" peaks [3.133] since there is no correspondence in the low-temperature electronic structure. The "extraordinary" peaks are equivalent to the "forbidden" magnon lines which have been observed in polarized neutron scattering on Ni [3.136].

Characteristic features of the changes are that, at the binding energy, where the SREDCs cross each other at low temperature, they also cross at high temperatures ("intensity invariance"). The total (spin-summed) intensity remains unchanged at any binding energy. Therefore, the extraordinary peaks can only be detected by *spin-resolved* photoelectron spectroscopy.

Changes in the *spin-summed* intensity are beyond this simple model. However, when the magnet becomes disordered, changes in the transport properties are also expected. They would manifest themselves generally in a decrease of the angle-resolved photoemission intensity if a state is observed which emits into a strongly confined direction at low temperatures. If changes in the spin-summed intensity are observed which are beyond "purely" thermal (i.e., non-magnetic) effects, this would be an indication that the photoelectron has been influenced by the magnetic disorder, i.e. that its coherence length is larger than the microscopic magnetic "domains". However, it might be difficult to separate the "purely" thermal effects (due to electron-phonon coupling) because of possible spin-phonon coupling.

3.8.2 Experimental Results for Fe and Ni and Their Interpretation

Ni was the first of the 3d-transition metals which had been investigated by spin-resolved photoelectron spectroscopy to detect changes which accompany the ferromagnetic to paramagnetic phase transition. The Ni Curie temperature is comparatively small, and it does not undergo structural phase transitions up to its melting point, unlike Fe and Co. However, because of the small value of the exchange splitting, it requires a very good experimental energy resolution to study its variation with temperature. A further complication is that the natural photoemission linewidths are already at room temperature comparable with the exchange splitting. We will therefore in the following first review data for Fe which display effects which might not be recognized easily for Ni.

a) Photoemission from Fe at Finite Temperatures

The exchange splitting of Fe at room temperature has been resolved for initial states near Γ, as shown in Fig. 3.16. We display in Fig. 3.32 the changes which occur upon heating to $T/T_c = 0.85$ [3.24]. The effect of temperature is a decrease of the peak heights of both the majority-spin state $\Gamma'^{\uparrow}_{25}$ and of its exchange-split counterpart, $\Gamma'^{\downarrow}_{25}$. The leading minority-spin peak near E_F decreases much more in intensity than the majority-spin peak at 2.6 eV binding energy. The peaks do not shift significantly.

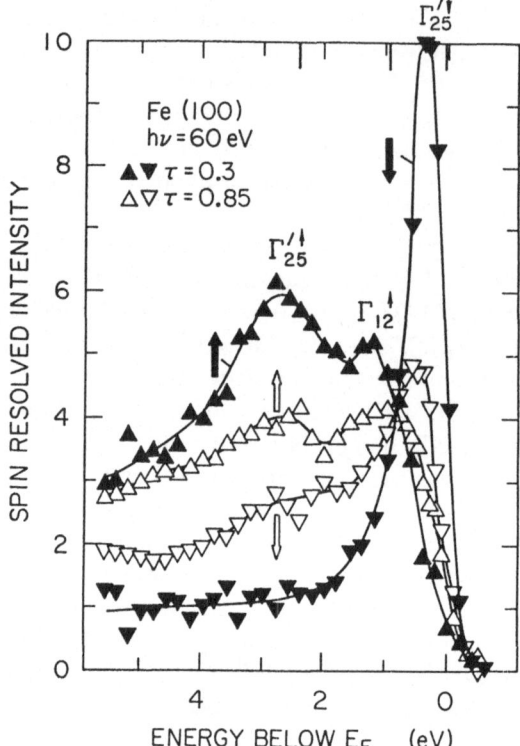

Fig. 3.32. Spin- and angle-resolved photoelectron energy distribution curves from Fe(100) for emission centered around the surface normal and predominantly s-polarized light, at two temperatures (T = 0.3 T$_C$ and T = 0.85 T$_C$). The photon energy is 60 eV and the initial states are near Γ [3.24]

An important observation is that at the position of $\Gamma'^{\uparrow}_{25}$, a new peak emerges in the minority-spin EDC. We observe here clearly one of the predictions of the model for transverse fluctuating magnetic moments (Fig. 3.31) and identify the new peak as an "extraordinary" peak [3.133]. The new peak is interpreted as originating from regions whose magnetization directions deviate from the long-range spontaneous magnetization direction. The other extraordinary peak (at the position of $\Gamma'^{\downarrow}_{25}$) is very weak and is hardly recognized. This results in a loss of the spin-summed intensity with increasing temperature at the leading peak energy near E$_F$ (see Fig. 3.33). It has been shown experimentally [3.103] that the loss of intensity is partially due to a broadening of the photoelectron emission cone, which indicates that with increased temperature, a broadening in the angular distribution and hence in wave vector occurs. We note here that with the more recently determined separation of the minority-spin peak from E$_F$ (0.07 eV) as discussed in Subsect. 3.5.2b, a strong decrease of intensity with temperature is also expected just from the Fermi function.

An EDC taken at about T$_C$ + 10 K is shown in Fig. 3.34. In order to find the peak positions, a fit with two Lorentzians and an inelastic background as the integral over the Lorentzians has been performed. It appears that the peak near 2.6 eV has not shifted. Considering the evolution of the SREDCs from

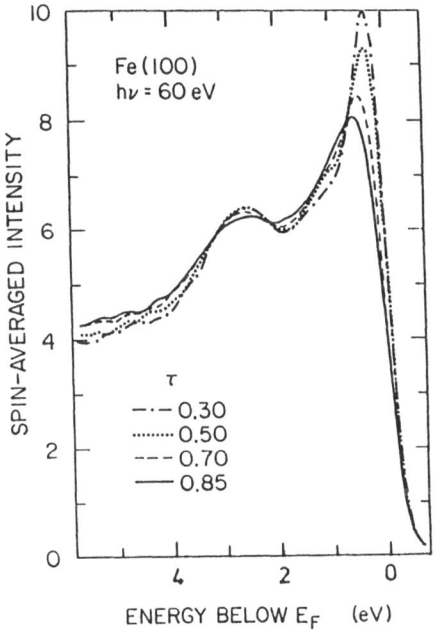

Fig. 3.33. Spin-averaged data as in Fig. 3.32 for a set of different temperatures [3.103]

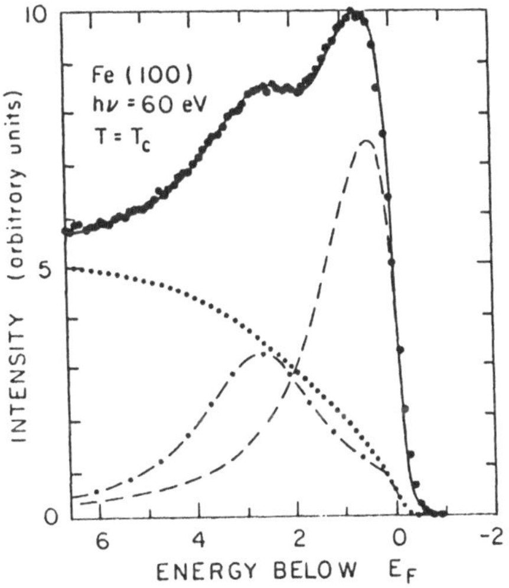

Fig. 3.34. As Fig. 3.33, but slightly above T_C [3.103]. The data are fitted by two Lorentzians and an inelastic background

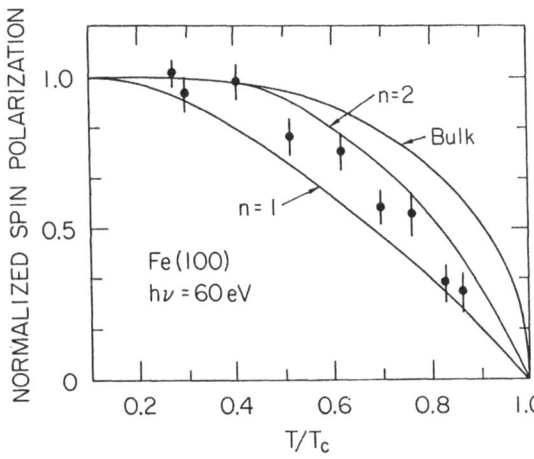

Fig. 3.35. Spin polarization of photoelectrons from Fe(100) as function of temperature and comparison with a layer-dependent mean-field calculation. The electron spectrometer is set fixed at 2.6 eV binding energy ($\Gamma'^{\uparrow}_{25}$) at 60 eV photon energy [3.103]

below T_C, it cannot be decided if the leading peak near E_F is derived from $\Gamma^{\uparrow}_{12}$ or from $\Gamma'^{\downarrow}_{25}$.

An important observation in Fig. 3.32 is that the separation of the exchange-split peaks decreases much less than the spontaneous magnetization (as judged from the decrease of the spin polarization, see Fig. 3.35). This demonstrates

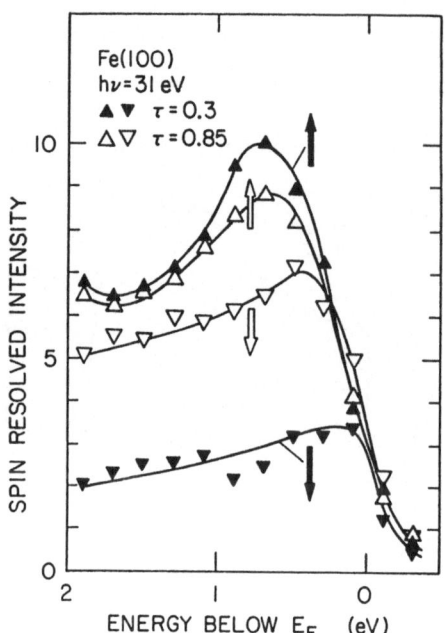

Fig. 3.36. Spin- and angle-resolved photoelectron energy distribution curves from Fe(100) at 31 eV photon energy [3.103]

that the Stoner model is not valid for Fe. Apparently, it also invalidates predictions in the disordered-local-moment picture that the $\Delta_5^{\uparrow,\downarrow}$-bands should coalesce above T_C [3.44,45]. Within the latter model, the $\Delta_5^{\uparrow,\downarrow}$-symmetry band have been predicted to "collapse" above T_C at about 0.7 eV above the room temperature binding energy of $\Gamma_{25}'^{\uparrow}$.

Effects which differ from those discussed above for the data taken at 60 eV photon energy have been observed at photon energies less than 35 eV. Rather than a *decrease* of spin-summed intensity near E_F, an *increase* was observed. The spin-resolved data revealed (Fig. 3.36) that the reason is an increase in minority-spin intensity which overcompensates the decrease of majority spin intensity in the binding energy range between E_F and about 3 eV below E_F. The increase in minority-spin intensity could be caused by a downwards shift of the unoccupied part of the minority-spin Δ_5-band at k vectors in the right half of the $\Gamma - H$ line in the Brillouin zone. This would be in agreement with predictions that it depends on the group velocity if the observed exchange splitting remains constant or decreases with increasing temperature [3.131]. Another reason could also be in phonon-assisted transitions from regions in k space where the minority-spin band is below E_F. Actually, the data at 31 eV photon energy differ from those at 60 eV in that the emission is not strongly peaked to the normal direction.

In accordance with the first interpretation of the $h\nu = 31$ eV data to indicate a decrease of the exchange splitting with temperature at these particular

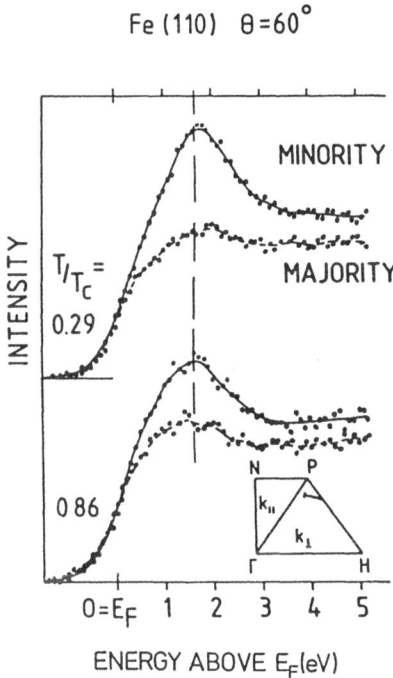

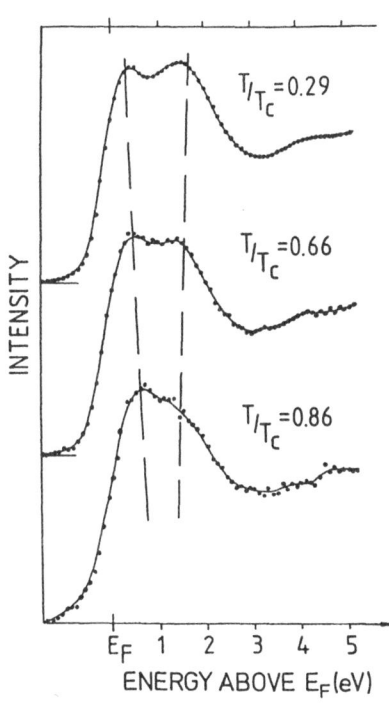

Fig. 3.37. Comparison of spin-resolved inverse photoemission spectra from Fe(110) at 60° angle of incidence at room temperature and near T_C [3.137]

Fig. 3.38. Spin-averaged inverse photoemission intensity spectra from Fe(100) at different temperatures [3.137]

wave vectors, "noncollapsing" and "collapsing" (empty) states have been inferred from spin polarized inverse photoemission data by *Kirschner* et al. [3.137] (Figs. 3.37, 38).

From the spin-resolved photoemission data taken at 60 eV photon energy (Fig. 3.32) the spin correlation length parameter has been determined by comparing with results from the cluster theory by *Haines* et al. [3.48]. Extraordinary peaks comparable to those experimentally observed are obtained when the correlation length parameter exceeds 4 Å, (Fig. 3.39). This value has been regarded as a lower limit. The data for initial state k-vectors in the right half of the Brillouin zone ($h\nu = 31$ eV, Fig. 3.36) are apparently not helpful to determine the correlation length parameter. Changes in the SREDCs similar as observed (Fig. 3.36) are obtained in the calculations independent of the degree of short range order due to the presence of the long-range magnetization below T_c [3.138].

b) Valence Band X-Ray Photoelectron Spectroscopy of Fe at Finite Temperatures

In view of the importance of the spin-resolved photoemission data for obtaining a microscopic picture of the ferromagnetic to paramagnetic phase transition,

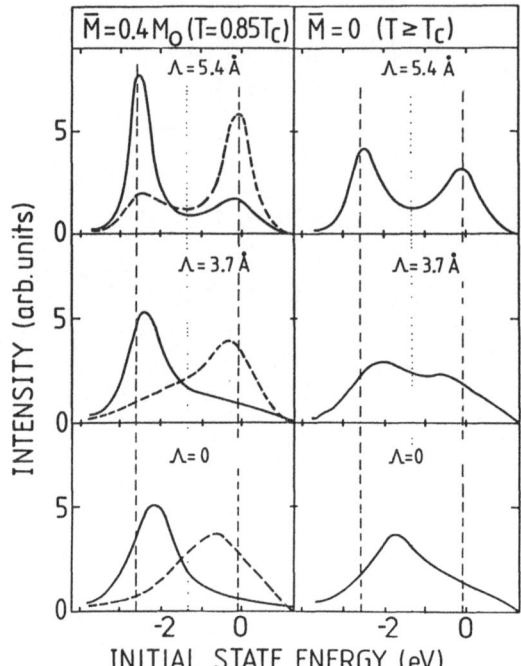

Fig. 3.39. Calculated spin-resolved photoelectron energy-distribution curves obtained by a cluster-calculation for initial states near Γ and two different temperatures ($T = 0.85\,T_C$ and $T > T_C$). The short-range order parameter λ is varied between 0 and 5.4 Å [3.48]

it appeared to be highly desirable to have an additional test by means of a different method. As such, valence band x-ray photoelectron spectroscopy offers itself since the DOS is probed rather than specific, differential electronic states as in angle-resolved photoemission. Any model on the electronic structure has to be tested finally *globally* by its predictions on the DOS. A merit of valence band XPS is that it is not as surface sensitive as ultraviolet photoemission (UPS). Rather than a mean probing depth of about 1–3 layers as deduced from Fig. 3.35 in UPS, the escape depth is about 15 Å or about 7 layers. Besides the practical advantage of reduced sensitivity to surface contamination which inevitably occurs at high temperatures, complications due to intrinsic surface effects (as, e.g., surface magnetism) are largely reduced, and the *bulk* electronic structure is tested.

As discussed in Sect. 3.2.2, predictions on the DOS above T_C have been made within a modified Stoner model [3.33], within the DLM model [3.43] (Fig. 3.2), and recently within the supercell model [3.49]. The result was that the DOS, as calculated by either of the above mentioned theories for the paramagnetic state above T_C, differs considerably from that at low temperatures. The changes occuring in the DOS for temperatures through T_C have been studied recently by *Kirby* et al. [3.101] by valence band x-ray photoelectron spectroscopy (XPS), employing a Mg K_α radiation source ($h\nu = 1254\,\text{eV}$). Only slight changes have been observed in the data between room temperature and $T = 1.034\,T_C$, (Fig. 3.40).

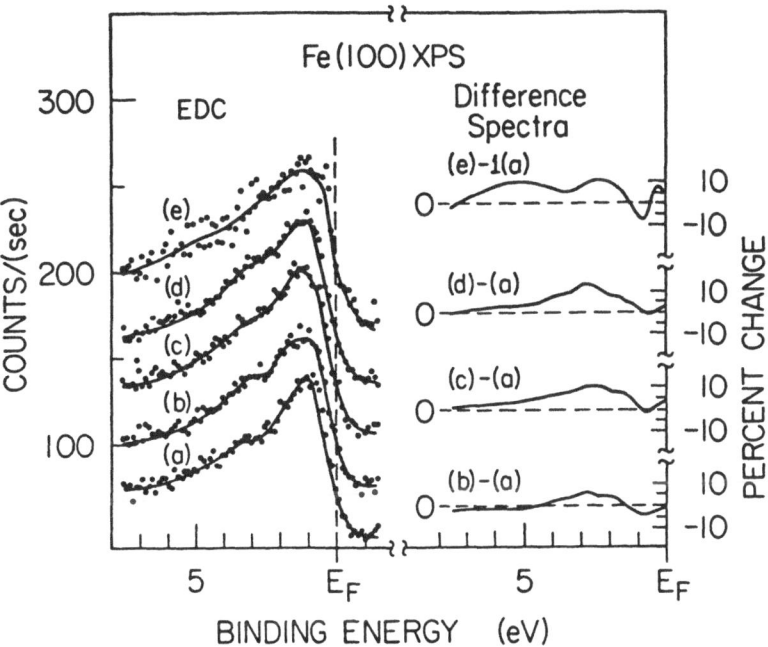

Fig. 3.40. X-ray photoelectron spectra from Fe(100) taken at different temperatures. The solid lines are averaging spline fits [3.101]. *(a)* 0.283 T_c, *(b)* 0.860 T_c, *(c)* 0.931 T_c, *(d)* 0.964 T_c *(e)* 1.034 T_c

From the XPS data on Fe at temperatures up to $T/T_C = 1.034$ [3.101], *Jarlborg* and *Peter*, by their supercell model, determined a value for the short range order parameter of about 5–6 Å [3.49]. *Clauberg* et al. [3.138] arrived at a similar value (6–8 Å) on the basis of the cluster model. This value is in fair agreement with the interpretation based on the spin-resolved photoemission results.

It is remarkable that similar results are obtained for such different photon energies (60 eV in the spin-resolved photoemission experiments, and 1254 eV in the XPS experiment) since the x-ray excited photoelectron spends much less time at a lattice site than the 60-eV photoelectron.

In conclusion, the spin-resolved photoemission data together with the XPS data indicate that the short-range order of Fe up to temperatures of $T = 1.034\,T_C$ is somewhere in between the DLM- and the fluctuating band model. The qualitative picture of Fe above T_C is similar to that of Fig. 3.41. It is interesting to note that a spin correlation length of about 5 Å at 1090 K had been determined as early as 1956 by small-angle neutron scattering on Fe [3.140]. The question remains open whether the critical fluctuations [3.141] with a diverging correlation length at T_C (which are responsible for the neutron-scattering data shown in Fig. 3.42) suffice to explain the photoemission data, or whether additional short-range order has to be invoked.

Fe

Fig. 3.41. The magnetic structure of Fe in the magnetic and
non-magnetic state (after *Swedenborgs* [3.139])

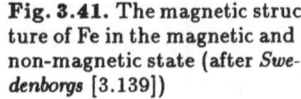

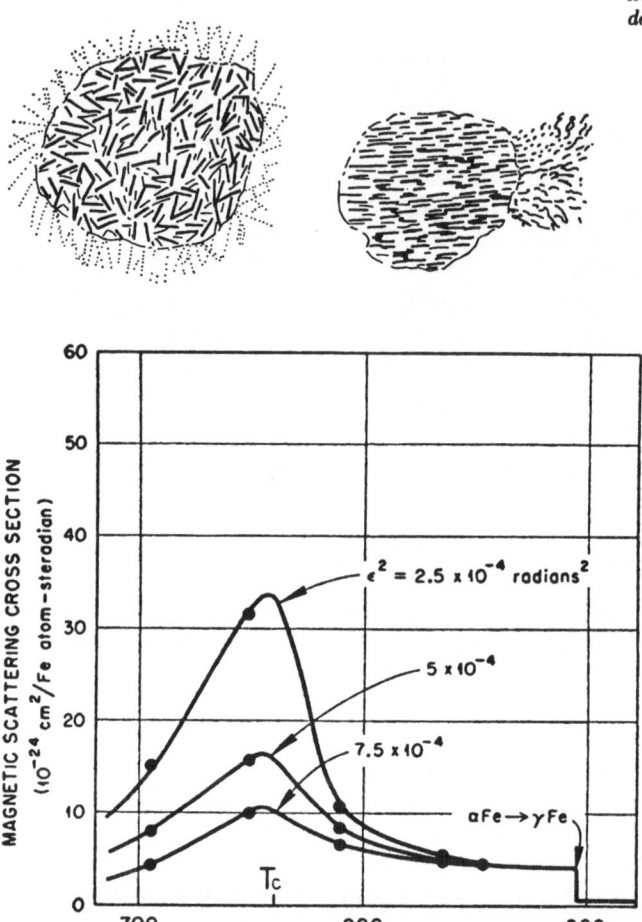

Fig. 3.42. Temperature dependence of magnetic scattering of 0.9 Å neutrons on Fe at fixed
scattering angles (after *Gersch* et al. [3.140])

The effect of critical fluctuations (pair interactions) has not yet been considered for interpreting photoemission data in the vicinity of T_C, and photoemission experiments have apparently not yet addressed this question. It might
be speculated that at high enough temperatures the short range order becomes
sufficiently broken-up to observe the DOS in the DLM limit. Asking for the temperature at which this might occur one might be guided by specific heat data
[3.142] and magnetic small-angle neutron scattering data [3.140] (Fig. 3.42)
which display a "tail" due to magnetic correlations up to $T = 1.1 T_C$ where
the $\alpha - \gamma$ phase transition occurs [3.143].

c) Points of Invariant Photoemission Intensities

The spin-resolved energy distribution curves as a function of temperature exhibit the characteristic feature of a point of invariant intensity at 1.2 eV binding energy (Figs. 3.32,33). At this energy, the SREDCs cross each other at room temperature (i.e., $P(1.2\,eV) = 0$). SREDCs taken at elevated temperatures at least up to $T/T_c = 0.85$ cross also at the same binding energy within the experimental errors: As compared to the changes in intensity of the leading peak, the changes at the invariant point energy are less by at least a factor of ten. An indication for the intensity invariance at binding energy where the spin polarization is zero has also been found in a recent finite-temperature valence band XPS study of Fe(100) [3.101].

Invariance of the spin-resolved intensities at the binding energy where the spin polarization is zero is a feature of the simple model calculation (Fig. 3.31) based on transverse fluctuation of magnetic moments. However, in the model calculation, the *spin-summed* intensity is conserved at any binding energy. We are therefore confronted with the question why in the experimental data the total intensity generally changes across the EDC, but hardly at the energy where the spin polarization vanishes.

Recent data on the *angular dependence* of EDCs from Fe(100) at 60 eV photon energy at room temperature revealed intensity-invariance near 1.2 eV binding energy, despite a strong decrease in intensity of the peak near E_F [3.144]. This supports the previously mentioned observation that the changes in intensity are consequence of an increased angular spread. This might indicate the existence of a spin-angle correlation in ferromagnets.

The intensity invariances might just be coincidental. Further data are certainly needed, especially at XPS energies. On the other hand, a point of invariant intensity at binding energy where the spin polarization is zero has originally been suggested by the author as existing also in spin-resolved photoemission data on Ni(110) [3.133].

3.8.3 Photoemission from Ni

Changes with temperature have been observed in spin-averaged EDCs from Ni(111) by *Himpsel* et al. [3.145] and by *Maetz* et al. [3.146]. It was found that the EDC becomes narrowed when approaching T_C. *Himpsel* et al. regarded this as evidence for a reduction of the exchange splitting by about 50 %. *Maetz* et al., however, considered similar data as evidence for a largely temperature-independent local exchange splitting and interpreted their data in the three-peak model of *Korenman* and *Prange* [3.42] by fitting the data with three Lorentzians. However, without spin analysis, it is impossible to determine the underlying spin structure, and the interpretation remains speculative.

Subsequently, a similar experiment has been performed with additional spin analysis by *Hopster* et al. [3.23] on a Ni(110) single-crystal using a laboratory NeI gas discharge lamp as the light source. The data are shown in Fig. 3.43a. At room temperature, the separation of the peaks in the SREDCs

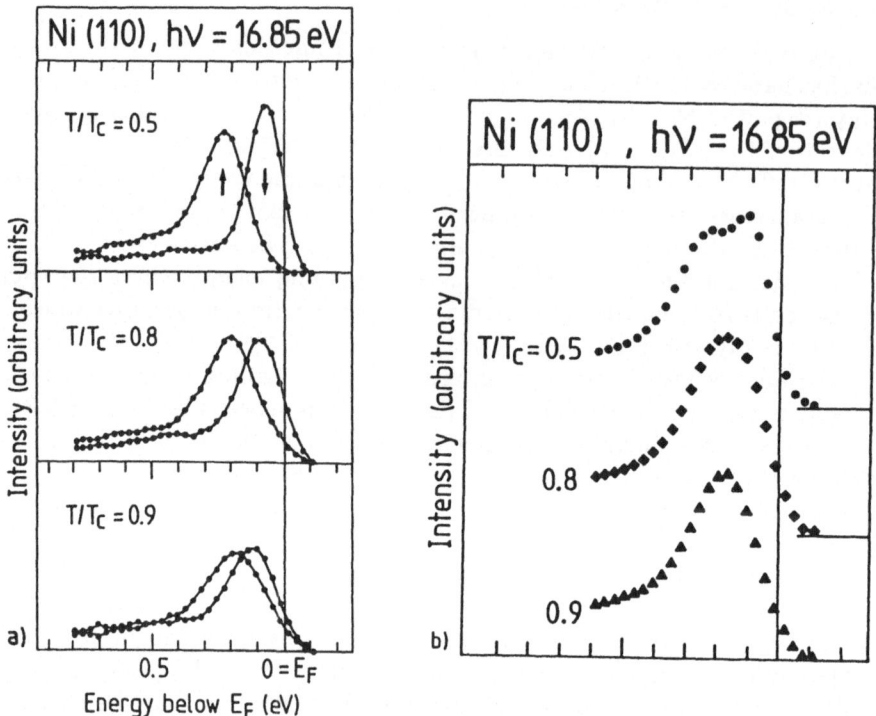

Fig. 3.43. (a) Spin- and angle-resolved energy distribution curves from Ni(110) at 16.8 eV photon energy (NeI) for normal emission and s-polarized light at three temperatures. (b) Spin-averaged data [3.23]

reveals an exchange splitting of about 0.18 eV. From an extrapolation to lower temperatures, it appears that the peak separation will increase furthermore to about 0.2 eV. Figure 3.43b shows the spin-summed data of Fig. 3.43a.

With increasing temperature, the exchange-split peak separation is reduced (see Fig. 3.44a) and the total EDC becomes narrowed as in the earlier spin-averaged data. The separation of the spin-split peaks tends to zero at T_C. This might lead to the conclusion that the Stoner model is valid which predicts a reduction of the exchange splitting proportional to the decrease of the spontaneous magnetization. However, the line widths are increasing with temperature, see Fig. 3.44b, and become constant near T_C. That the linewidths do not collapse to their room temperature values at T_C, might indicate that there is still magnetic disorder even above T_C, in disagreement with the Stoner model.

Figure 3.44 shows that the maximum linewidth is not obtained at T_C but rather at $0.9\,T_C$. It might be speculated that this is related to the observation that a sound velocity anomaly of Ni also occurs *below* T_C (see, e.g. *Kim* [3.148] and references therein).

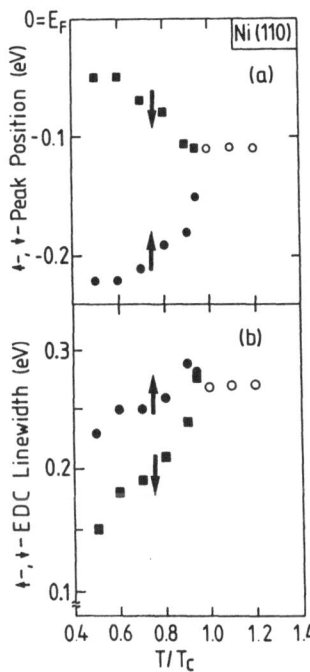

Fig. 3.44. (a) Initial state energies of spin-split states of Fig. 3.43 as a function of temperature. (b) Temperature-dependence of the width of the exchange-split lines [3.147]

The variation of the *spin-summed* EDCs with temperature (see Fig. 3.43b) is similar to that as observed by *Maetz* et al. [3.146] in that the EDC becomes narrower with increasing temperature. On the other hand, the strong intensity increase in the center of the EDC as observed by *Himpsel* et al. [3.145] in off-normal emission from Ni(111) has not been reported. Instead, the change in intensity in the spin-resolved data (Fig. 3.43a) is small at the binding energy E_0 where the spin polarization is zero [3.133]. In this respect the data are similar to the data on Fe(100) (Fig. 3.32). A difference from the Fe data is that the extraordinary peaks as defined above are certainly not observed in the Ni data (Fig. 3.43a). But it might be speculated that they are just not resolved. The possible overlap with the "ordinary" peaks might be a reason for the broadening of the spin-resolved lineshapes. The extraordinary peaks in spin-resolved photoemission EDCs would be analogous to forbidden magnon lines in polarized neutron scattering [3.136].

Certainly, more attention has to be paid to the normalization of Ni photoemission intensities in the future. If confirmed, the collaps of the peak separation would be caused by intensity losses of the ordinary peaks and a partial overlap with the increasing extraordinary peak intensities. However, this is still somewhat speculative and certainly deserves further experimental attention.

3.9 Concluding Remarks and Outlook

It has been demonstrated in this review that spin- and angle-resolved photoemission is capable of mapping the electronic structure of ferromagnets at temperatures up to the Curie temperature. The exchange splitting has directly been observed in the prototype materials Fe and Ni. The low-temperature spin- and angle-resolved photoemission data on Fe and Ni agree within about 0.5 eV with calculated bulk electronic structures. A discrepancy of this order is most significant for Ni since it compares with the value of the exchange splitting itself. The Ni problem has been discussed in the literature mostly in terms of correlation effects associated with the photoemission process. For Fe(100), data taken at 60 eV photon energy where surface sensitivity is largest, agree quantitatively with the surface electronic structure as calculated by *Ohnishi* et al. [3.108]. This might be regarded as evidence for the observation of the predicted enhanced surface exchange splitting of the Fe(100) surface.

The study of the electronic structure of Fe at elevated temperatures revealed by comparison with the cluster calculation by *Haines* et al. [3.48] that short-range magnetic order of more than 4 Å persists up to temperatures of at least $1.034\,T_C$. A key observation has been the emergence of an "extraordinary" peak when approaching T_C in the minority-spin energy distribution curve at the binding energy where the majority-spin EDC has its prominent peak. This is in disagreement with the Stoner model, and apparently also in disagreement with the disordered-local-moment picture which predicts that the splitting of the relevant states with Δ_5 wave function symmetry vanishs at T_C.

For Ni, it is still questionable if the observed collapsing of the separation of the exchange-split peaks when approaching T_C indicates a vanishing local exchange splitting. Further experimental data for different initial state k-vectors with emphasis on the intensity normalization should help to answer this question.

It has been found that the resonant photoemission intensity in the vicinity of the $3p - 3d$ excitation threshold of Fe is closely related to the MVV Auger transition. This observation might lead to better understanding of electron correlation effects also in other 3d transition metals as Ni and Co.

The study of the electronic structure of more complicated materials as Heusler- and Invar alloys is just in the beginning. Future work will center more on the investigation of compound materials, which is significantly more difficult to perform than the study of the pure 3d-metals.

The current interest in problems of surface and interface magnetism has been illustrated. Due to the small probing depths, photoemission is ideally suited for the study of magnetism of ultrathin films (in the monolayer range) of ferromagnets on non-magnetic substrates. First experiments have just been performed. Also, chemisorption on ferromagnetic surfaces is an interesting field which has not been fully explored.

The high spin polarization of secondary electrons has stimulated the development of theories on elementary (Stoner-) excitations, and accomplished operation of spin-polarized scanning microscopes with very high resolution.

Provisions will be made in the future to cool the samples to liquid Helium temperatures in order to investigate materials with low Curie temperatures. Anisotropy problems with ultrathin films demand a measurement of the full polarization vector. With the synchrotron radiation flux increased by orders of magnitude by employing wigglers, undulators, and free-electron lasers, it will be possible to increase the energy resolution in spin-resolved photoemission considerably. This is mandatory for investigating materials with small exchange splitting, as, e.g., Ni, and for the study of critical phenomena in the vicinity of the Curie temperature.

An important development will be spin-resolved x-ray photoelectron spectroscopy. This will allow direct test of the density-of states which is the relevant quantity determining the occurence of ferromagnetism. Rather than a point-by-point comparison with calculated band energies for selected k-vectors this allows a global determination of the DOS. The larger escape depth of the photoelectrons at x-ray energies is of advantage for testing bulk properties.

Considerable progress has been made recently in the development of such related techniques as spin-polarized inverse photoemission, spin-resolved energy loss spectroscopy, and spin-resolved Auger-electron spectroscopy. The interplay between the different spin-resolving methods will help to obtain a better understanding of problems related to the fascinating field of ferromagnetism.

References

3.1   H.C. Siegmann, F. Meier, M. Erbudak, M. Landolt: Adv. El. El. Phys. **62**, 2 (1984)
3.2   E.L. Garwin, D.T. Pierce, H.C. Siegmann: Helv. Phys. Acta **47**, 343 (1974);
       D.T. Pierce, F. Meier, P. Zürcher: Appl. Phys. Lett. **26**, 670 (1975)
3.3   A. Eyers, F. Schäfers, G. Schönhense, U. Heinzmann, H.P. Oepen, K. Hünlich,
       J.Kirschner: Phys. Rev. Lett. **52**, 1559 (1984)
3.4   E. Fues, H. Hellmann: Phys. Z. **31**, 465 (1930)
3.5   E.O. Kane, Phys. Rev. Lett. **12**, 97 (1964)
3.6   C.N. Berglund, W.E. Spicer: Phys. Rev. A**136**, 1030 and 1044 (1964)
3.7   U. Gerhardt, E. Dietz: Phys. Rev. Lett. **26**, 1477 (1971)
3.8.   N.V. Smith, M.M. Traum: Phys. Rev. B**11**, 2087 (1975)
3.9   P.J. Feibelman, D.E. Eastman: Phys. Rev. B**10**, 4932 (1974)
3.10   P. Thiry, D. Chandesris, Y. Lecante, C. Guillot, R. Pincheaux, Y. Petroff: Phys. Rev.
       Lett. **43**, 82 (1979)
3.11   R. Courths, S. Hüfner: Phys. Rep. **112**, 53 (1984)
3.12   F.J. Himpsel: Adv. Phys. **32**, 1 (1983)
3.13   N.F. Mott, H.S.W. Massey: *The Theory of Atomic Collisions* (Clarendon Press, Oxford 1965), Chapt. IX
3.14   J. Kessler: *Polarized Electrons*, 2nd ed., Springer Ser. Atoms Plasmas, Vol. 1, Heidelberg 1985
3.15   G. Busch, M. Campagna, P. Cotti, H.C. Siegmann: Phys. Rev. Lett. **22**, 597 (1969)
3.16   G. Chrobok, M. Hofmann: Phys. Lett. **57A**, 257 (1976)
3.17   E. Kisker, G. Baum, A.H. Mahan, W. Raith, B. Reihl: Phys. Rev. B**18**, 2256 (1978)
3.18   R.J. Celotta, D.T. Pierce, G.-C. Wang, S.D. Bader, G.P. Felcher: Phys. Rev. Lett
       43, 728 (1979)

3.19   E. Kisker, W. Gudat, E. Kuhlmann, R. Clauberg, M.Campagna: Phys. Rev. Lett. **45**, 2053 (1980)
3.20   E. Kisker, R. Clauberg, W. Gudat: Rev. Sci. Instr. **53**, 507 (1982)
3.21   J. Unguris, D.T. Pierce, A. Galejs, R. Celotta: Phys. Rev. Lett. **49**, 72 (1982)
3.22   R. Clauberg, W. Gudat, E. Kisker, E. Kuhlmann: G.M. Rothberg, Phys. Rev. Lett. **47**, 1314 (1981)
3.23   R. Raue, H. Hopster, R. Clauberg: Phys. Rev. Lett. **50**, 1623 (1983); H. Hopster, R. Raue, G. Güntherodt, E. Kisker, R. Clauberg, M. Campagna: Phys. Rev. Lett. **51**, 829 (1983)
3.24   E. Kisker, K. Schröder, M. Campagna, W. Gudat: Phys. Rev. Lett. **52**, 2285 (1984)
3.25   K.B. Hathaway, H.J.F. Jansen, A.J. Freeman: Phys. Rev. B**31**, 7603 (1985); C.S.Wang, B.M. Klein, H. Krakauer: Phys. Rev. Lett. **54**, 1852 (1985)
3.26   E.C. Stoner: Proc. Roy. Soc. A**154**, 656 (1936)
3.27   V.L. Moruzzi, J.F. Janak, A.R. Williams: *Calculated Electronic Properties of Metals,* (Pergamon, New York 1978)
3.28   V. Heine, J.H. Samson, C.M.M. Nex: J. Phys. F**11**, 2645 (1981); M.V. You, V. Heine, J. Phys. F**12**, 177 (1982)
3.29   see, e.g., *Selected Papers in Magnetism,* ed. by D.M. Edwards, J. Magn. Magn. Mat. **46** (1984)
3.30   see e.g., E.P. Wohlfarth, Inst. Phys. Conf. Ser. **55**, 161 (1980)
3.31   O. Gunnarsson: J. Phys. F**6**, **587** (1976)
3.32   E.P. Wohlfarth: Physica **119B**, 203 (1983)
3.33   T. Jarlborg, M. Peter: J. Magn. Magn. Mat. **42**, 89 (1984)
3.34   D.M. Edwards in: *Electron Correlation and Magnetism in Narrow-Band Systems,* Springer Ser. Sol. State Sci., Vol. 29, ed. by T. Moriya (Berlin: Springer, Heidelberg 1981), p.73
3.35   H. Capellmann: J. Magn. Magn. Mat. **28**, 250 (1982)
3.36   P.J. Brown, D. Deportes, D.Givord, K.R.A. Ziebeck: J. Magn. Magn. Mat. **31–34**, 295 (1983)
3.37   H. Capellmann: J. Phys. F**4**, 1466 (1974)
3.38   V. Korenman, J. Murray, R.E. Prange: Phys. Rev. B**16**, 4032 (1977)
3.39   D.M. Edwards: J. Magn. Magn. Mat. **15–18**, 262 (1980)
3.40   J.W. Lynn: Phys. Rev. B**11**, 2624 (1975)
3.41   O. Steinsvoll, C.F. Majkrzak, G. Shirane, J. Wicksted: Phys. Rev. Lett. **51**, 300 (1983)
3.42   V. Korenman, R.E. Prange: Phys. Rev. Lett. **44**, 1291 (1980)
3.43   A.J. Pindor, J. Staunton, G.M. Stocks, H. Winter: J. Phys. F**13**, 979 (1983); T. Oguchi, K. Terakura, N. Hamada: J. Phys. F**13**, 145 (1983); H. Hasegawa, J. Phys. F**13**, 1915 (1983)
3.44   J. Staunton, B.L. Gyorffy, A.J. Pindor, G.M. Stocks, H. Winter: J. Phys. F**15**, 1387 (1985)
3.45   P.J. Durham,J. Staunton, B.L. Gyorffy: J. Magn. Magn. Mat. **45**, 38 (1984)
3.46   T. Moriya, Y. Takahashi: J. Phys. Soc. Japan **45**, 397 (1978)
3.47   T. Moriya: J. Magn. Magn. Mat. **45**, 79 (1984)
3.48   E. Haines, R. Clauberg, R. Feder: Phys. Rev. Lett. l**54**, 932 (1985)
3.49   T. Jarlborg, M. Peter: Phys. Rev. B**32**, 5435 (1985)
3.50   see, e.g. A. Liebsch: Festkörperprobleme **19**, 209 (Vieweg, Braunschweig 1979)
3.51   see, e.g. N.V. Smith, in *Photoemission in Solids,* ed. by M. Cardona, L. Ley, Topics Appl. Phys., Vol. 26
3.52   see, e.g. J.C. Slater, *Quantum Theory of Molecules and Solids* (McGraw Hill, New York 1965)
3.53   E. Kisker, R.E. Kirby, E.L. Garwin, F.K. King, E. Tamura, R. Feder: J. Appl. Phys. **57**, 3021 (1985); E. Tamura, R. Feder, J. Krewer, R.E. Kirby, E. Kisker, E.L. Garwin, F.K. King: Sol. State Commun., **55**, 543 (1985)
3.54   M. Wöhlecke, A. Baalmann, M. Neumann: Sol. State Commun. **49**, 217 (1984)
3.55   J. Hermanson: Sol. State Commun. **22**, 9 (1977)
3.56   M.P. Seah, W.A. Dench, Surf. Interf. Anal. **1**, 2 (1979)
3.57   E. Kisker, M. Campagna, W. Gudat, E. Kuhlmann, in: Festkörperprobleme **19**, XIX, 259 (Vieweg, Braunschweig 1979)

3.58   K. Schröder, G.A. Prinz, K.-H. Walker, E. Kisker: J. Appl. Phys. **57**, 3669 (1985)
3.59   G.A. Prinz, E. Kisker, K.B. Hathaway, K. Schröder, K.-H. Walker: J. Appl. Phys. **57**, 3024 (1985)
3.60   W. Gudat, E. Kisker, G.M. Rothberg, C. Depautex: Nucl. Instr. Meth. **195**, 233 (1982)
3.61   V.W. Hughes, R.L. Long, M.S. Lubell, M. Posner, W. Raith: Phys. Rev. **A5**, 195 (1972)
3.62   R. Raue, H. Hopster, E. Kisker: Rev. Sci. Instr. **55**, 383 (1984)
3.63   M. Campagna, D.T. Pierce, F. Meie, K. Sattler, H.C. Siegmann: Adv. Electron. Electron Phys. **41**, 113 (1976)
3.64   E. Kisker, A.H. Mahan: unpublished
3.65   K.-H. Walker, K. Schröder, E. Kisker, G.A. Prinz: to be published
3.66   G.A. Prinz, G.T. Rado, J.J. Krebs: J. Appl. Phys. **53**, 2087 (1982)
3.67   E. Kisker: Rev. Sci. Instr. **53**, 507 (1982)
3.68   H.C. Siegmann, D.T. Pierce, R.J. Celotta: Phys. Rev. Lett. **46**, 452 (1981)
3.69   J. Kirschner, R. Feder: Phys. Rev. Lett. **42**, 1008 (1979)
3.70   E.P. Wohlfarth: Phys. Lett. **36A**, 131 (1971)
3.71   E.P. Wohlfarth: Phys. Rev. Lett. **38**, 524 (1977)
3.72   W. Eib, S.F. Alvarado: Phys. Rev. Lett. **37**, 444 (1976)
3.73   I.D. Moore, J.B. Pendry: J. Phys. **C11**, 4615 (1978)
3.74   C.S. Wang, J. Callaway: Phys. Rev. **B15**, 298 (1977)
3.75   E. Kisker, W. Gudat, M. Campagna, E. Kuhlmann, H. Hopster, I.D. Moore: Phys. Rev. Lett. **43**, 966 (1979)
3.76   P. Heimann, F.J. Himpsel, D.E. Eastman: Sol. State Commun. **39**, 219 (1981)
3.77   R. Clauberg: Phys. Rev. **B27**, 4644 (1983)
3.78   E. Kisker (unpublished)
3.79   W. Eib, B. Reihl: Phys. Rev. Lett. **40**, 1674 (1978)
3.80   F. Meier, D. Pescia, T. Schriber: Phys. Rev. Lett. **48**, 645 (1982)
3.81   M. Landolt, P. Niedermann, D. Mauri, Phys. Rev. Lett. **48**, 1632 (1982)
3.82   G. Allan: Surf. Sci. **74**, 79 (1978);
        D.R. Grempel: Phys. Rev. **B24**, 3929 (1981)
3.83   R.H. Victora, L.M. Falicov: Phys. Rev. **B31**, 7335 (1985)
3.84   C.Rau, S. Eichner: Phys. Rev. Lett. **47**, 939 (1981)
3.85   L.E. Klebanoff, R.H. Victora, L.M. Falicov, D.A. Shirley: Phys. Rev. **B32**, 1997 (1985)
3.86   G.L. Bona, F. Meier, M. Taborelli, E. Bucher, P.H. Schmidt: Sol. State Commun. **56**, 391 (1985)
3.87   R.A. de Groot, F.M. Müller, P.G. van Engen, K.H.J. Buschow: Phys. Rev. Lett. **50**, 2024 (1983)
3.88   J. Callaway: Inst. Phys. Conf. Ser. **55**, 1 (1980)
3.89   A. Liebsch: Phys. Rev. Lett. **43**, 1431 (1979);
        A. Liebsch: Phys. Rev. **B23**, 5203 (1981)
3.90   P. Oles, G. Stollhoff: Phys. Rev. **B29**, 314 (1984)
3.91   C.S.Wang: J. Magn. Magn. Mat. **31-34**, 95 (1983)
3.92   R.H. Victora, L.M. Falicov: Phys. Rev. Lett. **55**, 1140 (1985)
3.93   J. Unguris, A. Seiler, R.J. Celotta, D.T. Pierce, P.D. Johnson, N.V. Smith: Phys. Rev. Lett. **49**, 1047 (1982)
3.94   G. Borstel, G. Thörner, M. Donath, V. Dose, A. Goldmann: Sol. State Commun. **55**, 469 (1985)
3.95   H. Martensson, P.O. Nilsson: Phys. Rev. **B30**, 3047 (1984)
3.96   P. Heimann, N. Neddermeyer: Phys. Rev. **B18**, 3537 (1978)
3.97   A. Schultz, R. Courths, H. Schultz, S. Hüfner: J. Phys. **F9**, L41 (1979)
3.98   D.E.Eastman, F.J. Himpsel, J.A. Knapp: Phys. Rev. Lett. **44**, 95 (1980)
3.99   A.M. Turner, A.W. Donoho, J.L. Erskine: Phys. Rev. **B29**, 2986 (1984);
        Surface effects are discussed in: A.M. Turner, J.L. Erskine: Phys. Rev. **B30**, 6675 (1984). See also references therein
3.100   Y. Sakisaka, T. Rhodin, D. Mueller: Sol. State Commun. **53**, 793 (1985)
3.101   R.E. Kirby, E. Kisker, F.K. King, E.L. Garwin: Sol. State Commun. **56**, 425 (1985)
3.102   R. Feder, W. Gudat, E. Kisker, A. Rodriguez, K. Schröder: Sol. State Commun. **46**, 619 (1983)

3.103 E. Kisker, K. Schröder, W. Gudat, M. Campagna: Phys. Rev. B**31**, 329 (1985)
3.104 J. Callaway, C.S. Wang: Phys. Rev. B**16**, 2095 (1977)
3.105 R. Feder, A. Rodriguez, E. Baier, E. Kisker: Sol. State Commun. **52**, 57 (1984)
3.106 G. Borstel: Sol. State Commun. **53**, 87 (1985)
3.107 E. Kisker, K. Schröder: to be published
3.108 S. Onishi, A.J. Freeman, M. Weinert: Phys. Rev. B**28**, 6741 (1983)
3.109 R.H. Victora, L.M. Falicov, S. Ishida: Phys. Rev. B**30**, 3896 (1984)
3.110 S. Hüfner, G.K. Wertheim: Phys. Lett. **51A**, 299 (1975)
3.111 C. Guillot, Y. Ballu, J. Paigne, J. Lecante, R.P.Jain, P. Thiry, R. Pinchaux, Y. Petroff, L.M. Falicov: Phys. Rev. Lett. **39**, 1632 (1977)
3.112 D.R. Penn: Phys. Rev. Lett. **42**, 921 (1979)
3.113 N.V. Mott: Adv. Phys. **13**, 325 (1964)
3.114 L.A. Feldkamp. L.C. Davis: Phys. Rev. Lett. **43**, 151 (1979)
3.115 D. Chandesris, J. Lecante, Y. Petroff: Phys. Rev. B**27**, 2630 (1983)
3.116 K. Schröder, E. Kisker, A. Bringer: Sol. State Commun. **55**, 377 (1985)
3.117 H. Kato, T. Ishii, S. Masuda, Y. Harada, T. Miyano, T. Komeda, M. Onchi, Y. Sakisaka: Phys. Rev. B**32**, 1992 (1985)
3.118 M. Landolt, R. Allenspach, D. Mauri: J. Appl. Phys. **57**, 3626 (1985)
3.119 P.A. Wolff: Phys. Rev. **95**, 56 (1954)
3.120 J. Unguris, D.T. Pierce, A. Galejs, R.J. Celotta: Phys. Rev. Lett. **49**, 72 (1982)
3.121 E. Kisker, W. Gudat, K. Schröder: Sol. State Commun. **44**, 591 (1982)
3.122 H. Hopster, R. Raue, E. Kisker, M. Campagna, G. Güntherodt: Phys. Rev. Lett **50**, 70 (1983)
3.123 A. Allenspach, M. Taborelli, M. Landolt: Phys. Rev. Lett. **55**, 2599 (1985)
3.124 D.R. Penn, S.P. Appel, S.M. Girvin: Phys. Rev. Lett. **55**, 518 (1985)
3.125 J. Glazer, E. Tosatti: Sol. State Commun. **52**, 905 (1984)
3.126 K. Koike, K. Hayakawa: Appl. Phys. Lett. **45**, 585 (1984)
3.127 J. Unguris, G. Hembree, R.J. Celotta, D.T. Pierce: J. Magn. Magn. Mat. **54–57**, 1629 (1986)
3.128 J. Kirschner, D. Rebenstorff, H. Ibach: Phys. Rev. Lett. **53**, 698 (1984)
3.129 H. Hopster, R. Raue, R. Clauberg: Phys. Rev. Lett. **53**, 695 (1984)
3.130 J. Kirschner: Phys. Rev. Lett. **55**, 973 (1985)
3.131 E.M. Haines, V. Heine, A. Ziegler: J. Phys. F**15**, 661 (1985)
3.132 V. Korenmann, R.E. Prange: Phys. Rev. Lett. **53**, 186 (1984)
3.133 E. Kisker, J. Magn. Magn. Mat. **45**, 151 (1984)
3.134 D.M. Edwards: J. Phys. C**16**, L327 (1983)
3.135 D.M. Edwards: J. Magn. Magn. Mat. **45**, 151 (1984)
3.136 R.D. Lowde, R.M. Moon, B. Pagonis, C.H. Perry, J.B. Sokoloff, R.S. Vaughan-Watkins, M.C.K. Wiltshire, J. Crangle: J. Phys. F**13**, 249 (1983)
3.137 J. Kirschner, M. Glöbl, V.Dose, H. Scheidt: Phys. Rev. Lett. **53**, 612 (1984)
3.138 R. Clauberg, E. Haines, R. Feder: Z. Phys. B**62**, 31 (1985)
3.139 E. Swedenborgs, Principia Rerum Naturalia (1734)
3.140 H.A. Gersch, C.G.Shull, M.K. Wilkinson: Phys. Rev. **103**, 525 (1956)
3.141 L. Van Hove: Phys. Rev. **95**, 1374 (1954)
3.142 M. Braun, R. Kohlhaas, O. Vollmer: Z. Angew. Phys. **25**, 365 (1968)
3.143 H. Hasegawa, D. Pettifor: Proc. of the Workshop on 3d Metallic Magnetism (Inst. von Laue-Langevin, Grenoble 1983), p.203, and priv. commun.
3.144 E. Kisker, K.-H. Walker: to be published
3.145 F.J. Himpsel, J.A. Knapp, D.E. Eastman: Phys. Rev. B**19**, 2919 (1979)
3.146 C.J. Maetz, U. Gerhardt, E. Dietz, A. Ziegler, R.J. Jelitto: Phys. Rev. Lett. **48**, 1686 (1982)
3.147 R. Raue, H. Hopster, R. Clauberg: Z. Phys. B**54**, 121 (1984)
3.148 D.J. Kim: Physica **119B**, 30 (1983)

4. The Local-Band Theory

V. Korenman
With 9 Figures

This chapter is concerned with the temperature dependence of the electronic structure in itinerant electron ferromagnets, such as iron and nickel. It presents a single, uniform description of this dependence throughout a temperature range which includes the paramagnetic state above the Curie temperature, as well as the T = 0 state, even with domains. The work to be discussed is the local-band theory (LBT). It has been developed over a period of years in collaboration with Richard Prange and others, [4.1-17]. Similar work was done independently by *Capellmann* [4.18]. Here I give, for the first time, full details of the LBT analysis of electronic states. Other aspects of the LBT, in particular the analysis of thermodynamic properties and of spin wave spectra, are not treated in this chapter. These topics have been reviewed, and relevant papers cited, in [4.17]. My purpose being mainly exposition, I will also not discuss in any detail the various opposing viewpoints, some of which will be found in other chapters.

4.1 Background

The main formal result I find is a rotationally invariant self-energy expansion for the single-electron Green's function, involving the interaction of electrons with spin fluctuations. The Green's function has an unusual structure, which is a consequence of the rotational symmetry. The self-energy is also unusual, and very complex, but the lowest-order term has a familiar form. The physical picture behind LBT requires that keeping only the lowest term be a good approximation. This probably fails at high temperatures in some systems, Invar for example, but recent experiments [4.19-22] on spin-polarized angle-resolved photoemission near and above the Curie temperature, and on inverse photoemission, indicate that it is quite good for paramagnetic nickel [4.16], and excellent for iron [4.17,23].

At lower temperatures this first self-energy term includes effects of spin-wave excitations on electronic states which are well known [4.24] from Fermi-liquid-theory arguments, but which previously have been difficult to include in the Green's function in a systematic way. Ground state correlations due to virtual spin waves are also incorporated naturally. Such correlations were studied some years ago by *Hertz* and *Edwards* [4.25], who found that a good treatment of vertex corrections is vital for a satisfactory result. A result similar to theirs was found even earlier by *Roth* [4.26] using a decoupling scheme, and later by *Liebsch* [4.27] using a complicated summation of ladder diagrams. In the strong limit at $T = 0$, which is the case carried through by those researchers, our results are quite similar to theirs, although we do not explicitly include vertex effects or ladder sums in the derivation. We have avoided them by maintaining explicit rotational invariance at every step.

As we are mainly concerned here with the temperature dependence, I shall not examine too closely the detailed nature of the ground state, but rather will assume a model which has proved successful, and add temperature effects to it. Among other things, this means that I will not include zero-point magnon effects when discussing finite temperatures. I shall also discuss briefly formal functional integral treatments, in order to understand how the zero-point spin-wave effects are included, and how other correlation effects might be. There is no need in our work to make the static approximation to the functional integral. Zero-point effects enter naturally when periodicity in imaginary time is imposed.

4.1.1 General Considerations

The ground state of itinerant electron ferromagnets is reasonably well described [4.28] by the local-spin density functional (LSDF) version of mean-field theory, (although there are some striking apparent failures of this model [4.29]). Here one finds the eigenstates of single electrons in a potential which includes a lattice term, and also a mean-field term to simulate the effects of electron-electron interaction. The mean field has the lattice periodicity, and includes spin independent and dependent parts. The charge and spin densities are determined by filling the eigenstates according to the Fermi factor, and the mean field, in turn, is found from these densities. The procedure is meant to be iterated to self consistency. In simple mean-field theories the spin-dependent term is proportional to the local spin density. The LSDF version uses a more complex dependence, which more accurately includes exchange and correlation effects.

In the ferromagnetic ground state the spin density (averaged over the unit cell) is uniform in direction. Electron states with spin in this direction (majority states) are lower in energy than those with opposite spins (minority states), because of the spin-dependent mean field. This energy splitting induces the non-zero spin density, whose strength is determined by the self-consistency condition.

In a magnetically isotropic system, this uniform spin density may point in any direction. There is no energy cost, nor change in energy eigenvalues, but now the majority and minority states point along the new magnetization direction – their wave functions have changed. Similarly, there is little energy cost in forming a state with a slow spatial variation of the spin-density direction. In this case the majority and minority states also have a spatially varying spin direction, locally almost parallel and antiparallel to the spin density.

These slowly varying magnetization states appear statically in domain structures, but also dynamically as long-wavelength spin waves, the dominant low-energy excitations of ferromagnetic systems. To determine temperature dependences correctly we must include the contribution of these excitations as well as the direct effect of the temperature dependence of the Fermi factors. Standard techniques, based on perturbations of the ground state keeping a fixed direction of the mean field, do not do this well, as they miss the local adjustments of the minority and majority-state wave functions. They do not easily recover the result that electronic energies are barely affected by a long-wavelength spin wave. Accurate vertex corrections are needed to take account of the wave-function adjustment, and they have only been worked out correctly in special cases.

To incorporate rotational invariance correctly, we must allow the mean field the freedom to vary in its local direction. Since this variation averages to zero, we cannot compute electronic states using the average field, but must treat each configuration of spin density directions separately when finding the electronic structure, with an average over configurations as the final step. In this way the adjustments of the wave function are easily incorporated. The procedure can be justified formally by using functional integrals, as will be discussed below. One problem is that performing these averages is difficult, even conceptually. Fortunately, in the approximation we adopt, the average is used only for computing spin-density correlation functions, and these can be estimated from experimental data.

The crucial problem is how to find the electronic states and energies in a field whose direction varies in space and time. Although formally quite general, our approach really relies on a number of observations which suggest the persistence of short-range magnetic order in iron and nickel even above the Curie temperature. These observations include a "direct" measurement [4.30,31] of the spin correlation function using quasielastic neutron scattering, an "indirect" measurement [4.32,33,7] which exploits the existence of short-wavelength spin-wave-like excitations above T_C, and thermodynamic arguments [4.34,12,14] which suggest that the paramagnetic state is locally very much like the ground state. Some of these results are controversial, and they have all been discussed at great length elsewhere [4.17]. The self-energy expression we find is written as an expansion in powers of space and time gradients of the mean-field directions. Short-range order is needed for this expansioin to be useful, and to make it possible to describe the system with only the lowest-order terms. Physically, we believe that local magnetic order persists above T_C because it is needed to maintain the energy splitting of majority and minority states, and thus the free energy gain of being magnetic. Our perturbation theory relies on changes of energy levels being small. Wave functions change dramatically, but we take this into account directly. Of course, this is not the only way to treat electronic states in a directionally varying mean field. An alternative approach (in which short-range order is not needed, indeed is difficult to include) uses the Coherent Potential Approximation (CPA). This will be described by *Gyorffy* in [4.35].

4.1.2 Green's Function Structure

Before the more detailed analysis we discuss the general form of our results. For simplicity, consider a one-band model with momentum-independent exchange splitting Δ in the ground state. Ignoring ground state correlations, the up- and down-spin Green's functions at $T = 0$ are

$$g_\pm(k, E) = [E - \varepsilon(k) \pm \Delta/2]^{-1} \ . \tag{4.1}$$

The spin density is

$$2\langle s \rangle = (2\pi)^{-4} \int d^3k \, dE (-2) \, \mathrm{Im}/, \{g_+(k, E) - g_-(k, E)\} f_0(E - \mu) \tag{4.2}$$

where f_0 is the Fermi function at temperature $T = 0$. The magnetization is fixed by requiring Δ to be a given function of $\langle s \rangle$, ($\Delta = 2U\langle s \rangle$, for example). Allowing f to be temperature dependent gives the Stoner model at finite T.

Attempts to include spin-wave effects in this model generally replace (4.1) by

$$g_\pm \to (W \pm 1 - \Sigma_\pm)^{-1} \tag{4.3}$$

where $\Sigma_\pm$ are self-energy corrections. [We write $E - \varepsilon(k) \equiv W$, and measure energies in units of $\Delta/2$, for simplicity]. Equations such as (4.3) are most useful if Σ is a smooth function. It is easy to see, however, that Σ is singular in (4.3) if the finite temperature state averages over configurations of slowly varying magnetization direction.

In a mean field with varying direction, the up- and down-spin Green's functions are components of a 2×2 matrix in spin space. In the ground state, (4.1) follows from the Green's function matrix

$$g_0 = (1 + \sigma^z)/2(W + 1) + (1 - \sigma^z)/2(W - 1) \ . \tag{4.4}$$

Similarly, if the magnetization is uniform in direction $\hat{n}$,

$$g_0 = (W - \hat{n} \cdot \boldsymbol{\sigma})/(W^2 - 1) \quad \text{and} \tag{4.5}$$

$$g_\pm = \mathrm{Tr}\{g_0(1 \pm \sigma^z)\}/4 = (W \mp \hat{n} \cdot \hat{z})/2(W^2 - 1) \ . \tag{4.6}$$

This already gives a singular Σ if put into the form of (4.3). In addition, (4.2) is no longer useful, as it gives $\langle \sigma^z \rangle$, where $\langle \boldsymbol{\sigma} \rangle \cdot \hat{n}$ is what is needed.

Of course, this is just a bad choice of coordinate system. But suppose the material has broken up into domains. Then one must average over $\hat{n}$ to find

$$g_\pm = (W \mp \langle \hat{n} \cdot \hat{z} \rangle)/2(W^2 - 1) \ , \quad \langle \hat{n} \cdot \hat{z} \rangle < 1 \ , \tag{4.6a}$$

and there is no choice of coordinate system in which Σ is not singular (and in which (4.2) is useful).

In the domain case it is customary to treat each domain separately, using (4.2 and 3) in each, and to put the macroscopic pieces together at the end. If the effective domains are small, and fluctuate in time (because they represent long-wavelength spin-wave excitations) such a separation is not possible. It is also not desirable since, as the domains get smaller, there finally must be an effect on the average electronic energies, which could not easily be taken into account in this manner.

The result of our analysis in Sects. 4.3 and 4, in contrast, joins smoothly onto (4.5 and 6) rather than (4.1). This is

$$g = [W - \Sigma - \langle \hat{n} \cdot \boldsymbol{\sigma} \rangle (1 - \Sigma')]/(W^2 - W\Sigma - 1) \ , \tag{4.7}$$

where Σ and Σ' vanish when gradients of the spin density do. These self-energy functions are smooth and have a simple form to second order in gradients. This lowest approximation recovers all known results about spin-wave effects at weak excitation, and satisfies all the requirements of rotational invariance, without vertex corrections.

In choosing configurations we choose $\hat{n}(r, t)$, the local direction of the mean field. The field strength depends on the magnitude of the local-spin density, which is $\hat{n}(r, t) \cdot \langle \boldsymbol{s}(r, t) \rangle$. We expect to find a positive contribution to this density from states near the majority spin energy, and a negative contribution from minority states. In fact, we find in place of (4.2) the expression

$$2 \langle \hat{n} \cdot \boldsymbol{s} \rangle = (2\pi)^{-4} \int d^3 k \, dW \, 2W \, \text{Im}/, \{ \text{Tr}(g) f_T (W + \varepsilon - \mu \} \ , \tag{4.8}$$

which has complete rotational symmetry, reduces to (4.2) in the fully aligned ground state, and has the other desirable properties described just above.

This provides most of the elements for a complete theory of itinerant electron ferromagnetism at finite temperatures. Given a mean-field potential (Δ in this one-band case), the expressions for Σ and Σ', and the temperature dependence of the spin direction correlation function, we can evaluate g from (4.7), and then find the particle and spin densities using g and (4.8). An appropriate prescription (LSDF, say) gives a corrected value for the mean field, and this is iterated to self consistency. In principle, the spin correlation function can itself be determined by using the total electronic energy for a given configuration as a weighting factor in a functional average. So far the program is far from complete. The rather complex self energies have not been evaluated for a realistic system, and the spin-correlation functions have not yet been predicted successfully.

4.1.3 Functional Integrals

The functional-integral analysis (which provides some formal justification for the treatment above) is based on the following identity [4.36]. Suppose the system has the Hamiltonian $H = H_0 - UA^2$. Then the partition function Z at temperature $T = 1/\beta k_B$ can be written as

$$Z \equiv \text{Tr}\{e^{-\beta H}\} = C\Pi_\tau \int_{-\infty}^{\infty} d\mu(\tau) \exp\left(-U \int_0^{\beta} \mu^2(\tau') d\tau'\right) Z\{\mu(\tau)\} \tag{4.9a}$$

113

$$Z\{\mu(\tau)\} \equiv \mathrm{Tr} \left\{ \exp \left[- \int_0^\beta [H_0 - 2U\mu(\tau')A]d\tau' \right] \right\}_+ . \tag{4.9b}$$

Equation (4.9a) is to be understood as the limit as $N \to \infty$ of the product of N integrals, where the interval $[0, \beta]$ is divided into N steps labelled by τ. The constant C becomes singular in the infinite N limit, but in such a way as to leave Z finite. The $+$ label in the definition of Z is "imaginary time ordering": the operators in this expression are ordered so that those with a larger value of τ stand to the left.

Suppose A is a density operator. Then (4.9b) replaces a problem of mutually interacting particles with one where the particles interact only with a fixed external field μ. The cost of this drastic simplification is the average in (4.9a) over all possible "histories" $\mu(\tau)$ of the field. In a three-dimensional problem there are operators A defined at every lattice point, or on a very fine grained lattice which fills space in the limit of a continuum model. Equations (4.9) are generalized accordingly. Then the average is over independent "histories" $\mu(\mathbf{r}, \tau)$ at every spatial point.

Averages of electron operators, Green's functions for example, are found by inserting the operators within the ordering bracket in (4.9b). Physical quantities are evaluated, then, by first computing them in the presence of a given "mean field" $\mu(\mathbf{r}, \tau)$, and averaging over the function μ as the last step. This corresponds to the procedure we outlined in Sect. 4.1.1. Because of the form of $Z\{\mu\}$, in particular the time ordering, all correlation functions (which are defined with operators evaluated at imaginary times $i\tau$, $i\tau'$, etc.) are periodic functions of these arguments in the range $[0, \beta]$. Physical correlations are found by analytic continuation, as described, for example, in the text by *Fetter* and *Walecka* [4.37]. It is this periodicity which brings in the Fermi and Bose factors needed for the correct quantum statistics.

One commonly used simplification is the "saddle-point" approximation, where only the single value of μ which makes the exponent in (4.9) stationary is retained. The stationarity condition is that $\mu = \langle A \rangle$. This is the standard mean field theory, with field $U\langle A \rangle$. Since the potentials μ simulate the effects of A in general, a close connection between μ and A is maintained in more accurate approximations. Evaluating physical quantities requires correlation functions of the $\mu(\tau)$, and these functions are closely related to the corresponding correlations of A.

Another common simplification is the static approximation, where μ has no τ dependence, although it takes the full range of values at every spatial point. This leads to a Ginzburg-Landau like free energy expression, which is an improvement over mean-field theory. The periodicity in τ is not respected, however, so effects of the quantum properties of A are lost. In particular, zero-point fluctuations of A are omitted. Theories based on the CPA must make the static approximation, as the CPA is formulated to deal only with stationary potentials. An important feature of our work is that we have no need to do so. We are able to write an expression for the one-particle Green's function and formally manipulate it into the form of (4.7), where Σ and Σ' are expressed in terms of correlations of the μ potentials. The dependence on τ is retained at every step, and we impose periodicity in the final expressions. Then no quantum effects are lost.

To go one step farther, consider a one-band Hubbard model

$$H = \sum_k \epsilon(k)n(k) + U \sum_i (n_{i+}n_{i-})$$

where $n_{i\pm}$ are the up- and down-spin density operators at site i. This may be rewritten in a number of ways to bring it into a form where the identity can be applied. Some of these are

$$4n_+ n_- = n^2 - (s^z)^2 \tag{4.10a}$$

$$= n^2 - \boldsymbol{s}\cdot\boldsymbol{s}/3 \tag{4.10b}$$

$$= n^2 - (\boldsymbol{s}\cdot\hat{\mu})^2 \tag{4.10c}$$

where $n = n_+ + n_-$ and $\boldsymbol{s} = \sum_{\alpha\beta} \psi_\alpha^\dagger \boldsymbol{\sigma} \psi_\beta$.

Equations (4.10a and b) have, respectively, two and four operators corresponding to A at every site, so require two and four auxiliary fields per site in forming the functional integral. These are the basis for the "two-field" and "four-field" methods used in the literature. The four-field method has the advantage of explicit rotational invariance, but there are difficulties associated with the factor of $1/3$. We prefer (4.10c) which maintains rotational invariance, but avoids the $1/3$ "problem". Two fields are needed in forming the functional integral, but also we must specify the unit vector $\hat{\mu}$, *which is arbitrary for every value of τ and i.* We choose to *average over $\hat{\mu}$ at every space-time point.* This yields a mean field with space- and time-dependent directions, and shows how we may justify the averaging procedure we have proposed to use.

In what follows we present a complete description of our analysis. Since much of this is novel and previously unpublished, we provide considerable algebraic detail, the worst of it appearing in several appendices.

We start in Sect. 4.2 with a direct treatment of electron states in a field with varying direction, paralleling elementary treatments of the Stoner model. This gives a direct look at how wave functions and energies depend on the excitation of long-wavelength magnons, how the self-consistent magnetization strength varies, and how the total energy and free energy behave. We also introduce here the very useful computational device of a locally rotated frame of reference. Much of the material of Sect. 4.2 was already presented in [4.2a], but we provide some additional details.

In Sect. 4.3 we discuss more formal Green's function manipulations, and find an expression for the Green's functions close to the final result. This allows us to make connection with known Landau-Fermi liquid results at low temperatures, and provides us with exact sum rules from which we can understand many features of the recent spin- and angle-resolved photoemission experiments [4.19–22]. This section uses techniques discussed by *Capellmann* [4.18] and by *Capellmann* and *Prange* [4.10], but carries them much farther. Most of this material is new, but some of the results were written down earlier [4.16].

In Sect. 4.4 we complete the Green's function analysis by incorporating the device of locally rotated frames of reference. This allows better account to be taken of spin-direction correlations, leads to the results in (4.7,8), and naturally describes zero-point effects. This material appears here for the first time.

Finally, in Sect. 4.5, we describe some simple model calculations of the predicted temperature dependence of photoemission spectra, and compare them with experimental results.

4.2 Single-Particle States and Energies

In this section we consider the eigenstates and energies for single electrons, in an otherwise uniform medium, subject to a self-consistent exchange field whose direction varies with position and time.

4.2.1 The Rotated Reference Frame

If the field directions are static, the wave functions satisfy

$$(-\hbar^2/2m)\nabla^2\psi_i(r) - (\Delta/2)\hat{n}(r)\cdot\sigma\psi_i(r) = E_i\psi_i(r) \ . \tag{4.11}$$

The self-consistency condition (for a simple mean-field theory) or the saddle-point condition (for functional integrals) is

$$\Delta/U = 2\langle\hat{n}(r)\cdot s(r)\rangle = V^{-1}\sum_i f(E_i)\int d^3r\,\psi_i^*(r)\hat{n}\cdot\sigma\psi_i(r) \ . \tag{4.12}$$

Here s is the spin density, $\hat{n}(r)$ is a unit vector in the direction of the exchange field, U is the exchange constant and $f(E)$ is the Fermi function.

Since the exchange energy is an energy of interaction, the total energy density of the electrons in the system is

$$\varepsilon = \sum_i f(E_i)E_i + \Delta^2/4U \ . \tag{4.13}$$

For slow variations of $\hat{n}(r)$, the eigenstates will be spinors whose spin direction closely follows $\hat{n}$. Spinors which always "point along or opposite to $\hat{n}$" are $\alpha_\pm(r)$ where

$$\hat{n}(r)\cdot\sigma\alpha_\pm(r) = \pm\alpha_\pm(r) \ . \tag{4.14}$$

Let $R(r)$ be the 2×2 matrix which rotates the z-axis into $\hat{n}(r)$. Specifically

$$R(r)\sigma^z R^{-1}(r) = \hat{n}(r)\cdot\boldsymbol{\sigma} \; . \tag{4.15}$$

Then

$$\alpha_+(r) = R(r)\begin{pmatrix}1\\0\end{pmatrix} \; , \quad \alpha_-(r) = R(r)\begin{pmatrix}0\\1\end{pmatrix}$$

and R is explicitly

$$R = \begin{pmatrix} \cos(\theta/2)\,e^{-i(\varphi+b)/2} & -\sin(\theta/2)\,e^{-i(\varphi-b)/2} \\ \sin(\theta/2)\,e^{i(\varphi-b)/2} & \cos(\theta/2)\,e^{i(\varphi+b)/2} \end{pmatrix} \tag{4.16}$$

where $\theta(r)$ and $\varphi(r)$ are the polar and azimuthal angles of the unit vector $\hat{n}(r)$, while $b(r)$ is the third Euler angle (usually called ψ) completing the specification of the coordinate system rotation. In our case $b(r)$ is arbitrary. $R(r)$ may also be written as

$$R = e^{-i\varphi\sigma^z/2}\,e^{-i\theta\sigma^y/2}\,e^{-ib\sigma^z/2} \; . \tag{4.17}$$

We express $\psi(r)$ in terms of α_+ and α_-. Formally this means

$$\psi(r) \equiv R(r)\chi(r) \; , \tag{4.18}$$

where $\chi(r)$ is the "spinor wave function in the rotated frame". Note that if $\psi(r) = \alpha_+(r)$ then $\chi(r) = \begin{pmatrix}1\\0\end{pmatrix}$. For slowly varying $\hat{n}$ we expect χ to deviate only slightly from this form. Using (4.11), $\chi_i(r)$ satisfies

$$-(\hbar^2/2m)R^{-1}\nabla^2(R\chi_i) - (\Delta/2)R^{-1}(\hat{n}\cdot\sigma)R\chi_i = E_i\chi_i \; .$$

Using (4.15) this is

$$-(\hbar^2/2m)\nabla^2\chi_i - (\Delta\sigma^z/2)\chi_i - (\hbar^2/2m)R^{-1}[\nabla^2,R]\chi_i = E_i\chi_i \tag{4.19}$$

while the consistency condition (4.12) is

$$\Delta/U = V^{-1}\sum_i f(E_i)\int d^3 r\,\chi_i^*(r)\sigma^z\chi_i(r) \; . \tag{4.20}$$

There is no change in (4.13) for the energy density.

If R is a constant matrix, then $[\nabla^2,R] = 0$, and (4.19,20) are the usual Stoner mean-field theory of the uniformly magnetized state. Deviations from the Stoner theory depend on space (and time) gradients of $\hat{n}(r)$. For weak gradients we use perturbation theory as follows.

Using (4.16,17) we find, after some algebra,

$$-iR^{-1}(r)\nabla R(r) \equiv \boldsymbol{A}(r) = \boldsymbol{a}^*(r)\sigma^+ + \boldsymbol{a}(r)\sigma^- - \boldsymbol{g}(r)\sigma^z \; . \tag{4.21a}$$

where

$$2\boldsymbol{a}(\mathrm{r}) = (\sin\theta\nabla\varphi - i\nabla\theta)e^{-ib} ,$$ (4.22a)

$$2\boldsymbol{g}(\mathrm{r}) = (\cos\theta\nabla\varphi + \nabla b) .$$ (4.22b)

Similarly, if θ, φ, and b are time dependent, we find

$$-i\mathrm{R}^{-1}\partial_t\mathrm{R} \equiv \hat{\mathrm{A}}(\mathrm{r},t) = \hat{a}^*\sigma^+ + \hat{a}\sigma^- - \hat{g}\sigma^z$$ (4.21b)

with

$$2\hat{a}(\mathrm{r},t) = (\sin\theta\partial_t\varphi - i\partial_t\theta)e^{-ib} ,$$ (4.22c)

$$2\hat{g}(\mathrm{r},t) = (\cos\theta\partial_t\varphi + \partial_t b) .$$ (4.22d)

Here we define $\sigma^\pm \equiv (\sigma^x \pm i\sigma^y)/2$. (We use a different convention in Sect. 4.4).

The quantities $\boldsymbol{a}(\mathrm{r},t)$, $\hat{a}(\mathrm{r},t)$, $\boldsymbol{g}(\mathrm{r},t)$, and $\hat{g}(\mathrm{r},t)$ characterize the directional disorder, and play an important role in what follows. Related to gradients of the three Euler angles, they are not all independent. Some relations among them are worked out in Appendix 4.A.1.

The perturbation term in (4.19) is

$$\mathrm{R}^{-1}[\nabla^2, \mathrm{R}] = i\nabla\cdot\boldsymbol{A} + 2i\boldsymbol{A}\cdot\nabla - \boldsymbol{A}\cdot\boldsymbol{A}$$ (4.23)

which gives

$$-(\hbar^2/2m)(\nabla - i\boldsymbol{g}\sigma^z)^2\chi - (\Delta\sigma^z/2)\chi + (\hbar^2/2m)\boldsymbol{a}^*\cdot\boldsymbol{a}\chi$$
$$+ (\hbar^2/2mi)[(2\boldsymbol{a}^*\cdot\nabla + \nabla\cdot\boldsymbol{a}^*)\sigma^+ + (2\boldsymbol{a}\cdot\nabla + \nabla\cdot\boldsymbol{a})\sigma^-]\chi = \mathrm{E}\chi .$$ (4.24)

This corresponds to the effective Hamiltonian density

$$\chi^\dagger(\mathrm{r})\mathrm{H}_{\mathrm{eff}}\chi(\mathrm{r}) = (\hbar^2/2m)[\, |(\nabla - i\boldsymbol{g}\sigma^z)\chi|^2 + |a|^2 |\chi|^2]$$
$$+ (\hbar^2/2mi)[\boldsymbol{a}^*\cdot(\chi^\dagger\sigma^+\nabla\chi - \nabla\chi^\dagger\sigma^+\chi) - cc] - (\Delta/2)\chi^\dagger\sigma^z\chi .$$ (4.25a)

If we include the time dependence of $\hat{n}$, then we must use the time dependent Schrödinger equation. The right-hand side of (4.19) is replaced by $i\hbar\partial_t\chi - \hbar\hat{A}\chi$. This adds to the effective Hamiltonian the new term

$$\chi^\dagger\mathrm{H}_{\mathrm{eff}}^t\chi = -\hbar\hat{g}\chi^\dagger\sigma^z\chi + \hbar\hat{a}^*\chi^\dagger\sigma^+\chi + \hbar\hat{a}\chi^\dagger\sigma^-\chi .$$ (4.25b)

Equations (4.25) describe a spinor particle of charge q in a uniform exchange field, subject to "electric" and "magnetic" fields derived from the scalar and vector "potentials" $\mp\hbar\hat{g}/q$, $\pm\hbar c\boldsymbol{g}/q$. The "fields" depend on the spin state. There are also spin-flip interactions whose strength is given by $\hat{a}$ and $\boldsymbol{a}$. The effective fields are found in Appendix 4.A.1, see (A.1.1a) and (A.1.2a). They do not depend on the arbitrary Euler angle $b(\mathrm{r},t)$. The arbitrariness appears in this context as gauge invariance.

4.2.2 Perturbation Theory

The "magnetic" field can only contribute to the energy in second order, which is fourth order in $\boldsymbol{a}$. We will neglect it to the order considered in this section. As will be seen later, because the spin wave spectrum is quadratic, $\hat{a}$ is second order in gradients, so the "electric" field is third order. Except for the term proportional to $\hat{g}$ it can also be neglected.

Eigenstates and energies are useful in a time-dependent situation if we may use an adiabatic approximation. That is, the right-hand side of (4.19) is set equal to $E_i(t)\psi_i(r,t)$, while (4.20) becomes a time average involving $\chi_i(r,t)$ and $E_i(t)$, and similarly for (4.13). Then $E_i(t) = \chi^\dagger(H_{eff} - \hbar\hat{A})\sigma$ and the explicit terms in $\hat{g}$ and $\hat{a}$ drop out in evaluating the energy. (Of course, $\hat{a}$ and especially $\nabla\hat{g}$ can play a role in determining χ, and thus indirectly E. But this is an effect of order higher than we consider).

The zeroth-order eigenstates of (4.25) are

$$\chi^0_{k+}(r) = e^{ik\cdot r}\begin{pmatrix} 1 \\ 0 \end{pmatrix} \quad \chi^0_{k-}(r) = e^{ik\cdot r}\begin{pmatrix} 0 \\ 1 \end{pmatrix}$$

with energies $E^0_\pm(k) = \epsilon(k) \mp \Delta/2$. To second order the energies are

$$E_{k\pm} = E^0_{k\pm} + \hbar^2\langle |a|^2\rangle/2m + \int \frac{d^3q}{(2\pi)^3} \frac{\langle\langle |\hbar^2/m)(k\mp q/2)\cdot a(r)|^2\rangle}{\epsilon(k) - \epsilon(k-q)\mp\Delta} \quad (4.26)$$

where the angular brackets are a space average.

To lowest order in gradients of $\hat{n}$, set $k + q \approx k + q/2 \approx k$ in the last term of (4.26). Then

$$E_{k\pm} \approx \epsilon(k) \mp \Delta/2 + \hbar^2\langle |a|^2\rangle/2m \mp \hbar^2\langle |v\cdot a|^2\rangle/\Delta . \quad (4.27)$$

For Δ we need the self-consistent value. The normalized "up" spin wave function is, to first order,

$$\chi^1_+(r) \approx e^{ik\cdot r}\begin{pmatrix} 1 \\ -\hbar v\cdot a(r)/\Delta \end{pmatrix} \{1 + \hbar^2\langle |v\cdot a|^2\rangle/\Delta^2\}^{-1/2} \quad (4.28)$$

and similarly for χ^1_-. Then (4.20) gives

$$\Delta/U \approx (V^{-1})\sum_k [f(E_{k+}) - f(E_{k-})][1 - 2\hbar^2\langle |v\cdot a|^2\rangle/\Delta^2] \quad (4.29a)$$

$$\Delta = \Delta_0 - 2\langle |v\cdot a|^2\rangle_{avg}/\Delta . \quad (4.29b)$$

We define $\Delta_0 = U(N^+ - N^-)/V$ in terms of the occupation number difference of "majority" and "minority" band eigenstates. The local exchange-field strength Δ is reduced from Δ_0 due to an average "tilt" of the eigenstates from the local net magnetization direction. The avg on the "tilt" term denotes an average over "singly occupied" band states, as implicitly defined in (4.29a). Then the eigen energies are

$$E_{k\pm} \approx \epsilon(k) \mp \Delta_0/2 + \hbar^2\langle |a|^2\rangle/2m \mp \hbar^2[\langle |v\cdot a|^2\rangle - \langle |v\cdot a|^2\rangle_{avg}]/\Delta . \quad (4.30)$$

Note that the average over the singly occupied states of the exchange splitting is Δ_0.

To get the total energy change due to magnetic disorder, we proceed in two steps. First find the change for fixed Δ_0, and then add to it the change in energy due to level repopulation, corresponding to the change from the ground state value Δ_{00} to the actual value Δ_0. Using (4.13) the first contribution is

$$\delta\varepsilon_1 = \sum_i f^0(E_i)\delta E_i + \Delta_0(\Delta - \Delta_0)/2U + \sum_i \delta f(E_i)E_i^0 \ . \tag{4.31a}$$

Here $\delta E_i \equiv \delta E_{k\pm}$ is the difference between (4.27) and the ground state value. For this term the majority and minority state chemical potentials are adjusted to keep the total majority and minority occupations fixed. Inserting δE_i we find

$$\delta\varepsilon_1 = (2\pi)^{-3} \int d^3k\{[f_+^0(k) + f_-^0(k)]\langle |a|^2\rangle/2m + [f_+^0(k) - f_-^0(k)]$$
$$\times \langle |v \cdot a|^2\rangle/\Delta_0\} + \sum_i E_i^0 \delta f(E_i) \ . \tag{4.31b}$$

The first term is the random-phase-approximation (RPA) expression for $4A \langle |a|^2\rangle$ where A is the Bloch wall constant. $[A = (N_+ - N_-)D/4$, D being the spin-wave stiffness and $N_\pm$ the majority and minority occupation numbers.] It is a simple exercise to show (for fixed N_+ and N_-) that the second term is $-T\partial/\partial T$ of the first, which identifies $4A\langle |a|^2\rangle$ as the contribution to the *free energy* due to directional disorder at fixed occupations. The relation $4a^* \cdot a = (\nabla\hat{M})^2$ puts this in a more familiar form.

The effect of population change is computed for vanishing directional disorder, and is a standard Stoner model calculation. Adding the two terms we find the free energy for a given configuration of direction unit vectors

$$F = 4A \int d^3r \, |a(r)|^2 + X_S^{-1}(\Delta_0 - \Delta_{00})^2/2 \tag{4.32}$$

where X_S is proportional to the Stoner susceptibility *in the fully ordered state*.

We can find the equilibrium population change induced by directional disorder by minimizing F with respect to $\Delta_0 - \Delta_{00}$ then

$$(\Delta_0 - \Delta_{00}) = -4\langle |a|^2\rangle X_S(\partial A/\partial \Delta_0) \ . \tag{4.33}$$

Explicit expressions for the Stoner $\partial A/\partial \Delta_0$ and X_S are in [Ref. 4.2a, Sect. III.C] and will not be reproduced here. In fact, (4.32) is more general than the RPA. It is still correct when the physical values of A and X are used. Since X is extremely small in both iron and nickel, and A is not particularly sensitive to Δ_0, we conclude that disorder gives very little population change, and that the average exchange splitting is a very weak function of temperature, in these materials.

4.2.3 Discussion

The results of this section provide support for the conceptual picture of the local-band theory. An itinerant electron system can sustain a large loss of magnetization, yet retain its band structure essentially intact, if the magnetization loss is due to slowly varying directional disorder. In this case the energy changes (4.30) can be very small. Rather than the single variable of the Stoner model, which determined both the magnetization and the band energies, we now have separate variables, which are only weakly coupled. There is a natural separation of magnetic fluctuations into longitudinal ones – changes of Δ_0 – against which the system is very stiff, and transverse ones – directional fluctuations – which are easily established. The latter control the bulk magnetization of the system, as in the Heisenberg model. The former control the strength of local magnetism, and seem to play a relatively small role until well above the Curie temperature, at least in nickel and iron.

While this is conceptually very satisfactory, the results of this section are not useful for comparison with experiments such as photoelectron spectroscopy, since the eigenstates we consider are not directly observable. Experiments determine the spectrum of states of a given momentum and spin, which are complex combinations of the eigenstates. In the next two sections we discuss calculations of the average Green's functions of a spatially disordered magnetic system.

4.3 Green's Functions

In this section I reformulate the LBT in terms of measurable quantities, the thermodynamic single-particle Green's functions. I carry out the formal analysis in considerable generality, for arbitrary bands, allowing for time dependence and, initially, allowing for fluctuations in the local magnetization magnitude. The result found here, although correct, is completely satisfactory only in the static approximation, and I will present a "better" (including more correlation effects, but also more complex) analysis in Sect. 4.4. Nevertheless, the expressions found here are particularly useful for finding moments of the Green's functions, and I will discuss implications of these moments for the interpretation of photoelectron line shapes. These Green's functions are also convenient, in the low-excitation limit, to establish that they correctly incorporate physical effects which are known to be important on other grounds.

4.3.1 Formal Analysis

Consider an electronic system described by the single-particle Hamiltonian

$$H_{mag} = H - M(r, t) \qquad (4.34a)$$

where H contains the kinetic energy and a local, spin-independent potential $V(r)$, while M is a local exchange potential with arbitrary orientation. In what follows we will use a matrix notation, where the product of operators AB stands for

$$AB \equiv \sum_{\sigma''} \int dr'' dt'' A_{\sigma\sigma''}(rt, r''t'') B_{\sigma''\sigma'}(r''t'', r't') \ .$$

Then the Hamiltonian contains

$$H = [- (\hbar^2/2m)\nabla^2 \delta(r - r') + V(r)\delta(r - r')]\delta(t - t')\delta_{\sigma\sigma'} \ , \qquad (4.34b)$$

$$M = \boldsymbol{\Delta}(r, t) \cdot \boldsymbol{\sigma} \delta(r - r') \delta(t - t')/2 \ , \tag{4.34c}$$

$$\boldsymbol{\Delta}(r, t) \equiv \hat{n}(r, t) \Delta(rt) \ . \tag{4.34d}$$

We think of $V(r)$ and $\Delta(rt)$ as the local potentials in an LSDF calculation, with directional fluctuations added. Then they are to be determined self-consistently in terms of the derived charge and spin density. Alternatively, V and Δ are the auxiliary fields in a functional integral scheme. Then physical properties are found by averaging over these fields, and the analysis proceeds in imaginary time. In either case we must find the single-particle Green's function for the given functions $V(r)$ and $\Delta(rt)$, and average over the directions of the latter, at least. For simplicity, we omit fluctuations in $V(r)$ from the outset, since the LSDF scheme works quite well in normal metals. Then the Green's function is the average

$$G_{\sigma\sigma'}(rt, r't') \equiv \langle \hat{g}_{\sigma\sigma'}(rt, r't') \rangle \ , \tag{4.35}$$

$$\hat{g} = \{T - H + M\}^{-1} \ , \tag{4.36}$$

$$T \equiv i\hbar(\partial/\partial t)\delta(t - t')\delta(r - r')\delta_{\sigma\sigma'} \ . \tag{4.37}$$

Write

$$\begin{aligned} \hat{g} &= (T - H - M)[(T - H + M)(T - H - M)]^{-1} \\ &= (T - H - M)\{(T - H)^2 - M^2 - [T - H, M]\}^{-1} \ . \end{aligned} \tag{4.38}$$

Define

$$(T - H)^2 - M^2 \equiv \gamma^{-1} \ , \tag{4.39}$$

$$[T - H, M] \equiv K \ . \tag{4.40}$$

It is K which contains information about the space and time gradients of $\boldsymbol{\Delta}$. Note that

$$M^2 = (\Delta^2/4)\delta_{\sigma\sigma'} \tag{4.41}$$

proportional to the unit matrix in spin space. Then

$$\begin{aligned} \hat{g} &= (T - H - M)\gamma\{1 - K\gamma\}^{-1} \\ &= (T - H - M)\gamma(1 + K\gamma)\{1 - K\gamma K\gamma\}^{-1} \\ &= \{(T - H - M\gamma K) + [(T - H)\gamma K - M]\}(\gamma^{-1} - K\gamma K)^{-1} \ . \end{aligned} \tag{4.42}$$

We have broken $\hat{g}$ into two terms in which only even or only odd powers of M appear. In Appendix 4.A.3 we manipulate (4.42) to rewrite these two expressions as

$$\hat{g}_e = [T - H - (\Delta/2)(T - H - \delta\hat{\Sigma})^{-1}(\Delta/2)]^{-1} \ , \tag{4.43a}$$

$$\hat{g}_o = -\hat{n}\cdot\boldsymbol{\sigma}(\Delta/2)[(T-H)(2/\Delta)(T-H-\delta\hat{\Sigma})(\Delta/2) - \Delta^2/4]^{-1} , \quad (4.43b)$$

$$\delta\hat{\Sigma} = -\hat{n}\cdot\boldsymbol{\sigma}(T-H,\hat{n}\cdot\boldsymbol{\sigma}) . \quad (4.43c)$$

So far this is purely formal. The quantity $\delta\hat{\Sigma}$ depends on the fluctuating quantity $\hat{n}(rt)$, and averaging (4.43) is no easier than averaging (4.36). The advantage is that the physically relevant *gradients* of $\hat{n}$ appear here explicitly. To get a useful result, we expand $\hat{g}$ in powers of $\delta\hat{\Sigma}$ and formally average over fluctuations, to find expressions similar to (4.43) for the *averaged* Green's functions.

$$G_e = \langle\hat{g}_e\rangle = [T-H-(\Delta/2)(T-H-\Sigma)^{-1}(\Delta/2)]^{-1} , \quad (4.44a)$$

$$G_o = \langle\hat{g}_o\rangle = -\overline{\Sigma}[(T-H)(2/\Delta)(T-H-\Sigma)(\Delta/2) - \Delta^2/4]^{-1} . \quad (4.44b)$$

Details are in Appendix 4.A.4. Henceforth we no longer allow fluctuations in the magnitude $\Delta(r)$, and consider only the effects of directional fluctuations. As discussed briefly in Appendix 4.A.4, including both effects would require the appearance of many-particle ground state correlations in place of the single-particle ground state Green's functions which are in Σ. The quantities Σ and $\overline{\Sigma}$ are the fluctuation averages

$$\Sigma = \langle\delta\hat{\Sigma} + \delta\hat{\Sigma}\Gamma\delta\hat{\Sigma} + \ldots\rangle_{1-\text{irred}} \quad (4.44c)$$

$$\overline{\Sigma} = \langle\hat{n}\cdot\boldsymbol{\sigma} + \hat{n}\cdot\boldsymbol{\sigma}\Gamma\delta\hat{\Sigma} + \ldots\rangle_{1-\text{irred}}(\Delta/2) \quad (4.44d)$$

where

$$2\Gamma = \{(T-H+\Delta/2)^{-1} + (T-H-\Delta/2)^{-1}\}\delta_{\sigma\sigma'} \equiv (g_+ + g_-)1 \quad (4.44e)$$

and the notation $1 - \text{irred}$ is defined after (A.4.7).

As averages over fluctuations, Σ and $\overline{\Sigma}$ have the translational symmetry of the lattice and full time translational invariance, while Δ and H are lattice periodic. If the net magnetization $\langle\hat{n}\rangle$ defines the z-direction, then Σ and $\overline{\Sigma}$ each include only terms proportional to σ^z and 1, and are thus diagonal in the spin space, as are G_e and G_o. Given the correlation functions of the direction vectors $\hat{n}(rt)$, (4.44a–e) provide an explicit scheme for evaluating the disordered state Green's functions as a perturbation expansion in directional gradients. The self-consistency condition for the exchange field, or the saddle-point condition for the functional integral, is

$$\Delta(r) = U \int (dE/2\pi)f(E)(-2\,\text{Im})G_n(rrE) , \quad (4.45)$$

where $G_n(E)$ is the Fourier transform of the component of the spin density in its local direction,

$$G_n(rt, r't') = \text{tr}\langle\hat{n}(rt)\cdot\boldsymbol{\sigma}\hat{g}(rt, r't')\rangle , \quad (4.46)$$

123

In these equations $\hat{g}(E)$ is the limit as E approaches the real energy axis from the upper half plane. In the imaginary time case the usual connection between time ordered and retarded Green's functions has to be made.

From (4.34) and (4.36) we write

$$G_n(rr'E) = (2/\Delta(r))[2 - (E - H)\text{tr}G] \qquad (4.46a)$$

since $E - H$ is proportional to the unit spin matrix. Then (4.45) is

$$\Delta(r)/U = (2/\Delta(r)) \int (dE/\pi)f(E)(E - H) \text{ Im tr}G(rrE) . \qquad (4.47)$$

Similarly, using a standard trick, the total energy density is

$$\varepsilon = \int (dE/\pi)f(E)E(- \text{ Im })\text{tr}G(rrE) + \Delta(r)^2/4U \qquad (4.48)$$

4.3.2 Limiting Cases

We discussed in Sect. 4.1 the Green's functions in the limit of weak directional gradients. In that limit $\Sigma \rightarrow 0$ while $\overline{\Sigma} \rightarrow \langle n \cdot \sigma \rangle \Delta/2$. Then the sum of (4.44a and b) reduce to (4.5) while (4.47) becomes (4.8), Equation (4.48) reduces to the Stoner-model energy, independent of domain effects which could even reduce the net magnetization to zero. We do have the correct behavior in this case.

The next "check" of (4.44) is more involved. We analyze the Green's functions when $\delta\hat{\Sigma}$ and Σ are small but not negligible, and compare them with a Fermi-liquid analysis of a system of interacting electrons and spin waves. For this we require Σ and $\overline{\Sigma}$ to second order in directional gradients.

Define

$$m \equiv \hat{n} \cdot \sigma , \quad k \equiv [T - H, m] . \qquad (4.49)$$

We show in Appendix 4.A.5 that, to second order,

$$\Sigma = \Sigma_+ + \Sigma_- , \qquad (4.50)$$

$$2\overline{\Sigma}/\Delta = \langle m \rangle (1 - g_+\Sigma_+ - g_-\Sigma_-) , \qquad (4.51)$$

where

$$\Sigma_\pm \equiv - (\langle mk \rangle + \langle kg_\pm k \rangle)/2 . \qquad (4.52)$$

To evaluate $\Sigma_\pm$, use (4.37) for T and write H as $H(r, r')\delta(t - t')1$. Then

$$k = \sigma \cdot \{i\hbar\partial_t \hat{n}(rt)\delta(r - r') - H(r, r')[\hat{n}(r't) - \hat{n}(rt)]\} . \qquad (4.53)$$

Define

$$X(rt, r't') = \langle \hat{n}(rt) \cdot \sigma \hat{n}(r't') \cdot \sigma \rangle - \langle \hat{n}(rt) \rangle \cdot \sigma \langle \hat{n}(r't') \rangle \cdot \sigma$$

$$= \langle \delta\hat{n}(rt) \cdot \delta\hat{n}(r't') \rangle + i\sigma \cdot \langle \delta\hat{n}(rt) \times \delta\hat{n}(r't') \rangle \qquad (4.54)$$

where $\delta\hat{n}$ is the deviation from the average.

124

Then, for example,

$$-\langle mk \rangle = \{H(r, r')[X(rr') - X(rr)] - i\hbar\partial_{t'}X(rt, rt')\delta(r - r')\}\delta(t - t') .$$

This and other terms can be expanded in terms of band eigenstates to find a general, but unwieldy expression. We only write it down for a one-band case, with constant Δ, where X is a function only of coordinate differences. Then

$$2\Sigma_{\pm} = (2\pi)^{-4} \int d^3q \, d\Omega X(q, \Omega)[V(k, q, \Omega)$$
$$+ V(k, q, \Omega)^2 g_{\pm}(k - q, E - \hbar\Omega)] \tag{4.55}$$

with

$$V(k, q, \Omega) \equiv \epsilon(k - q) - \epsilon(k) + \hbar\Omega \tag{4.56}$$

and $\epsilon(k)$ the band energy.

The first term in (4.55) is just $-\langle mk \rangle$. If we expand $V(k, q, \Omega)$ in powers of q, the linear term gives no contribution to the integral if X has reflection symmetry, which we assume. Since the spin-wave spectrum is quadratic, $\hbar\Omega$ is also effectively quadratic in q. Then $\langle mk \rangle$ is second order in gradients of $\hat{n}$, as claimed before (A.5.5).

In this one-band model, (4.44a,b) are

$$G_e(k, E) = \{E - \epsilon(k) - \Delta^2/4[E - \epsilon(k) - \Sigma(k, E)]\}^{-1} , \tag{4.57a}$$

$$G_o(k, E) = -\overline{\Sigma}(k, E)\{[E - \epsilon(k)][E - \epsilon(k) - \Sigma(k, E)] - \Delta^2/4\}^{-1} . \tag{4.57b}$$

When $\Sigma = 0$, $G(k, E)$ has poles at $E = \epsilon \mp \Delta/2$, with residues $(1 \pm \hat{n}\cdot\boldsymbol{\sigma})/2$. At small, but finite Σ the poles are shifted to

$$E \approx \epsilon(k) \mp \Delta/2 + \Sigma(k, \epsilon(k) \mp \Delta/2)/2 \tag{4.58}$$

with residues

$$R \approx (1 \pm \Sigma/\Delta \pm 2\overline{\Sigma}/\Delta)/(2 - \Sigma') \tag{4.59}$$

where $\Sigma' = d\Sigma/dE$ and Σ, $\overline{\Sigma}$, and Σ' are to be evaluated at $E = \epsilon(k) \mp \Delta/2$. In addition to these poles, G also has spectral weight wherever Σ or $\overline{\Sigma}$ has a non-vanishing imaginary part.

[A complication of these formulas is that Σ, $\overline{\Sigma}$, G, and the residues in (4.59) are (diagonal) matrices in spin space. This comes from the σ^z component of X, seen in (4.54). From (4.50,51,55) we see that Σ contains a σ^z part, while $\overline{\Sigma}$ is dominantly proportional to σ^z, but contains a part proportional to the unit matrix as well. In what follows we treat separately the two components G^{++} and G^{--}, and the corresponding self energies, poles and residues. However, we will only write the component labels where needed for clarity.]

To evaluate G write

$$2\Sigma_{\pm}(E) = \langle\langle V + V^2/[E - \epsilon(k)\pm\Delta/2 - V]\rangle\rangle \quad \text{where} \tag{4.60}$$

$$\langle\langle f\rangle\rangle \equiv (2\pi)^{-4} \int d^3q\, d\Omega\, X(q,\Omega) f \ . \tag{4.61}$$

Keeping terms to order V^2 we find

$$2\Sigma(\epsilon\mp\Delta/2) = \langle\langle V\mp V^2/\Delta\rangle\rangle \ , \tag{4.62a}$$

$$2\Sigma'(\epsilon\mp\Delta/2) = -\langle\langle 1 + V^2/\Delta^2\rangle\rangle \ , \tag{4.62b}$$

$$2\overline{\Sigma}(\epsilon\mp\Delta/2)/\Delta = \langle m\rangle[1 + \langle\langle 1\pm V/\Delta - V^2/\Delta^2\rangle\rangle/2] \ . \tag{4.62c}$$

The shifted energies are

$$E\approx\epsilon(k)\mp\Delta/2 + (2\pi)^{-4} \int d^3q\, d\Omega[X(q,\Omega)/4]$$
$$\times[q^2/2m + \hbar\Omega\mp(\boldsymbol{v}\cdot\boldsymbol{q})^2/\Delta] \ . \tag{4.63}$$

With the identification, see (4.22a),

$$\int d^3q\, d\Omega\, q^2 X(q,\Omega)/(2\pi)^4 = 4\hbar^2\langle|a|^2\rangle \ . \tag{4.64}$$

Equation (4.63) is the same as (4.27) *except for the term in* Ω. We will return to discuss this additional term in what follows.

From (4.59 and 62) we evaluate the residues as

$$R\approx(1\pm\langle m\rangle)[1\pm\langle\langle V\rangle\rangle/2\Delta - \langle\langle V^2\rangle\rangle/\Delta^2]/2 - \langle\langle 1\rangle\rangle/4 \ . \tag{4.65}$$

For the moment ignore the terms $\langle\langle V\rangle\rangle$ and $\langle\langle V^2\rangle\rangle$ which depend on gradients. Suppose the net magnetization defines the direction z, so that $\langle m\rangle = \mu\sigma^z$, μ being the fractional polarization. Equations (4.54 and 61) give the evaluation

$$\langle\langle 1\rangle\rangle = (2\pi)^{-4} \int d^3q d\Omega X(q,\Omega) = (1 - \mu^2) \ . \tag{4.66}$$

Let N be the density of "up" spins in a fully polarized system. If ν spins per unit volume are turned over, then $\mu = (N - 2\nu)/N$. thus

$$1 - \mu^2\approx(4\nu/N) \tag{4.67}$$

where ν is the "number of excited spinwaves" and N the "number of magnetic electrons" per unit volume. (Since spin waves in a disordered state do not each result in one turned over spin, this is a useful identification only in an almost fully polarized system).

The Green's function G^{++} for a particle with spin up in the laboratory frame of reference is

$$G^{++} = \text{Tr}\{(1 + \sigma^z)G/2\} \ . \tag{4.68}$$

To get the residues, set $\langle m \rangle = +\mu$ in (4.65). Then

$$R^+(\epsilon - \Delta/2) = (1 + \mu)/2 - (1 - \mu^2)/4 \approx 1 - 2\nu/N \ ,$$

$$R^+(\epsilon + \Delta/2) = (1 - \mu)/2 - (1 - \mu^2)/4 \approx (\nu/N)^2$$

with similar results for G^{--}. The occupations of up- and down-spin electron states are appropriately related to the *number* of spin waves excited, although only the *gradient* correlations appear in the self energies.

The total spectral weight of G^{++} is 1, so the weight absent from the poles reappears in subsidiary peaks. Since Σ has a non-vanishing imaginary part near $E = \epsilon \pm \Delta/2$, this is where the peaks are located. By examining the origins of the various terms in (4.65), we conclude that the weight associated with $\langle\langle 1 \rangle\rangle$ reappears in the immediate vicinity of the pole it is associated with, while the weights $\langle\langle V \rangle\rangle$ and $\langle\langle V^2 \rangle\rangle$ move to the neighborhood of the opposite spin pole. Considering (4.47), then, the self consistent Δ is reduced not by the number of spin waves excited, but by the direction fluctuations associated with them. More quantitatively, write (4.47) as

$$\Delta/U = (2/\Delta) \sum_i \{(\Delta/2 - \delta\epsilon)[1 + \langle\langle V \rangle\rangle/2\Delta - \langle\langle V^2 \rangle\rangle/\Delta^2]\} \ .$$

The sum is over states such that the up-spin energy is below the Fermi energy and the down-spin energy above. The square bracket is the weight in $\text{Tr}\{G\}$ located near the up-spin energy. This includes the weight in the peak which stays near the pole. The term $\delta\epsilon$ is the shift of the pole as required by (4.47). In this notation, (4.63) is

$$\delta\epsilon = \langle\langle V \rangle\rangle/4 - \langle\langle V^2 \rangle\rangle/4\Delta \ .$$

Then

$$\Delta/U = \sum(1 - \langle\langle V^2 \rangle\rangle/2\Delta^2)$$

which, taking account of (4.64), is the same as (4.29). We find the same change of Δ as in Sect. 4.2. Equation (4.63) shows that the poles of G are displaced from the eigenenergies found in that subsection only by a term related to the average frequency in the disorder correlations.

4.3.3 Fermi-Liquid Theory

In Fermi-liquid theory the poles of the Green's functions are associated with quasi-particles, of energy $\epsilon_\pm(k)$, interacting with spin waves whose energy is $\Omega(q)$. Symmetry considerations require $\Omega \propto q^2$ and we write $\Omega = Dq^2$. The free energy is related to electron and spin-wave excitations by

$$\delta F = \sum_{k,\sigma} \epsilon_\sigma(k)\delta n_\sigma(k) + \sum_q Dq^2 \delta N(q) \ .$$

127

Taking mixed second derivatives we find

$$q^2 \delta D / \delta n_\sigma(k) = \delta \epsilon_\sigma(k) / \delta N(q) \ . \tag{4.69}$$

Using the relationship $4A = (N_+ - N_-)D$, – see remarks after (4.31) – we write

$$\delta \epsilon_\sigma(k) = \sum_q q^2 \delta N(q)[\delta(4A)/\delta n_\sigma(k) \mp D]/(N_+ - N_-) \ . \tag{4.70}$$

Then, using (4.31b) for $4A$, taking care to differentiate the Δ_0 in the denominator, we find

$$\delta \epsilon_\sigma(k) = \sum_q \delta N(q)\{q^2/2m \mp [(\boldsymbol{v}\cdot\boldsymbol{q})^2 - (\boldsymbol{v}\cdot\boldsymbol{q})^2_{\text{avg}}]/\Delta_0 \mp Dq^2\}/(N_+ - N_-) \ . \tag{4.71}$$

It is easy to see from (4.54,64,66, and 67) that

$$\delta N(q)/(N_+ - N_-) = (2\pi)^{-1} \int d\Omega \, X(q, \Omega)/4 \ . \tag{4.72}$$

Further, from (4.54),

$$X(rt, r't') = \langle \delta n^z(rt)\delta n^z(r't')\rangle$$
$$+ (1 + \sigma^z)\langle \delta n^- \delta n^+\rangle/2 + (1 - \sigma^z)\langle \delta n^+ \delta n^-\rangle/2 \ . \tag{4.73}$$

Then, since it is n^+ which evolves with frequency $+Dq^2$,

$$(2\pi)^{-1} \int d\Omega \, \Omega X(q, \Omega) = \mp 4Dq^2 \delta N(q)/(N_+ - N_-) \ , \tag{4.74}$$

where the signs correspond to the $(++)$ and $(--)$ components of G. Putting these together in (4.63), the position of the pole we find is precisely that required by Fermi-liquid theory, using the RPA evaluation of D, which is consistent with the other approximations we have made.

I comment on the difference between the eigenenergy and the position of the pole in G. As we saw in Sect. 4.2, the system eigenstates accounted for all the energy in the system – there are only electrons after all. In Fermi-liquid theory, however, the poles represent the contribution of only the electronic quasi-particles. There is also energy separately associated with the spin waves. The energy discrepancy $\pm \delta N D q^2/(N_+ - N_-)$ is just enough, summed over occupied states, to balance the spin-wave contribution. The shift is needed for Fermi-liquid theory to work. An alternate point of view is that the shift represents a simple Doppler shift associated with the precession of the electron as it follows the directional fluctuations. In any case, having the full Green's function allows the unambiguous evaluation of physical properties, so the precise interpretation is not important.

4.3.4 Green's Function Moments

Electronic spectrum measurements, such as angle-resolved photoemission, (ideally) measure the imaginary part of $G^{\sigma\sigma}(k, E)$ as a function of E for fixed k and σ. To predict the detailed shape of this spectral function requires a careful analysis of approximations to G, such as we have been discussing. It is very useful, however, first to examine some exact results for the energy moments of the spectrum. At least in simple cases these allow us to relate gross features of the spectrum (mean energy, line width) to particular properties of the underlying system.

To find the moments, we use an alternative form of the (unaveraged) Green's function

$$\hat{g}_e = \Gamma - \Gamma'(mk + k\Gamma k)(1 - \Gamma'k\Gamma'k)^{-1}\Gamma' , \tag{4.75a}$$

$$\hat{g}_o = (m - \Gamma k)(1 - \Gamma'k\Gamma'k)^{-1}\Gamma' , \tag{4.75b}$$

which are derived in Appendix 4.A.6. Note that Γ' is defined in (A.4.4) and Γ in (4.44e).

Moments of $\mathrm{Im}\{G\}$ are found by expanding G in inverse powers of E. The relation is

$$G(E) \sim \sum_{n=1}^{\infty} \langle E^{n-1} \rangle / E^n \tag{4.76a}$$

where

$$\langle E^n \rangle = \int (dE'/\pi)[- \mathrm{Im}\{G(E')\}](E')^n . \tag{4.76b}$$

Of course, G, and the moments, depend on the particular band state and spin projection measured.

The convenience of (4.75) follows from the observation that $\Gamma'(E) \sim 1/E^2$ while $\Gamma(E) \sim 1/E$. Then the denominator of (4.75a) first contributes to the expansion in order $1/E^8$. To order $1/E^9$ the denominator in (4.75b) can be expanded to lowest order. Then all the moments up to $\langle E^6 \rangle$ are precisely the corresponding moments of

$$G_6 \equiv \Gamma + \langle m \rangle \Gamma' - \Gamma' \langle mk + k\Gamma k \rangle \Gamma' + \langle m\Gamma'k\Gamma'k \rangle \Gamma' - \Gamma \langle k\Gamma'k\Gamma'k \rangle \Gamma' . \tag{4.77}$$

The first two terms of (4.77) are simply the domain result, say (4.6a). Since this is accurate to order $1/E^3$, the mean energy and line width reflect *only* the net magnetization and local exchange potential strength. In the one-band model used earlier

$$\langle E(k) \rangle^{\pm} = \epsilon(k) \mp \mu\Delta/2 , \tag{4.78a}$$

$$\langle [E(k) - \epsilon(k)]^2 \rangle^{\pm} = \Delta^2/4 . \tag{4.78b}$$

In principle, just these two moments allow one to distinguish between the Stoner model and some version of a fluctuating band model. In a Stoner model the up- and down-spin spectra will move together as the temperature increases, since Δ reduces to zero at the Curie point. Equation (4.78b) then requires lines with no magnetic component to their width. The fluctuating band models may also have up- and down-spin spectra which remain single lines, as

T increases. If so they will merge, but now because μ is decreasing to zero at T_C, while Δ retains a substantial fraction of its low-temperature value. Then (4.78b) requires a magnetic line broadening, and a measured broadening will provide an estimate for Δ. Of course, (4.78) is also consistent with the domain prediction, a two-peak structure for each spin at finite temperature, with a constant splitting equal to Δ. The "merging line" situation seems to describe recent experiments on nickel [4.19], while the "domain" case is apparently seen [4.20] at the Γ point of iron. As a practical matter, however, it is hard to reach conclusions based on just a few low moments, since the spectral functions may distort, as the temperature rises, and there may be mixing with other spectral features. The line wings contribute heavily to moments, and a measurement of the line width may well be very inaccurate. There are also other sources of line broadening which must be considered.

The moment analysis also suggests when the domain behavior or the single-line behavior is likely to occur. Corrections to the domain Green's function first appear in the third and fourth moments, due to the third term in (4.77). Comparing with (4.52), this is proportional to the second-order expression for Σ. In the one-band model, (4.55 and 61) show that the relevant terms are $\langle\langle V \rangle\rangle$ and $\langle\langle V^2 \rangle\rangle$. In particular,

$$\langle [E(k) - \epsilon(k)]^3 \rangle^\pm = (\Delta/2)^3 [\mp\mu + 2\langle\langle V \rangle\rangle/\Delta] \ , \tag{4.78c}$$

$$\langle [E(k) - \epsilon(k)]^4 \rangle^\pm = (\Delta/2)^4 [1 + 4\langle\langle V^2 \rangle\rangle/\Delta^2] \ . \tag{4.78d}$$

Using (4.56 and 61) we see that the corrections depend on the range of band energies mixed with the state of interest when absorbing or emitting a typical spin excitation in the thermal fluctuations. If this range is small compared to $\Delta/2$, corrections to the domain situation are small, and a two-peak line will be seen. If the range is comparable to Δ there will be considerable distortion, and a one-peak spectrum may result. Estimates for the parameters in nickel and iron support [4.16,17] this interpretation, as will be detailed in Sect. 4.5.

4.4 Green's Functions Revisited: The Rotated Frame

The expressions found for G in Sect. 4.3, are somewhat unsatisfactory, in two respects. The minor problem is that they would be easier to evaluate if Σ and $\overline{\Sigma}$ were spin scalars rather than matrices. The more significant problem can be seen in (4.55).

The final term in (4.55) has the form of a Feynman self-energy diagram describing the absorption or emission of a spin fluctuation by an electron. However, as it appears, *both* absorption *and* emission are allowed for *either* minority- or majority-spin electron line. This is unphysical since emission of a spin fluctuation requires that a minority electron "flip" into the majority state to conserve the local spin density. Our expansion of Σ to second order omits those fluctuation correlations which would enforce that rule.

By using the transformation to a locally rotated reference frame, as in Sect. 4.2, we solve both these problems, albeit at the expense of other complications. The results are very close to those of Sect. 4.3 (differing only when the fluctuations have a time dependence, so emission and absorption are distinguished). An extra advantage is that zero-point spin fluctuations are now incorporated in a natural way.

4.4.1 Formal Analysis

It is convenient in this section to think of the quantities $\psi_\alpha(\mathrm{rt})$ as Fermion field operators, rather than as wave functions.

As in (4.18), we expand the field operators in a basis of eigenspinors of the local exchange field direction. However, the final Green's functions are in the laboratory frame, and are expressed in terms of eigenspinors of σ^z. This leads us to define *projected* operators

$$\psi_\alpha^\mu(\mathrm{rt}) = [R(\mathrm{rt})\sigma^{\mu\dagger}R^{-1}(\mathrm{rt})]_{\alpha\beta}\psi_\beta(\mathrm{rt}) \tag{4.79}$$

where α and β are the usual spin indices. Note that

$$\psi_\alpha^0(\mathrm{rt}) = \psi_\alpha(\mathrm{rt}) \tag{4.80}$$

so we can recover properties of the usual operators from the set ψ^μ.

As in Sect. 4.3, using definitions (4.34 and 37), write

$$(T - H + M)\psi = 0 \ . \tag{4.81}$$

Then

$$(T - H)\psi^\mu = [T - H, R\sigma^{\mu\dagger}R^{-1}]\psi - R\sigma^{\mu\dagger}R^{-1}\Delta\cdot\sigma\psi/2 \ .$$

Now use (4.15) to choose the rotation matrix R. This gives

$$(T - H)\psi^\mu + (\Delta/2)(R\sigma^{\mu\dagger}\sigma^z R^{-1})\psi = [T - H, R\sigma^{\mu\dagger}R^{-1}]\psi \ . \tag{4.82}$$

Any 2×2 matrix is a linear combination of Pauli matrices and the identity. Using the convention

$$\sigma^\pm = (\sigma^+\pm i\sigma^y)/\sqrt{2} \ , \quad \sigma^0 = 1 \tag{4.83}$$

the coefficients for an arbitrary matrix U are

$$U = \sum_\mu \sigma^\mu \mathrm{Tr}\{U\sigma^{\mu\dagger}\}/2 \tag{4.84}$$

where μ runs over $0, z, +$, and $-$. Then (4.82) is

$$[(T - H)\delta^{\mu\nu} + (\Delta/2)\mathrm{d}^{\mu\nu} - K^{\mu\nu}]\psi^\nu \equiv Z^{\mu\nu}\psi^\nu = 0 \tag{4.85}$$

where

$$2d^{\mu\nu} = \text{Tr}\{\sigma^{\mu\dagger}\sigma^z\sigma^\nu\} \ , \tag{4.86}$$

$$2K^{\mu\nu} = \text{Tr}\{[T - H, R\sigma^{\mu\dagger}R^{-1}]R\sigma^\nu R^{-1}\} \ . \tag{4.87}$$

We are still using the space-time matrix notation introduced in Sect. 4.3. In a commutator, for example, (if x stands for rt)

$$[T, R] \equiv T(x - x')[R(x') - R(x)] \ , \tag{4.87a}$$

since T is a spin scalar. *The trace, however, is only over spin indices.* It follows from (4.86) that $d^{\mu\nu}$ is a real symmetric matrix, whose only non-vanishing components are

$$d^{0z} = d^{z0} = d^{++} = -d^{--} = 1 \ . \tag{4.88}$$

Furthermore, $K^{\mu\nu}$ is a Hermitian matrix as

$$K^{\mu\nu}(x, x')^* = K^{\nu\mu}(x', x) \tag{4.89a}$$

while, for all ν

$$K^{0\nu} = K^{\nu 0} = 0 \ . \tag{4.89b}$$

Finally, $Z^{\mu\nu}$ in (4.85) is a unit matrix in spin space. That is, (4.85) holds for the set $\psi_\alpha^\nu(x)$, *for each fixed* α.

Now define the Green's functions

$$\hat{g}_{\alpha\beta}^{\mu\nu}(x, x') = -i\langle\{\psi_\alpha^\mu(x), \psi_\beta^{\nu\dagger}(x')\}\rangle\eta(t - t') \tag{4.90}$$

where the angular bracket is a quantum mechanical expectation value and the curly bracket an anticommutator. As in Sect. 4.3 we defer the average over fields Δ. To be definite we write the retarded Green's function. A time-ordered function is more appropriate in the imaginary time case, but nothing essential changes.

From (4.85) the Green's function satisfies

$$Z^{\mu\nu}\hat{g}_{\alpha\beta}^{\nu\tau}(x, x') = \delta(x - x')(R\sigma^{\mu\dagger}\sigma^\tau R^{-1})_{\alpha\beta} \ . \tag{4.91}$$

To find the physical up- and down-spin functions it is sufficient to find

$$\hat{g}_e \equiv \text{Tr}\{\hat{g}^{00}\}/2 = (\hat{g}_+ + \hat{g}_-)/2 \ , \tag{4.92a}$$

$$\hat{g}_o \equiv \text{Tr}\{\hat{g}^{00}\sigma^z\}/2 = (\hat{g}_+ - \hat{g}_-)/2 \ , \tag{4.92b}$$

while the local spin magnitude is

$$\langle\psi^\dagger\hat{n}\cdot\boldsymbol{\sigma}\psi\rangle = \langle\psi^\dagger R\sigma^z R^{-1}\psi\rangle \tag{4.92c}$$

which is related to $\text{Tr}\{\hat{g}^{z0}\}$ by thermal factors. Then we restrict our attention to

$$\hat{g}_e^\nu \equiv \text{Tr}\{\hat{g}^{\nu 0}\}/2 \ , \quad \hat{g}_e = \hat{g}_e^0 \ , \tag{4.93a}$$

$$\hat{g}_o^\nu \equiv \text{Tr}\{(\hat{g}^{\nu 0}\sigma^z)\}/2 \ , \quad \hat{g}_o = \hat{g}_o^0 \ . \tag{4.93b}$$

These satisfy

$$Z^{\mu\nu}\hat{g}_e^\nu = \delta^{\mu 0}\delta(x - x') \tag{4.94a}$$

$$Z^{\mu\nu}\hat{g}_o^\nu = f^{\mu z}(x)\delta(x - x') \tag{4.94b}$$

where

$$2f^{\mu\nu} \equiv \text{Tr}\{R\sigma^{\mu\dagger}R^{-1}\sigma^\nu\} \ . \tag{4.94c}$$

It follows that

$$f^{0\nu} = f^{\nu 0} = 0 \ , \quad \nu \neq 0 \ : \quad f^{00} = 1 \ . \tag{4.94d}$$

In Appendix 4.A.7 we invert these equations to find

$$\hat{g}_e = \{T - H - (\Delta/2)[T - H - \delta\hat{\Sigma}]^{-1}(\Delta/2)\}^{-1} \ , \tag{4.95a}$$

$$\hat{g}_o = -(\Delta/2)[(T - H - \delta\hat{\Sigma})(2/\Delta)(T - H)(\Delta/2) - \Delta^2/4]^{-1}\hat{F} \tag{4.95b}$$

where

$$\delta\hat{\Sigma} = K^{zz} + \sum_{\alpha,\beta=+,-} (K^{z\alpha}\hat{\Gamma}_{\alpha\beta}K^{\beta z}) \tag{4.96a}$$

$$\hat{F} = f^{zz} + \sum_{\alpha,\beta=+,-} (K^{z\alpha}\hat{\Gamma}_{\alpha\beta}f^{\beta z}) \tag{4.96b}$$

and $\hat{\Gamma}$ is the inverse of Z_4, defined in (A.7.4d). Equation (A.7.7c) is

$$(\Delta/2)\hat{g}_e^z = 1 - (T - H)\hat{g}_e \ . \tag{4.95c}$$

Equations (4.95a,b) are written to be almost identical to (4.43a,b). Following the steps in Appendix 4.A.4 we can write

$$G_e = \langle\hat{g}_e\rangle = \{T - H - (\Delta/2)[T - H - \Sigma]^{-1}(\Delta/2)\}^{-1} \ , \tag{4.97a}$$

$$G_o = \langle\hat{g}_o\rangle = -(\Delta/2)[(T - H - \Sigma)(2/\Delta)(T - H)(\Delta/2) - \Delta^2/4]^{-1}\overline{\Sigma} \ , \tag{4.97b}$$

$$\Sigma = \langle\delta\hat{\Sigma} + \delta\hat{\Sigma}\Gamma\delta\hat{\Sigma} + \ldots\rangle_{1-\text{irred}} \ , \tag{4.97c}$$

$$\overline{\Sigma} = \langle\hat{F} + \delta\hat{\Sigma}\Gamma\hat{F} + \ldots\rangle_{1-\text{irred}} \ . \tag{4.97d}$$

Note that (4.95c) (and its configuration average) plays the role of (4.46a) in providing a connection between the Green's functions and the local spin density magnitude. The Green's functions and self energies in (4.97) are all spin scalars, removing the first of the "problems" referred to at the start of this subsection. The complication is that $\delta\hat{\Sigma}$ and $\hat{F}$ are now harder to evaluate than the corresponding quantities in Sect. 4.3.

In Appendix 4.A.8 we find explicit expressions for the components of G. The matrix $\hat{\Gamma}$ (which appears in (4.96) for $\delta\hat{\Sigma}$ and $\hat{F}$) is the inverse of

$$\hat{\Gamma}^{-1} = \begin{pmatrix} \hat{\Gamma}_+^{-1} & a^*\cdot a^*/m \\ a\cdot a/m & \hat{\Gamma}_-^{-1} \end{pmatrix} \tag{4.98a}$$

where

$$\hat{\Gamma}_\pm^{-1} = T \pm 2\hat{g} - V + (\nabla \mp 2ig)^2/2m - |a|^2/2m \pm \Delta/2 . \tag{4.98b}$$

Important components of $K^{\mu\nu}$ are

$$K^{zz} = 2a^*\cdot a/m , \tag{4.99a}$$

$$K^{z-} = \sqrt{2}[\hat{a}^* - i(\nabla - 2ig)\cdot a^*/2m - ia^*\cdot(\nabla + 2ig)/m] , \tag{4.99b}$$

$$K^{-z} = \sqrt{2}[\hat{a} - i(\nabla + 2ig)\cdot a/2m - ia\cdot\nabla/m] , \tag{4.99c}$$

$$K^{z+} = -\sqrt{2}[\hat{a} - i(\nabla + 2ig)\cdot a/2m - ia\cdot(\nabla - 2ig)/m] , \tag{4.99d}$$

$$K^{+z} = -\sqrt{2}[\hat{a}^* - i(\nabla - 2ig)\cdot a^*/2m - ia^*\cdot\nabla/m] . \tag{4.99e}$$

As in Sect. 4.2, the construction of $\delta\hat{\Sigma}$ and $\hat{F}$ requires solving the problem of electrons moving in effective "electric" and "magnetic" fields, now given by the "potentials" $2g$ and $2\hat{g}$. The up- and down-spin electrons carry opposite "charge", and the spin-flip potentials $a\cdot a$ and its conjugate carry the needed compensating charge. There is gauge invariance again, associated with the Euler angle b, and $\delta\hat{\Sigma}$ and $\hat{F}$ are explicitly gauge invariant. Although more complex than in Sect. 4.3, the formal evaluation of the self energy is well defined. Further, comparing (4.96,98 and 99), we see that the "majority" Green's function g_+ is predominately associated with the factors $a(x)a^*(x')$, while g_- couples with the complex conjugate of this product. Now in Appendix 4.A.2 we show that $a(rt)$ contains mostly positive frequencies, and a^* negative ones, at least for slowly varying disorder, *but even in the paramagnetic state*. Then in the self energy a majority spin is coupled with the *creation* of a spin collective excitation, while a minority spin couples with the destruction of one, except for corrections depending on the degree of disorder. This is what is required physically, and corrects the serious flaw in the previous analysis, mentioned at the beginning of this section.

4.4.2 Approximate Evaluation

Although the problem of evaluating Σ and $\bar{\Sigma}$ is well formulated, it is, as usual, difficult to find a systematic perturbation expansion in the potentials g and $\hat{g}$. We restrict our attention to the lowest-order terms, which already contain most of the physics. In Appendix 4.A.9 we use the classical path approximation for electromagnetic potentials to find

$$\Sigma(x, x') = \Sigma_+(x, x') + \Sigma_-(x, x') \ , \tag{4.100a}$$

$$\begin{aligned} \Sigma_+(x, x') =& 2\{\langle [T - H, d]d^*\rangle\delta(x - x') \\ &+ \langle [T - H, d]g_+[d^*, T - H)]\} \ , \end{aligned} \tag{4.100b}$$

$$\begin{aligned} \Sigma_-(x, x') =& 2\{\langle [T - H, d^*]d\rangle\delta(x - x') \\ &+ \langle [T - H, d^*]g_-[d, T - H])\} \ . \end{aligned} \tag{4.100c}$$

In the same approximation we have

$$\bar{\Sigma} = \langle \cos \theta\rangle(1 - \Sigma_+g_+ - \Sigma_-g_-) \ . \tag{4.101}$$

These expressions are again almost identical to expressions in Sect. 4.3, now (4.49–52). The differences are, first, that the new Σ and $\bar{\Sigma}$ are scalars in spin space, and second, that the correlation function $D = \langle dd^*\rangle$ appears, rather than $X = \langle mm\rangle$. The function D is an unsymmetric function of frequency, even in the paramagnetic state, which ensures that Σ and $\bar{\Sigma}$ depend only on the physically relevant excitation processes. Under the same approximation as in Sect. 4.3, expressions such as (4.55 and 56) are readily found in the present case. In particular, corresponding to (4.55)

$$\Sigma_\pm(k, E) = 2(2\pi)^{-4} \int d^3q d\Omega \, D(q, \Omega)[V_\pm + V_\pm^2 g_\pm(k \mp q, E \mp \Omega)] \ , \tag{4.102a}$$

$$V_\pm = \epsilon(k \mp q) - \epsilon(k) \pm \Omega \ . \tag{4.102b}$$

The $\pm$ signs reflect that fact that $D(x, x')$ appears in $\Sigma_+(x, x')$ while $D^*(x, x') = D(x', x)$ is in $\Sigma_-(x, x')$. At low excitation $D(q, \Omega)$ has most of its weight at the spin-wave pole $\Omega = Dq^2$. The intermediate state in (4.102a) has an extra spin excitation associated with a majority spin electron, and a reduced excitation associated with a minority spin electron, as is required.

For the Green's functions to be useful, Σ and $\bar{\Sigma}$ should be relatively smooth functions. There is no problem with Σ which has a familiar form, but $\bar{\Sigma}$ is apparently singular at the poles of g_+ and g_-. In fact there is no singularity. Formally, from (4.100b)

$$\begin{aligned} \Sigma_+/2 &= \langle [T - H, d]\{d^* + (T - H + \Delta/2)^{-1}[d^*, T - H + \Delta/2]\} \\ &= \langle [T - H, d]g_+d^*\rangle g_+^{-1} \end{aligned} \tag{4.103}$$

and similarly for Σ_-. This removes the possible singularity. It is also useful to carry this rewriting one step further. Then

$$\Sigma_+/2 = g_+^{-1}\langle dg_+ d^*\rangle g_+^{-1} - \langle dd^*\rangle g_+^{-1} \tag{4.104a}$$

$$\Sigma_\pm(k, E) = 2(2\pi)^{-4} \int d^3q d\Omega \, D(q, \Omega)[(E - \epsilon(k)\pm\Delta/2)^2 g_\pm(k\mp q, E\mp\Omega)$$
$$- (E - \epsilon(k)\pm\Delta/2)] \tag{4.104b}$$

where the second form is appropriate for a one-band model.

There is one caveat necessary when using these second-order expressions. Since $\Sigma_\pm$ vanish near one or the other ground-state pole, higher-order terms may actually be important there. In particular, the approximations may not be equally good for G_e and G_0, and their sum or difference (which give G_+ and G_-, see (4.102a,b)) may show an incorrect sign of imaginary part over a small energy range. This has no bearing on the overall energetics, the local charge and spin density, or the self consistency. These all depend only on G_e, which has the correct analytic properties by construction. The problem appears (when it does) only in one of the single-spin spectral functions (but not in the sum of the two). It is eliminated by adding a small imaginary part to all the self energies, to simulate higher-order terms.

4.4.3 Zero-Point Effects

With (4.101 and 104) we can now treat zero-point spin fluctuations. This is done formally in the functional integral method by letting $\langle dd^*\rangle$ and $g_\pm$ be periodic functions in imaginary time. More directly, the terms in (4.104b) involving $g_\pm$ are familiar self-energy expressions for boson emission (Σ_+) and absorption (Σ_-) if the factors $D(q, \Omega)$ are interpreted as (some care is necessary to get factors of two correct)

$$D(q, \Omega) \rightarrow N(\Omega) A(q, \Omega)/N^2 \tag{4.105a}$$

with $N(\Omega)$ the Bose statistical distribution function, $A(q, \Omega)$ the spin wave spectral function, and N the number of magnetic electrons, as in Sect. 4.3. (For small q and low temperatures A is proportional to $N\delta(\Omega - Dq^2)$, and $N = N_+ - N_-$.) To include zero-point effects, we simply make the replacements

$$N(\Omega) \rightarrow (1 + N)(1 - f) + Nf = N(\Omega) + 1 - f_+(\epsilon(k - q)) \quad (\text{in } \Sigma_+), \tag{4.105b}$$

$$N(\Omega) \rightarrow N(\Omega) + f_-(\epsilon(k + q)) \quad (\text{in } \Sigma_-) . \tag{4.105c}$$

Note that in the final term of (4.104) $N(\Omega)$ is not modified.

Because of the novel form of (4.97), we have to see if (4.105) give sensible results. In particular, in the strong limit ground state, the spin-up Green's function is simply g_+, since no real spin excitations are present to be absorbed.

The spin-down states, however, are modified by the interaction. Since both self energies appear in both G_e and G_o, it is not obvious that this asymmetry will hold.

In the ground state $N(\Omega)$ is zero for positive Ω. But also $A(q, \Omega)$ has weight only at positive frequency, so $N(\Omega)$ may simply be set to zero. In the strong limit $f_- = 0$ at $T = 0$, so Σ_- vanishes in this case. Using (4.104a), write $\Sigma_+ \equiv \Lambda' g_+^{-1}$. Also, in the ground state $\langle \cos \theta \rangle = 1$ so $\overline{\Sigma} \to 1 - \Lambda'$. Putting these together into (4.97) it is easy to show that

$$G_+ = G_e + G_o = (T - H + \Delta/2)^{-1} = g_+ . \tag{4.106}$$

Then the up-spin Green's function is indeed inert in this situation.

In considering the spin-down function, note that the term $\langle dd^* \rangle$ in (4.104a) vanishes in the ground state $(N(\Omega)A(q, \Omega) = 0)$ so that $\Sigma_+ = g_+^{-1} \Lambda g_+^{-1}$. For simplicity, consider the one-band case. Write $E - \epsilon(k) \equiv z$. Then

$$G_- = G_e - G_o = (z - \Sigma + \overline{\Sigma}\Delta/2)[z(z - \Sigma) - \Delta^2/4]^{-1} ,$$

$$G_- = (z + \Delta/2 - \Sigma_{\text{eff}})^{-1} \quad \text{where} \qquad . \tag{4.107a}$$

$$\Sigma_{\text{eff}} = \Delta/\{1 - (\Lambda\Delta/2)/[1 - \Lambda(z + \Delta/2)]\} \tag{4.107b}$$

Equation (4.107a) is written for ready comparison with the work of *Hertz* and *Edwards* [4.25]. These researchers also considered the strong-limit ground state correlation problem, and found an expression for G_- in just this form (when account is taken of a shift in origin of the energy scale). Their expression for Σ_{eff} is similar to (4.107b), but they have Σ_0/Δ in place of $(\Lambda\Delta/2)/[1 - \Lambda(z + \Delta/2)]$. Now if we use (4.105a), then it follows that $\Lambda\Delta^2 = 2\Sigma_0$ as defined in [4.25], so except for the denominator the expressions are then the same. If Λ falls off fast enough at large z, the qualitative features of this expression do not change when the extra denominator is restored. (Otherwise, a better approximation for the self energy is required for a satisfactory $T = 0$ Green's function.) The caveat discussed at the end of Sect. 4.4.2 does apply here as well, but not in the strong limit.

We have found, for the first time, temperature-dependent Green's functions for itinerant ferromagnets which fully respect the rotation invariance of the system, incorporate the effects of spin fluctuations, yet allow for strong magnetic behavior even in the paramagnetic state. Zero-point fluctuations are included, as are substantial vertex corrections. The up-spin states are unaffected by zero-point fluctuations in the strong limit, as they must be. If we ignore ground state fluctuations, then at finite temperature the spectral functions develop satellite structure, while the quasi-particle poles shift and change residue, as required by Fermi-liquid theory. As magnetic disorder increases, there is an admixture of minority spin weight in the up-spin Green's function, reflecting real local fluctuations in the magnetization direction. This all describes physically correct behavior. The price paid for it is the appearance in the self energy of a correlation function which is difficult to evaluate, and an expression for the self energy which is hard to expand beyond the lowest order.

4.5 Model Analysis

Except for some corrections which are expected to be small, the spectrum of photoemission from a particular band state is proportional to the imaginary part of the Green's function for a hole in that state. In this section I present a simple model analysis of the temperature dependence of the photoemission spectrum, using parameters appropriate for two bands of nickel, and I compare with the recent SARPES studies of nickel [4.19] and iron [4.20]. The predicted behavior agrees quite well with the nickel measurements, which show an apparent disappearance of the spin splitting as the temperature is raised to the Curie point. I will show that the theory nevertheless allows an almost temperature-independent local magnetization strength and intrinsic splitting parameter. The experimental results in iron are also what is expected, although no quantitative comparison is given here. The expected effects of direction fluctuations are quite small in this case, so details of the experimental observations may be dominated by extraneous factors. Some of the material in this section has already appeared in [4.16 and 17], but I will provide more details, and some additional information.

4.5.1 Temperature-Dependent Parameters

In the Green's function analysis of Sects. 4.3 and 4, we can distinguish three separate sources of temperature dependence. Equations (4.44 or 97) show that G changes if $\Delta(r)$, $V(r)$ (which is contained in H), or Σ depends on temperature. Now we have dropped consideration of *fluctuations* in V and Δ, but the mean values vary because of temperature-induced changes in the charge and magnetization densities. Changes in $V(r)$ are likely to be small, and are generally ignored. I ignore them here as well. In the Stoner model Δ drops to zero at T_C, since this is the only means available to drive the net magnetization to zero at the Curie temperature. For rather compelling theoretical and experimental reasons, however, it is now widely believed that $\Delta(r)$ is only weakly temperature dependent, so the temperature evolution of the photoemission spectrum (and the electronic structure) is mainly in that of Σ. I will mention briefly three lines of argument for the approximate constancy of Δ, and then will consider Σ in more detail.

The first concerns energy changes which would be associated with changes in Δ. *Janak* and *Williams* [4.34a], and *Poulson* et al. [4.34b] have shown that the increase in volume of the magnetic transition metals over the trend of their non-magnetic neighbours is quantitatively explained by the energy associated with the exchange splitting of the ferromagnetic energy bands. If the splitting disappears at T_C, a volume change on the order of 5 % is predicted in iron.

The actual change seen is two orders of magnitude smaller, implying that the splitting, and thus Δ, have not been reduced. Using similar reasoning, *Holden* et al. [4.39] compared the numerical values of volume change and total magnetic energy associated with the phase transition, and concluded that changes in Δ were no more than a few percent in iron and nickel.

Somewhat different is an argument which actually accounts for the observed energy and volume changes. In our discussion of total energy in Sect. 4.2, we found that the free energy *separated* into two terms, one associated with directional disorder $[A(\nabla \hat{M})^2]$ and the other with the amplitude of the local magnetic moment. We find [4.2a,12] that the magnetic energy changes through T_C in iron and nickel are given quite closely by the first term alone, with known values of A and values of the disorder scale found by other means. In addition, the volume change at T_C has a part related to the pressure dependence of A. This pressure dependence has been measured in nickel [4.40], and again the agreement [4.41] with the measured volume change is quite good. Values of $\partial T_C / \partial P$ are also predicted successfully in this way [4.14]. We conclude that observed magnetic energy changes are predominantly given by the directional disorder term, so changes in Δ must be small.

Finally, neutron scattering experiments [4.30,31] show considerable magnetic scattering in the paramagnetic state of iron and nickel, so there must be a substantial moment. The precise value is hard to extract, but it is comparable to the ground state value. Of course, the large Curie-Weiss like paramagnetic susceptibility has long been taken as evidence of the presence of a large magnetic moment above T_C.

In light of these arguments we concentrate on the effects of directional disorder on the electronic states and energies, and pay less attention to changes in the self-consistent potentials. The latter enter automatically if the the full self-consistent calculation proposed in the earlier sections is carried out, but this is still too ambitious a program, and only simple single-band models have been worked out so far. Even where the simple model closely resembles part of the spectrum of a real system it is not correct to apply self consistency. The potential Δ has contributions from many bands, and self consistency can be invoked only when all are included. As we will see, different bands behave differently under the effects of fluctuations, so this is not a trivial point.

The expressions for Green's functions in Sects. 4.3 and 4 require the temperature-dependent direction correlation functions, $X(rr'\omega)$ or $D(rr'\omega)$, for r and r′ arbitrary positions in the same or different unit cells, and for all ω. Except at the lowest temperatures, however, where a spin-wave expansion may be appropriate, these correlations are not well understood theoretically. There important data [4.30,31] from neutron-scattering studies, but this gives only qualitative guidance, because of resolution problems and of the difficulty of getting polarized scattering information below the Curie point. In what follows, for simplicity, we choose a form for the correlation functions which is analytically convenient, and depends on three parameters. Of these, two are temperature dependent, and give the overall strength of direction fluctuations and the range of their spatial frequency content. These parameters are determined from the temperature dependence of the net magnetization and of the magnetic energy. The third parameter determines the frequency scale of temporal fluctuations. For want of information we keep this parameter constant, at a value given by the ground state spin-wave stiffness. In any case, the results do not depend sensitively on this value.

If X is the direction correlation function of (4.54), then (4.64 and 66) give its normalization and second q moment

$$(2\pi)^{-4} \int d^3q \, d\Omega \, X(q,\Omega) = 1 - \mu^2 \ , \tag{4.108a}$$

$$(2\pi)^{-4} \int d^3q \, d\Omega \, q^2 X(q,\Omega) = \langle (\nabla \hat{M})^2 \rangle = E_{mag}/A \ , \tag{4.108b}$$

where the identification with the thermodynamically measured magnetic energy follows from the free-energy expression in (4.32). Above T_C (4.108a) must be supplemented by a condition relating $X(0,0)$ to the suceptibility.

The function $D(x,x')$ is defined in (A.9.4). Using (4.22a) we see that it satisfies

$$(2\pi)^{-4} \int d^3q \, d\Omega \, q^2 D(q,\Omega) = \langle (\nabla \hat{M})^2 \rangle / 4 \ . \tag{4.108c}$$

In fact we have already used the identity of X and 4D at low excitation in Sect. 4.4. This identity does not persist at larger disorder, and there is no normalization requirement corresponding to (4.108a). Rather we use a condition relating the fourth q moments of X and D. In order for the third and fourth energy moments of the Green's functions, discussed at the end of Sect. 4.3, to be correct, we require the condition

$$(2\pi)^{-4} \int d^3q \, d\Omega \, 4q^4 D(q,\Omega) =$$
$$(2\pi)^{-4} \int d^3q \, d\Omega \, q^4 X(q,\Omega) - \langle (\nabla \hat{M})^2 \rangle^2 \ . \tag{4.108d}$$

In fact it follows easily from the properties of the rotation matrix (4.15 and 4.21) that

$$\nabla^2 \nabla'^2 \hat{n}(x) \cdot \sigma \hat{n}(x') \cdot \sigma \, |_{x' \to x} = 16(a^* \cdot a)^2 + 4 \, |(\nabla + 2ig) \cdot a \, |^2 \ .$$

Taking account of (4.54, A.9.3d and A.9.4a), and assuming that $|a|^2$, the energy density, does not have large fluctuations, we arrive at (4.108d).

We approximate X and D by the cutoff flat distributions

$$X(q) = X(0)\eta(Q_1 - q) \ , \tag{4.109a}$$

$$D(q) = D(0)\eta(Q_0 - q) \ . \tag{4.109b}$$

Then the parameters $D(0)$ and Q_0 are easily found in terms of the physical quantities in (4.108). Finally, we use the frequency dependence

$$D(q, \Omega) = D(q)\delta(\Omega - Dq^2) \qquad (4.109c)$$

with D the ground state value of the spin-wave stiffness.

We consider a single parabolic band with effective mass m and constant spin splitting Δ. It is convenient to measure energies in units of $\Delta/2$ and momenta in units of $\sqrt{m\Delta}$. Then (4.108 and 109) give

$$Q_0^2/m\Delta \equiv Q^2 = (5/3)[\langle(\nabla\hat{M})^2\rangle/m\Delta](0.16 + 0.84\mu^2)/(1 - \mu^2) , \qquad (4.110a)$$

$$(m\Delta)^{3/2}D(0) = (5\pi^2/2)[\langle(\nabla\hat{M})^2\rangle/m\Delta]/Q^5 . \qquad (4.110b)$$

We write the temperature dependence of the magnetization and magnetic energy using a formula suggested by A.S. Arrott, which interpolates between known behaviors near $T = 0$ and the Curie point. For nickel we find for the magnetization

$$\mu = M/M_0 = (1 - t)^{0.378}/\{1 - 0.378t + 0.12t^{3/2} - 0.039t^{7/2}\} \qquad (4.111)$$

which matches to the fit of *Kouvel* and *Comly* [4.42] near T_C and that of *Riedi* [4.43] near $T = 0$. (We have defined $t = T/T_C$.) For the magnetic energy

$$E(t)/E(1) = 1 - [(1 - t) - 0.821(1 - t)^{1.1}]/g(t) , \qquad (4.112a)$$

$$g(t) = 0.179(1 - 0.540t - 0.253t^2 + 0.157t^{5/2} + 0.169t^{9/2}) , \qquad (4.112b)$$

$$E(1) = 225\,\text{cal/mole} . \qquad (4.112c)$$

Equations (4.112) match the critical specific heat fit of *Connelly* et al. [4.44], and a low temperature estimate of the spin-wave contribution to the energy based on the magnetization data of *Riedi* [4.43]. They also take account of the measurement of *Pawel* and *Stansbury* [4.45], who found a total magnetic energy of 285 cal/mole far above T_C, and our estimate from their curves that about 80 % of this is realized at the Curie temperature.

The evaluation of $\langle(\nabla\hat{M})^2\rangle$ is completed by our estimate [4.2a,7,12] that it reaches a value of about $0.2\,\text{Å}^{-2}$ well above T_C in nickel.

Using this approximation for $D(q, \Omega)$, $\Sigma_\pm$ are found using (4.104b), $\overline{\Sigma}$ using (4.101), G_e and G_o using (4.97a,b), and G_n from the configuration average of (4.95c). (In the single-band approximation $T - H$ in these equations becomes $E - \epsilon(k)$.) As discussed at the end of Sect. 4.4.2, it may be necessary to add an extra small imaginary part to the self energies when computing the spin-up or -down Green's functions.

4.5.2 Model Nickel Spectra

The figures illustrate the temperature evolution of single-particle Green's functions for nickel, computed as described above. Calculations are shown for a "light" band and a "heavy" band. The light band has the effective mass

$m^* = 3m$, and the splitting $\Delta = 0.2\,\mathrm{eV}$, values corresponding to the band of S4 symmetry at the nickel X-point. This is the only nickel band for which temperature-dependent SARPES results are so far available [4.19]. It is a convenient band to study experimentally since all spin-up and spin-down states are filled, so peaks of both polarizations can be seen. For the same reason, however, this band makes no contribution to the magnetic moment of nickel, and only indirect information about the magnetic state is obtained. The heavy band has $m^* = 5m$ and $\Delta = 0.3\,\mathrm{eV}$, and it represents states which do contribute to the magnetization.

Figure 4.1 shows the spectral weight function $(-\operatorname{Im}/,\{G_e\})$ and its resolution into up- and down-spin components for three temperatures near the bottom of the light band. A temperature-independent Lorentzian broadening has been added to simulate the experimental conditions. Aside from a temperature rescaling required by the extreme surface sensitivity of the photoemission probe [4.46], Fig. 4.1 is quite similar to the experimental results of [4.19], which measure a corresponding state.

At first sight the temperature dependence in Fig. 4.1 seems to verify the Stoner model, where the up- and down-spin states converge as the intrinsic splitting Δ reduces with increasing temperature. This figure was computed with constant Δ, however, so the observed behavior does not require that Δ reduce.

One indication that the intrinsic splitting does not decrease as much as it seems to is that the single-spin peaks broaden as they move together. The theoretical connection between the broadening and the value of Δ can be seen from the moment relations (4.78) and the discussion which follows them. This broadening was remarked in the experiments [4.19], where it was correctly ascribed to magnetic behavior continuing above T_C. The amount of broadening seen is consistent with a value of Δ which does not vary significantly with temperature.

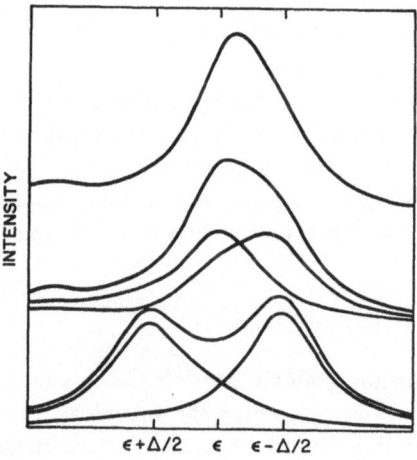

Fig. 4.1. Spectral function and its resolution into up- and down-spin components, near the bottom of the "light" band of nickel (see text). These are hole bands so excitation energies are plotted increasing to the left. The three points marked on the energy axis are the band energy, and the positions of the up- and down-spin peaks in the ground state. A temperature-independent Lorentzian broadening has been added to the computed curves. From the top down, the curves correspond to relative magnetization 0.1, 0.37, and 0.91. The individual spin components are omitted from the top curve for clarity: they are essentially identical to the total spectral function in this case

While the detailed analysis of Sect. 4.4 is thus seen to be consistent with experiment, one can ask what remains of the more heuristic discussion of Sect. 4.2, based on sharp, well separated instantaneous energy eigenstates. In fact, these states are not directly probed by the experiments, nor are they addressed by the theoretical Green's function. The instantaneous eigenstates are not momentum eigenstates, while the experiments (and theory) study the energy spectrum at fixed momentum. Then inhomogeneous broadening gives a complex spectrum. It is only when this broadening can be made small that the two-separated-peak energy spectrum of Sect. 4.2 can be expected to be seen.

Some of the complexity resulting from the inhomogeneous broadening is seen in Fig. 4.2, which shows the temperature dependence of the same spectral function as in Fig. 4.1, at higher resolution. Note the deep satellite which appears as the temperature increases. For a better look at the "local" situation, Fig. 4.3 shows the corresponding curves of G_n. G_n, which is defined in (4.46), and computed as the configuration average of (4.95c) gives the spectral function for the projection of the Green's function along the local magnetization direction. Integrated to the Fermi energy, G_n yields the contribution to the local magnetization of the particular momentum eigenstate, see (4.45). A substantial energy splitting between parallel and antiparallel spins is evident in the figure, but the effects of inhomogeneous broadening are still considerable. From Fig. 4.3 we see that the deep satellite is associated with the locally-minority spin direction.

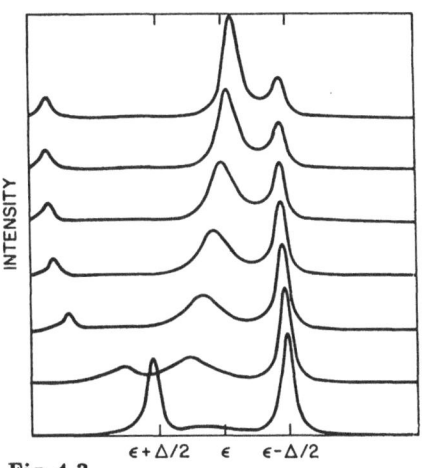 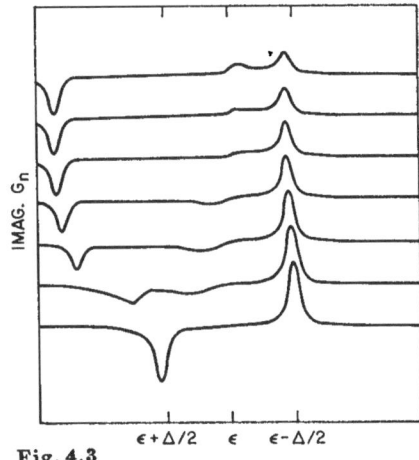

$\epsilon + \Delta/2 \quad \epsilon \quad \epsilon - \Delta/2$

Fig. 4.2

$\epsilon + \Delta/2 \quad \epsilon \quad \epsilon - \Delta/2$

Fig. 4.3

Fig. 4.2. Spectral function near the bottom of the "light" band of nickel. Same band state and conventions as in Fig. 4.1, but added broadening is reduced. From the top down the curves correspond to relative magnetization 0.1, 0.25, 0.37, 0.50, 0.65, 0.80, and 0.91

Fig. 4.3. Imaginary part of G_n (see text) for the same conditions as in Fig. 4.2. A positive value means that the local magnetization strength increases if the state is excited. That is, it is a locally-minority electron state

The temperature evolution of G_n, in conjunction with (4.45), raises the question of the consistency of assuming a constant Δ, since the contribution of this state to the magnetic moment apparently reduces substantially, as the temperature is raised. Indeed, the integral over positive G_n in Fig. 4.3 reduces at T_C to only 60% of its low temperature value. Self consistency would lead to dramatic changes in Δ if these states were, in fact, responsible for the magnetization.

But, as already noted, it is the heavy band which determines M. Fig. 4.4 shows the temperature evolution of the spectral function near the bottom of this band, while Figs. 4.5 and 6 show, respectively, the corresponding G_n and the spectral function including the experimental broadening. In contrast with the Stoner model we have substantially different behavior for the two different bands. This could be seen experimentally, except that inverse photoemission is

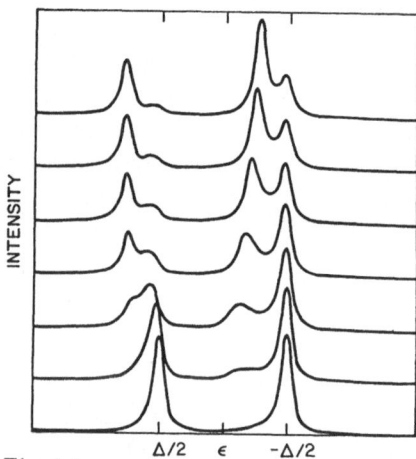

Fig. 4.4

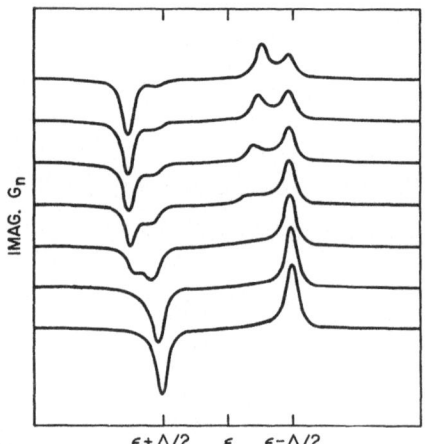

Fig. 4.5

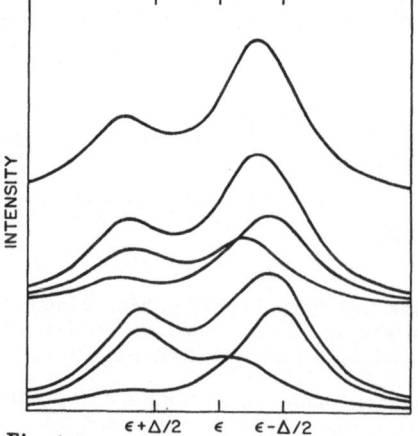

Fig. 4.6

Fig. 4.4. Same as Fig. 4.2, but near the bottom of the "heavy" band of nickel. From the top down the curves correspond to relative magnetization 0, 0.25, 0.37, 0.50, 0.65, 0.80, and 0.91

Fig. 4.5. Same as Fig. 4.3, but near the bottom of the "heavy" band of nickel. Magnetizations as in Fig. 4.4

Fig. 4.6. Same as Fig. 4.1, but near the bottom of the "heavy" band of nickel. From the top down the curves correspond to relative magnetization 0.1, 0.37, and 0.65. A curve at 0.91 would be indistinguishable from the lowest curve in Fig. 4.1

required to look at the empty minority spin state, and the required resolution has not yet been attained for this probe. The potential contribution of this state to M at T_C, found by integrating the curves in Fig. 4.5 over the region where they are positive, is about 90% of the low temperature value, so a substantial Δ may be maintained in the paramagnetic state. Of course, this model is far too crude to give a more quantitative estimate.

The reason for these varied behaviors can be understood from the moment relations in (4.78c,d). In the approximation we are using the needed quantities are

$$2\langle\langle V\rangle\rangle/\Delta = (1\mp 2mD)\langle(\bigtriangledown\hat{M})^2\rangle/m\Delta \ , \qquad (4.113a)$$

$$4\langle\langle V^2\rangle\rangle/\Delta^2 = (1.19)(2\langle\langle V\rangle\rangle/\Delta)^2/(1-\mu^2) + (4k^2/m\Delta)\langle(\bigtriangledown\hat{M})^2\rangle/m\Delta \ . \qquad (4.113b)$$

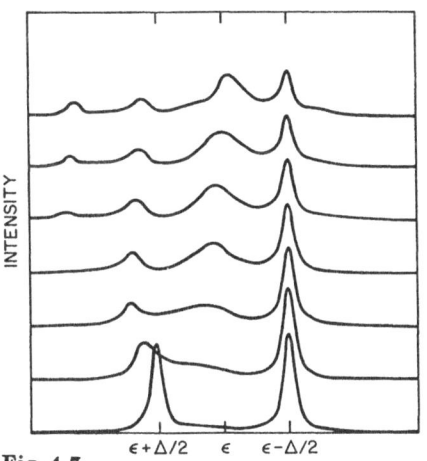

Fig. 4.7

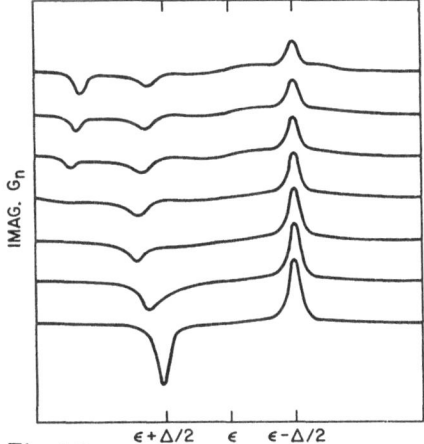

Fig. 4.8

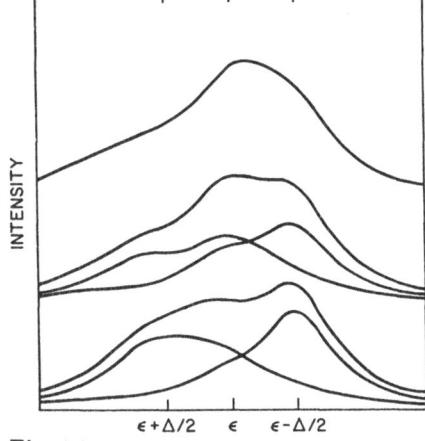

Fig. 4.9

Fig. 4.7. Same as Fig. 4.2, but at higher momentum in the "heavy" band of nickel. Magnetizations as in Fig. 4.4

Fig. 4.8. Same as Fig. 4.3. Conditions as in Fig. 4.7

Fig. 4.9. Same as Fig. 4.1, but at higher momentum in the "heavy" band of nickel. Magnetizations as in Fig. 4.6

145

We are using the value $2mD = 0.4$. Then (4.113) is $2(1\pm0.4)$ for the light band at T_C, and two and a half times smaller for the heavy band. The light band has substantial inhomogeneous broadening, while the heavy band shows more intrinsic structure, because the disorder contribution to the higher moments is much less.

Equation (4.113b) shows that the line shape also depends on the momentum value within a band, there being more line distortion where the band is more dispersive. Figures 4.7,8, and 9 show the temperature evolution for a state with $k^2 = m\Delta$, for the heavy band. The situation is intermediate between those of Figs. 4.2 and 4, both of which have $k^2 = 0.1\,m\Delta$. This higher momentum state corresponds to one studied [4.47] in spin unresolved photoemission some time ago. The experiments were analyzed very convincingly in terms of a three-peak structure, and Fig. 4.7 shows that this has a sound theoretical basis. Unfortunately, there is no justification for using a three-peak analysis, or any other simple ansatz, in the general case.

4.5.3 Iron Results

Because of its small spin splitting relative to available experimental resolution, and the rather large value of $2\langle\langle V\rangle\rangle/\Delta$ even in favorable cases, nickel does not provide a sharp experimental test of our analysis. The experiments [4.19] are consistent with theory, but do not prove it. However, since the disorder scale $\langle(\nabla\hat{M})^2\rangle$ is estimated [4.2a,7,12] to be about the same in iron as in nickel, but the splitting Δ is so much larger, iron provides a much better laboratory for this purpose.

In fact, the available experimental results are in good qualitative agreement with our analysis. In particular, SARPES experiments near the Γ point [4.20], where the bands are quite flat, and $2\langle\langle V\rangle\rangle/\Delta$ correspondingly small, show two peaks which do not move, as the temperature is raised to T_C. Photoelectrons in the two peaks are strongly polarized at low temperature, becoming less so as the temperature rises. The behavior is that of a domain situation, as we expect. Close to the H point, however, where the band dispersion is considerable, there is substantial temperature dependence of the peak position. Similarly, in inverse photoemission [4.21], temperature dependence was seen where the band has large dispersion, but not where it is flat.

It is difficult to make a quantitative comparison between theory and experiment without elaborate calculation. A one-band model is inappropriate in this case, and the magnetization fluctuations cause considerable mixing among bands. The quantities determining the energy moments of the Green's functions corresponding to $\langle\langle V\rangle\rangle$ and $\langle\langle V^2\rangle\rangle$, depend strongly on matrix elements, and are hard to estimate reliably. However, a quantitative comparison has been made [4.23] between the Γ point experiments and a quite different calculation based on the same model we use. *Haines* et al. [4.23] computed photoemission for a Monte Carlo ensemble of local-field-direction configurations, with a given correlation range. By comparing results obtained for several different ranges, they concluded that a correlation range of at least 4 Å is required, since a smaller ordering scale gives too large a change with temperature of the theoretical spectrum. The value we have used for the short-range order scale in iron is about 6 Å, in their notation. The quantitative agreement is very gratifying.

4.6 Conclusion

What I have presented in this chapter is a rotationally invariant generalization of electron-spin wave perturbation theory. The generalization is in that the perturbation is no longer about the fully aligned ground state, but about states with slowly varying magnetization direction. By choosing zero-order wave functions which follow this variation (using a locally rotated reference frame) I make the perturbation small even if the global magnetization has changed greatly from the ground state value. Experiments indicate that this perturbtion theory works reasonably well in nickel, and very well in iron, even in the paramagnetic state.

In conventional theories vertex corrections are used to account for these adjustments of wave functions to local conditions. We make the adjustments directly, and find agreement with conventional analyses of the ground state [4.25,27] which make vertex corrections, and which have had considerable success. We also match Landau-Fermi liquid results at low but finite temperatures, a result which has eluded most conventional approaches (but see [4.48]). Our predictions for the vicinity of the Curie temperature match those of no other spin-wave theory, but do agree with experiments on nickel and iron.

The work described has its difficulties and complexities. The self-energy functions are hard to evaluate beyond the lowest-order, and depend on a magnetization correlation function which is hard either to compute or to measure as a function of temperature. Important effects may have been left out of the model. Much more complete analysis must be carried out before a full comparison with experiment can be made.

4.A Appendix

4.A.1

Equations (4.21) can be written

$$\partial_\mu R = iRA_\mu$$

where μ runs over x, y, z, and t. Then

$$\partial_\mu \partial_\nu R = iR\partial_\mu A_\nu - RA_\mu A_\nu.$$

It follows that

$$(\partial_\mu A_\nu - \partial_\nu A_\mu) = -i(A_\mu A_\nu - A_\nu A_\mu) \ .$$

Using the explicit expressions in (4.21), matching coefficients of each σ matrix, and letting μ and ν be spatial variables only, we find

$$\nabla \times g = i a^* \times a = \nabla \cos(\theta) \times \nabla \varphi , \qquad (A.1.1a)$$

$$(\nabla - 2ig) \times a^* = 0 , \qquad (A.1.1b)$$

$$(\nabla + 2ig) \times a = 0 . \qquad (A.1.1.c)$$

If one variable is space and the other time we have

$$(\partial_t g - \nabla \hat{g}) = i(\hat{a}^* a - a^* \hat{a}) , \qquad (A.1.2a)$$

$$(\partial_t - 2i\hat{g}) a^* = (\nabla - 2ig)\hat{a}^* , \qquad (A.1.2b)$$

$$(\partial_t + 2i\hat{g}) a = (\nabla - 2ig)\hat{a} . \qquad (A.1.2c)$$

4.A.2

We suppose that the spin density satisfies the Landau Lifshitz equation (LLE)

$$\partial \hat{n}/\partial t = D\hat{n} \times \nabla^2 \hat{n} \qquad (A.2.1)$$

[We have shown in [4.4] that a slight generalization of the LLE is *implied* by the time-dependent versions of (4.11 and 12).] From (A.2.1) write

$$i(\partial/\partial t)(\hat{n} \cdot \boldsymbol{\sigma}) = D(\hat{n} \cdot \boldsymbol{\sigma})\nabla^2(\hat{n} \cdot \boldsymbol{\sigma})$$

where we understand that we must drop any term on the right-hand side proportional to the unit matrix in spin space.

Using (4.15,21) write this as

$$iR[\hat{A}, \sigma^z]R^{-1} = DR\sigma^z R^{-1}\{R[\nabla \cdot A, \sigma^z]R^{-1} + iR[A, [A, \sigma^z]]R^{-1}\}$$

or

$$i[\hat{A}, \sigma^z] = D\{\sigma^z[\nabla \cdot A, \sigma^z] + i\sigma^z[A, [A, \sigma^z]]\} . \qquad (A.2.2)$$

Using the explicit expressions for A and $\hat{A}$, and matching coefficients of σ^+ and σ^-, we find

$$i\hat{a} = -D(\nabla + 2ig) \cdot a , \quad i\hat{a}^* = D(\nabla - 2ig) \cdot a^* . \qquad (A.2.3)$$

Equations of motion for θ and φ follow easily from these.

For our purposes, we combine (A.2.3) with (A.1.2). Then, for example, from (A.1.2c)

$$(i\partial_t - 2\hat{g})a = -D(\nabla + 2ig)(\nabla + 2ig) \cdot a . \qquad (A.2.4)$$

Now a simple extension of a fimiliar identity is (writing $\tilde{\nabla} \equiv \nabla + 2ig$)

$$\tilde{\nabla} \times \tilde{\nabla} \times a = \tilde{\nabla}\tilde{\nabla} \cdot a - \tilde{\nabla}^2 a - 2ia \times \nabla \times g .$$

In view of (A.1.1), Eq. (A.2.4) is

$$[(i\partial_t - 2\hat{g}) + D(\nabla + 2i\boldsymbol{g})^2]\boldsymbol{a} = 2D\boldsymbol{a}\times(\boldsymbol{a}^*\times\boldsymbol{a}) \ , \tag{A.2.5a}$$

$$[(i\partial_t + 2\hat{g}) + D(\nabla - 2i\boldsymbol{g})^2]\boldsymbol{a}^* = 2D\boldsymbol{a}^*\times(\boldsymbol{a}^*\times\boldsymbol{a}) \ . \tag{A.2.5b}$$

Clearly, if we can ignore $\boldsymbol{g}$, $\hat{g}$, and $\nabla\times\boldsymbol{g}$, then $\boldsymbol{a}$ has weight at $\omega = Dq^2$ while $\boldsymbol{a}^*$ has weight at $\omega = -Dq^2$. These play the role of local-spin wave annihilation and creation "operators", even in the paramagnetic state, as long as the gradients are not too large. This justifies the discussion after (4.99).

4.A.3

Rewrite (4.42) as N/D where

$$D = (T - H)(T - H - M\gamma K) - (M^2 - (T - H)M\gamma K + K\gamma K)$$

Using (4.40) the second term is

$$-[M^2 - M(T - H)\gamma K] = -(\Delta^2/4)[1 - (4/\Delta^2)M(T - H)\gamma K] \ .$$

Similarly, the second term in N is $- M[1 - (4/\Delta^2)M(T - H)\gamma K]$. Then

$$\hat{g}_e = \{(T - H) - (\Delta^2/4)[1 - (4/\Delta^2)M(T - H)\gamma K]/[T - H - M\gamma K]\}^{-1} \ ,$$

$$\hat{g}_o = - M\{(T - H)(T - H - M\gamma K)/[1 - (4/\Delta^2)$$
$$M(T - H)\gamma K] - (\Delta^2/4)\}^{-1} \ . \tag{A.3.1}$$

Write

$$(T - H - M\gamma K)/[1 - (4/\Delta^2)M(T - H)\gamma K] \equiv T - H - N/D \ . \tag{A.3.2}$$

where

$$N = M\gamma K - (T - H)(4/\Delta^2)M(T - H)\gamma K$$
$$= M\gamma K - M(4/\Delta^2)(T - H)^2\gamma K - [T - H, 4M/\Delta^2](T - H)\gamma K \ .$$

But

$$(4/\Delta^2)(T - H)^2\gamma = (4/\Delta^2)(1 + M^2\gamma) = 4/\Delta^2 + \gamma$$

and

$$[T - H, 4M/\Delta^2] = 4K/\Delta^2 - (4M/\Delta^2)[T - H, M^2](4/\Delta^2)$$
$$= 4K/\Delta^2 - (4M/\Delta^2)(MK + KM)(4/\Delta^2)$$
$$= -(4/\Delta^2)MKM(4/\Delta^2) \ .$$

Then

$$N = -(4/\Delta^2)MK\{1 - (4/\Delta^2)M(T - H)\gamma K\} = -(4/\Delta^2)MKD$$

Inserting this into (A.3.2), and then into (A.3.1), gives

$$\hat{g}_e = [T - H - (\Delta^2/4)(T - H - \hat{\Sigma})^{-1}]^{-1} \ , \tag{A.3.3a}$$

$$\hat{g}_o = -M[(T - H)(T - H - \hat{\Sigma}) - \Delta^2/4]^{-1} \tag{A.3.3b}$$

where

$$\hat{\Sigma} = -(4/\Delta^2)MK = -(2/\Delta)\hat{n}\cdot\boldsymbol{\sigma}[T - H, \hat{n}\cdot\boldsymbol{\sigma}\Delta/2]$$
$$= -(2/\Delta)\hat{n}\cdot\boldsymbol{\sigma}[T - H, \hat{n}\cdot\boldsymbol{\sigma}]\Delta/2 - (2/\Delta)[T - H, \Delta/2] ,$$

$$\hat{\Sigma} \equiv (2/\Delta)\delta\hat{\Sigma}(\Delta/2) + \hat{\Sigma}_0 . \tag{A.3.4}$$

Inserting this last form into (A.3.3), after some simple manipulations, we find (4.43).

4.A.4

Write $\hat{\Sigma}$ in (A.3.4) as $\hat{\Sigma}_0 + \hat{\Sigma}_1$. Define a new γ by

$$\gamma^{-1} \equiv (T - H)(T - H - \hat{\Sigma}_0) - \Delta^2/4 . \tag{A.4.1}$$

Then (A.3.3a) is

$$\hat{g}_e = (T - H - \hat{\Sigma}_0)\gamma + [(T - H - \hat{\Sigma}_0)\gamma(T - H) - 1]$$
$$\times\hat{\Sigma}_1[1 - \gamma(T - H)\hat{\Sigma}_1]^{-1}\gamma . \tag{A.4.2}$$

Now
$$\gamma^{-1} = (T - H\pm\Delta/2)(T - H - \hat{\Sigma}_0\mp\Delta/2)\mp(\Delta/2)$$
$$(T - H - \hat{\Sigma}_0)\pm(T - H)\Delta/2 .$$

The last two terms cancel from the definition of $\hat{\Sigma}_0$. Then

$$2(T - H - \hat{\Sigma}_0)\gamma = (T - H + \Delta/2)^{-1} + (T - H - \Delta/2)^{-1}$$
$$= g_+ + g_- \equiv 2\Gamma . \tag{A.4.3}$$

Also
$$(T - H - \hat{\Sigma}_0)\gamma(T - H) - 1 = -\Gamma'\Delta/2 ,$$
$$2\Gamma' \equiv g_+ - g_- , \tag{A.4.4a}$$
$$\gamma(T - H) = (2/\Delta)\Gamma(\Delta/2) ; \quad (\Delta/2)\gamma = -\Gamma' . \tag{A.4.4b}$$

Then (A.4.2) is

$$\hat{g}_e = \Gamma + \Gamma'\delta\hat{\Sigma}(1 - \Gamma\delta\hat{\Sigma})^{-1}\Gamma' \tag{A.4.5a}$$

where we rewrite $\hat{\Sigma}_1$ as $(2/\Delta)\delta\hat{\Sigma}(\Delta/2)$. Similarly, (A.3.3b) is

$$\hat{g}_o = -M[1 - \gamma(T - H)\hat{\Sigma}_1]^{-1}\gamma = \hat{n}\cdot\boldsymbol{\sigma}(1 - \Gamma\delta\hat{\Sigma})^{-1}\Gamma' . \tag{A.4.5b}$$

In (A.4.5) $\delta\hat{\Sigma}$ is a fluctuating quantity. To perform the average of $\hat{g}_e$ we expand in powers of $\delta\hat{\Sigma}$

$$\langle\hat{g}_e\rangle = \langle\Gamma\rangle + \langle\Gamma'(\delta\hat{\Sigma} + \delta\hat{\Sigma}\Gamma\delta\hat{\Sigma} + \ldots)\Gamma'\rangle . \tag{A.4.6a}$$

To this point our only approximation is in neglecting fluctuations of H. Now we invoke the stiffness of the system against longitudinal fluctuations and drop fluctuations in $\Delta(r)$ as well. We take $\Delta(r)$ to be the self-consistent (spatially varying) value. Then Γ and Γ' may be treated as constants in the averaging process.

Expand each product of $\delta\hat{\Sigma}$'s in cumulants. Call two $\delta\hat{\Sigma}$'s correlated in a particular term if they appear in the same cumulant. Then we can write

$$\langle \hat{g}_e \rangle = \Gamma + \Gamma'\{\Sigma + \Sigma\Gamma\Sigma + \Sigma\Gamma\Sigma\Gamma\Sigma + \ldots\}\Gamma' \tag{A.4.6b}$$

where

$$\Sigma \equiv \langle \delta\hat{\Sigma} + \delta\hat{\Sigma}\Gamma\delta\hat{\Sigma} + \ldots \rangle_{1-\text{irred}} \;. \tag{A.4.7}$$

The label $1 - \text{irred}$ means that for any Γ in this expression, there is at least one pair of correlated $\delta\hat{\Sigma}$, such that one member of the pair is on each side of Γ. Comparing (A.4.6b) with (A.4.6a), we see that we can retrace our steps to the equivalent of (A.3.3a), and find (4.44a).

Similarly, we expand and resum (A.4.5b), and find (4.44b) with (4.44d) for $\overline{\Sigma}$.

It is worth noting that (A.4.7) can be replaced by a 2-irreducible sum, involving a renormalized Γ. This gives a self-consistent theory which, however, we do not pursue at this time.

Finally, note that *retaining* fluctuations in Δ would lead to extremely complex results. The functions Γ could not be treated as constant in the averaging process of (A.4.6a). Products of Γ under the average would have to be replaced by true multiparticle correlations, rather than by products of zero-order Green's functions. A self-consistent theory of one-particle Green's functions, taking these fluctuations of Δ fully into account, would be very difficult.

4.A.5

From (4.43c and 44c)

$$\Sigma = \langle -mk + mk\Gamma mk \rangle_{1-\text{irred}} = -\langle mk \rangle + \langle mk\Gamma mk \rangle - \langle mk \rangle\Gamma\langle mk \rangle \tag{A.5.1}$$

Equation (4.49) implies

$$km + mk = 0 \;. \tag{A.5.2}$$

Then

$$\langle mk\Gamma mk \rangle = -\langle k\Gamma k \rangle - \langle k[m,\Gamma]mk \rangle \;. \tag{A.5.3}$$

But using (4.44e)

$$2[m,\Gamma] = -\{g_+[m,g_+^{-1}]g_+ + g_-[m,g_-^{-1}]g_-\} = g_+kg_+ + g_-kg_- \;. \tag{A.5.4}$$

Then the second term in (A.5.3) is third order in gradients. We will see later that $\langle mk \rangle$ is second order. Then, to the order of interest

$$\Sigma = -\langle mk \rangle - \langle k\Gamma k \rangle \;. \tag{A.5.5}$$

To find $\overline{\Sigma}$ to this order, (4.44d) is

$$2\overline{\Sigma}/\Delta \approx \langle m - m\Gamma mk + m\Gamma mk\Gamma mk\rangle_{1-\text{irred}}$$
$$= \langle m\rangle(1 + \Gamma\langle mk\rangle) - \langle m\Gamma mk\rangle - \langle m\rangle\Gamma\langle mk\Gamma mk\rangle$$
$$- \langle m\rangle\Gamma\langle mk\rangle\Gamma\langle mk\rangle - \langle m\Gamma mk\rangle\Gamma\langle mk\rangle + \langle m\Gamma mk\Gamma mk\rangle \ .$$

Using (A.5.2,4), terms of second order are

$$2\overline{\Sigma}/\Delta \approx \langle m\rangle\{1 + \Gamma\langle mk\rangle + \Gamma\langle k\Gamma k\rangle\} + \langle m\Gamma'k\Gamma'k\rangle \qquad (A.5.6)$$

where Γ' is defined in (A.4.4). The fluctuating part of m in the last term of (A.5.6) gives a contribution of third order. Replacing m by its average in this term, we find

$$2\overline{\Sigma}/\Delta \approx \langle m\rangle(1 + \Gamma\langle mk\rangle + g_+\langle kg_+k\rangle/2 + g_-\langle kg_-k\rangle/2) \ .$$

This leads directly to (4.50–52).

4.A.6

Start with (A.4.5a) and use (A.5.2), (4.43 and 49) to write

$$\hat{g}_e = \Gamma - \Gamma'mk\{1 + \Gamma mk\}^{-1}\Gamma'$$
$$= \Gamma - \Gamma'mk(1 - m\Gamma k)\{(1 + \Gamma mk)(1 - m\Gamma k)\}^{-1}\Gamma'$$
$$= \Gamma - \{\Gamma'mk + \Gamma'k\Gamma k\}\{1 + [\Gamma, m]k + \Gamma k\Gamma k\}^{-1}\Gamma' \ . \qquad (A.6.1)$$

Then using (A.5.4) we find (4.75a). Similarly, (A.4.5b) is

$$\hat{g}_o = m\{1 + \Gamma mk\}^{-1}\Gamma' = m\{1 - m\Gamma k\}\{1 - \Gamma'k\Gamma'k\}^{-1}\Gamma' \qquad (A.6.2)$$

which gives (4.75b).

4.A.7

Write $Z^{\mu\nu}$ and its inverse in the 2×2 block form

$$Z \equiv \begin{pmatrix} Z_1 & Z_2 \\ Z_3 & Z_4 \end{pmatrix} \quad Z^{-1} \equiv \begin{pmatrix} U_1 & U_2 \\ U_3 & U_4 \end{pmatrix} \qquad (A.7.1)$$

where the order of the rows and columns in $Z^{\mu\nu}$ is $0, z, +, -$. Then

$$U_1 = \{Z_1 - Z_2 Z_4^{-1} Z_3\}^{-1} \ , \quad U_2 = -U_1 Z_2 Z_4^{-1} \ . \qquad (A.7.2)$$

The solutions to (4.94a,b,d) are

$$\hat{g}_e^0 = (U_1)^{11} \ , \qquad (A.7.3a)$$

$$\hat{g}_e^z = (U_1)^{21} \, , \tag{A.7.3b}$$

$$\hat{g}_o^0 = (U_1)^{12} f^{zz} + (U_2)^{11} f^{+z} + (U_2)^{12} f^{-z} \, , \tag{A.7.3c}$$

where the rows and columns of 2×2 matrices are labelled conventionally as 1 and 2.

From the definition of $Z^{\mu\nu}$ in (4.85), using (4.88 and 89), we have

$$Z_1 = \begin{pmatrix} T-H & \Delta/2 \\ \Delta/2 & T-H-K^{zz} \end{pmatrix} \, , \quad Z_2 = \begin{pmatrix} 0 & 0 \\ -K^{z+} & -K^{z-} \end{pmatrix} \, ,$$

$$Z_3 = \begin{pmatrix} 0 & -K^{+z} \\ 0 & -K^{-z} \end{pmatrix} \, ,$$

$$Z_4 = \begin{pmatrix} T-H+\Delta/2-K^{++} & -K^{+-} \\ -K^{-+} & T-H-\Delta/2-K^{--} \end{pmatrix} \, . \tag{A.7.4}$$

Then

$$U_1^{-1} = \begin{pmatrix} T-H & \Delta/2 \\ \Delta/2 & T-H-\delta\hat{\Sigma} \end{pmatrix} \, , \tag{A.7.5}$$

$$\delta\hat{\Sigma} \equiv K^{zz} + \sum_{\alpha,\beta=+,-} K^{z\alpha}(Z_4^{-1})^{\alpha\beta} K^{\beta z} \, , \tag{A.7.6a}$$

$$(U_2)^{1\beta} = (U_1)^{12} \sum_{\alpha=+,-} K^{z\alpha}(Z_4^{-1})^{\alpha\beta} \, , \tag{A.7.6b}$$

Inverting (A.7.5) gives

$$(U_1)^{11} = [T-H-(\Delta/2)(T-H-\delta\hat{\Sigma})^{-1}(\Delta/2)]^{-1} \, , \tag{A.7.7a}$$

$$(U_1)^{12} = [(T-H-\delta\hat{\Sigma})(2/\Delta)(T-H)-\Delta/2]^{-1} \, , \tag{A.7.7b}$$

$$(T-H)(U_1)^{11} + (\Delta/2)(U_1)^{21} = 1 \, . \tag{A.7.7c}$$

Putting these together with (A.7.3,6) and (4.92) we find (4.95).

4.A.8

To find $f^{\mu\nu}$, simply apply the definition (4.94c), using R as given in (4.16) and the conventions of (4.83). The values needed are

$$f^{zz} = \cos\theta \, ; \quad f^{+z} = (f^{-z})^* = -\sin\theta e^{ib}/\sqrt{2} \, . \tag{A.8.1}$$

Another useful result is that $f^{\mu\nu}$ is *unitary* as a 4×4 matrix, or, considering (4.94d), as a 3×3 matrix on omitting the 0 components. To prove this, first note

$$f^{*\nu\lambda} = \text{Tr}\{R\sigma^\nu R^{-1}\sigma^{\lambda\dagger}\}/2 \, .$$

153

Then, using (4.84)

$$\sum_\lambda f^{\mu\lambda} f^{*\nu\lambda} = \text{Tr}\{R\sigma^{\mu\dagger}R^{-1}R\sigma^\nu R^{-1}\}/2 = \delta^{\mu\nu} \tag{A.8.2}$$

which is what is required.

A similar argument, with the definition (4.87), gives

$$K^{\mu\nu} = \sum_\lambda [T - H, f^{\mu\lambda}] f^{\dagger\lambda\nu} \tag{A.8.3}$$

which is useful in what follows.

There are several ways to find $K^{\mu\nu}$, including use of (A.8.3). We use (4.87) directly, with the definitions (4.21 and 22) *modified* to reflect the difference betwen the conventions for $\sigma^\pm$ used in Sects. 4.2 and 4. Then

$$[T - H, R\sigma^{\mu\dagger}R^{-1}] = R\{[\sigma^{\mu\dagger}, \hat{A}] - \mathbf{A}\cdot[A, \sigma^{\mu\dagger}]/2m - [\sigma^{\mu\dagger}, A]\cdot\mathbf{A}/2m$$
$$- [\sigma^{\mu\dagger}, i\nabla\cdot A]/2m\}R^{-1} - R[\sigma^{\mu\dagger}, A]R^{-1}\cdot i\nabla/m .$$

Potential terms in H play no role in the commutator. Completing (4.87) this is

$$2K^{\mu\nu} = \text{Tr}\{[\sigma^{\mu\dagger}, \hat{A}]\sigma^\nu - [\sigma^{\mu\dagger}, A]\cdot[\sigma^\nu, A]/2m$$
$$- [\sigma^{\mu\dagger}, i\nabla\cdot A]\sigma^\nu - [\sigma^{\mu\dagger}, A]\sigma^\nu\cdot i\nabla/m\}$$
$$= \text{Tr}\{[\sigma^\nu, \sigma^{\mu\dagger}](\hat{A} - i\nabla\cdot A/2m - \mathbf{A}\cdot i\nabla/m) - [\sigma^{\mu\dagger}, A]\cdot[\sigma^\nu, A]/2m\} .$$

The various pieces of this expression can be read off, using (4.21 and 22). After some algebra we find

$$K^{--} = 2(\hat{g} - i\nabla\cdot\mathbf{g}/2m - i\mathbf{g}\cdot\nabla/m + |\mathbf{a}|^2/2m + |\mathbf{g}|^2/m) ,$$
$$K^{++} = 2(-\hat{g} + i\nabla\cdot\mathbf{g}/2m + i\mathbf{g}\cdot\nabla/m + |\mathbf{a}|^2/2m + |\mathbf{g}|^2/m) ,$$
$$K^{+-*} = K^{-+} = -\mathbf{a}\cdot\mathbf{a}/m ,$$

which leads to (4.98). The remaining components of $K^{\mu\nu}$ are written as (4.99).

4.A.9

From (4.96 and 99) $\delta\hat{\Sigma}$ is second-order in gradients. Also $\hat{F}$ contains all the second order terms of $\overline{\Sigma}$. Then we approximate (4.97c,d) as

$$\Sigma \approx \langle\delta\hat{\Sigma}\rangle ; \quad \overline{\Sigma} \approx \langle\hat{F}\rangle \tag{A.9.1}$$

where, for consistency, the terms quadratic in $\mathbf{a}$ and $\mathbf{a}^*$ are to be omitted in (4.98). Then

$$\Sigma \approx \langle K^{zz}\rangle + \langle K^{z+}\Gamma_+ K^{+z}\rangle + \langle K^{z-}\Gamma_- K^{-z}\rangle , \tag{A.9.2a}$$
$$\Gamma_\pm^{-1} \equiv T \pm 2\hat{g} - V + (\nabla \mp 2i\mathbf{g})^2/2m \pm \Delta/2 . \tag{A.9.2b}$$

Note that $\Gamma_\pm$ is the Green's function for a single particle in the lattice effective potential $V(r) \mp \Delta(r)/2$, and also subject to electric and magnetic fields determined by the potentials $2g$ and $2\hat{g}$. To the order of interest this Green's function is approximated by

$$\Gamma_\pm(x, x') \approx g_\pm(x, x') e^{\pm 2iI(x,x')} , \qquad (A.9.3a)$$

$$I(x, x') \equiv \int_{x'}^{x} [g(x_1) \cdot d\mathbf{r}_1 + \hat{g}(x_1) dt_1] \qquad (A.9.3b)$$

where the integral is along the straight line path in the four dimensional $x = (\mathbf{r}, t)$ space between x' and x. Here

$$g_\pm = (T - H \pm \Delta/2)^{-1} \qquad (A.9.3c)$$

the Green's fucntion in the absence of disorder. Equation (A.9.3) constitutes the classical path approximation for the effects of electromagnetic potentials on charged-particle propagation.

With this approximation Σ will involve the quantity $\langle K^{z\pm}(x) \exp[\pm 2iI(x, x')] K^{\pm z}(x') \rangle$. But (A.9.3a) implies the relations

$$\partial_\mu \{ f(x) e^{\pm iI(x,x')} \} \approx e^{\pm iI(x,x')} (\partial_\mu \pm 2ig_\mu) f(x) \qquad (A.9.3d)$$

for $\mu = x, y, z, t$. Comparing with (A.1.1,2), this implies that products

$$\hat{D}_{\mu\nu}(x, x') \equiv a_\mu(x) e^{iI(x,x')} a_\nu^*(x') \qquad (A.9.4a)$$

(with $\hat{a} \equiv a_0$), can be written approximately as gradients of a function $\hat{D}$

$$\hat{D}_{\mu\nu}(x, x') \approx \partial_\mu \partial_\nu' \hat{D}(x, x') . \qquad (A.9.4b)$$

With the expressions in (4.99) this gives an evaluation of Σ in terms of $g_\pm$ and $D = \langle \hat{D} \rangle$.

For later manipulations it is useful (but changes no results) to pretend there is a function $d(x)$ such that

$$\hat{D}(x, x') = d(x) d^*(x') . \qquad (A.9.4c)$$

Then, for example, $K^{z+}(x) e^{iI} \approx i\sqrt{2}[T - H, d]$, and (A.9.2a) is

$$\Sigma(x, x') \approx 4\langle [-H, d] d^* \rangle \delta(x - x') + 2\langle [T - H, d] g_+ [d^*, T - H] \rangle$$
$$+ 2\langle [T - H, d^*] g_- [d, T - H] \rangle . \qquad (A.9.5)$$

By time translation invariance (for $x = x'$)

$$\langle T dd^* \rangle = 0 = \langle d^*[T, d] \rangle + \langle d[T, d^*] \rangle \quad \text{while}$$
$$2m \langle dH d^* \rangle = \langle \nabla d \cdot \nabla d^* \rangle = 2m \langle d^* H d \rangle .$$

Then Σ can be written as (4.100).

To find $\overline{\Sigma}$, first use (A.8.3) to get the exact relation

$$(T - H + \Delta\sigma^z/2 - K)^{\mu\beta}f^{\beta z} = f^{\beta z}(T - H + \Delta\sigma^z/2)^{\mu\beta} + k^{\mu z}f^{zz} \qquad (A.9.6)$$

where μ and β run over + and −. The first parenthesis in (A.9.6) is $\hat{\Gamma}^{-1}$ while the parenthesis on the right-hand side is $g_\mu^{-1}\delta^{\mu\beta}$. Then

$$\hat{\Gamma}_{\mu\lambda}(f^{\lambda z}g_\lambda^{-1}) = f^{\mu z} - \hat{\Gamma}_{\mu\lambda}K^{\lambda z}f^{zz} \ .$$

Simple manipulations give

$$\hat{\Gamma}_{\mu\lambda}f^{\lambda z} = (f^{\mu z} - \hat{\Gamma}_{\mu\lambda}K^{\lambda z}f^{zz})g_\mu + \hat{\Gamma}_{\mu,-\mu}f^{-\mu,z}(g_{-\mu}^{-1} - g_\mu^{-1})g_\mu \ . \qquad (A.9.7)$$

This is exact and, with (4.96) gives an expression for $\hat{F}$. To the order of interest we drop off-diagonal terms in $\hat{\Gamma}$, and thus the second term in (A.9.7). Then

$$\hat{F} \approx f^{zz} + \sum_\mu (K^{z\mu}f^{\mu z} - K^{z\mu}\hat{\Gamma}_\mu K^{\mu z}f^{zz})g_\mu \ . \qquad (A.9.8)$$

Now from (A.8.1) explicit calculation gives

$$(\nabla + 2i\boldsymbol{g})f^{-z} = -i\sqrt{2}\,\cos\theta\,\boldsymbol{a} = -i\sqrt{2}\boldsymbol{a}f^{zz} \qquad (A.9.9)$$

and the complex conjugate for f^{+z}. With (4.99) we find

$$\langle K^{z+}f^{+z}\rangle = -\langle \boldsymbol{a}\cdot\boldsymbol{a}^*\cos\theta\rangle/m - \sqrt{2}\langle \hat{a}f^{+z}\rangle \ ,$$
$$\langle K^{z-}f^{-z}\rangle = -\langle \boldsymbol{a}^*\cdot\boldsymbol{a}\cos\theta\rangle/m + \sqrt{2}\langle \hat{a}^*f^{-z}\rangle \ . \qquad (A.9.10)$$

Now

$$\langle \hat{a}(x)f^{+z}(x')\rangle = \langle \hat{a}(x)e^{2iI(x,x')}f^{+z}(x')\rangle\delta(x - x') \ .$$

Using the time derivative analogue of (A.9.9 and 3d) we have

$$\partial_0'\langle \hat{a}(x)e^{2iI(x,x')}f^{+z}(x')\rangle_{x'=x} = i\sqrt{2}\langle \hat{a}\hat{a}^*\cos\theta\rangle \ .$$

We expect that $|\hat{a}|^2$ is only weakly correlated with $\cos\theta$, so replace $\cos\theta$ by its average. The function $\langle \hat{a}\hat{a}^*\rangle$ is given by $\partial_0\partial_0'D$ so we conclude that

$$-\sqrt{2}\langle \hat{a}f^{+z}\rangle \approx 2i\langle \cos\theta\rangle\langle d\partial_0 d^*\rangle = 2\langle \cos\theta\rangle\langle dTd^*\rangle \ ,$$
$$\sqrt{2}\langle \hat{a}^*f^{-z}\rangle \approx 2\langle \cos\theta\rangle\langle d^*Td\rangle \ . \qquad (A.9.11)$$

Then (A.9.10) is

$$\langle K^{z+}f^{+z}\rangle \approx 2\langle \cos\theta\rangle\langle d(T - H)d^*\rangle \ ,$$

$$\langle K^{z-}f^{-z}\rangle \approx 2\langle \cos\theta\rangle\langle d^*(T - H)d\rangle \ . \qquad (A.9.12)$$

Using (A.9.5, 12, and 8), we find (4.100).

Acknowledgements. The local-band theory described has evolved during a long collaboration with Richard Prange. He shares responsibility for the ideas presented, but bears none for any deficiencies of presentation.
This work was supported by the National Science Foundation under Grant No. DMR82–13768. Computer time and facilities were provided by the University of Maryland Computer Science Center.

References

4.1   R.E. Prange, V. Korenman: In "Magnetism and Magnetic Materials 1974", AIP Conf. Proc. **24**, 325 (1975);
V. Korenman, J.L. Murray, R.E. Prange: In "Magnetism and Magnetic Materials 1975", AIP Conf. Proc. **29**, 321 (1976)

4.2   V. Korenman, J.L. Murray, R.E. Prange: Phys. Rev. **B16**, 4032, 4048, 4058 (1977)

4.3   V. Korenman, R.E. Prange: In "Transition Metals 1976", Inst. Phys. Conf. Ser. **39**, 562 (1978);
R.E. Prange, V. Korenman, Ibid. 567 (1978)

4.4   V. Korenman, R.E. Prange: J. Appl. Phys. **50**, 1779 (1979)

4.5   R.E. Prange: J. Appl. Phys. **50**, 7445 (1979)

4.6   R.E. Prange, V. Korenman: Phys. Rev. **B19**, 4691, 4698 (1979)

4.7   V. Korenman; R.E. Prange: Solid State Commun. **31**, 909 (1979)

4.8   V. Korenman, R.E. Prange: Phys. Rev.Lett. **44**, 1291 (1980)

4.9   V. Korenman: In "Transition Metals 1980", Inst. Phys. Conf. Ser. **55**, 195 (1981)

4.10 H. Capellmann, R.E. Prange: Phys. Rev. **B23**, 4709 (1981).

4.11 R.E. Prange: In "Electron Correlation and Magnetism in Narrow-Band Systems", ed. by T. Moriya, Springer Ser. Solid-State Sci., Vol. 29 (Springer, Berlin, Heidelberg 1981) ps. 55, 63, 66, 69

4.12 V. Korenman, B. Wyman: Phys. Rev. **B24**, 5413 (1981)

4.13 C.S. Wang, R.E. Prange, V. Korenman: Phys. Rev. **B25**, 5766 (1982)

4.14 V. Korenman: Physica **119B**, 21 (1983)

4.15 V. Korenman: J. Magn. Magn. Mater. **31–34**, 301 (1983)

4.16 V. Korenman, R.E. Prange: Phys. Rev. Lett. **53**, 186 (1984)

4.17 V. Korenman: J. Appl. Phys. **57**, 3000 (1985)

4.18 H. Capellmann, J. Phys. **F4**, 1966 (1974), Solid State Commun. 30, 7 (1979), Z. Physik **B34**, 29 (1979); and **B35**, 269 (1979)

4.19 H. Hopster, R. Raue, G. Guntherodt, E. Kisker, R. Clauberg, M. Campagna: Phys. Rev. Lett. **51**, 829 (1983)

4.20 E. Kisker, K. Schröder, M. Campagna, W. Gudat: Phys. Rev. Lett. **52**, 2285 (1984)

4.21 J. Kirschner, M. Globl, V. Dose, H. Scheidt: Phys. Rev. Lett. **53**, 612 (1984)

4.22 E. Kisker: J. Magn. Magn. Mat. **45**, 23 (1984)

4.23 E.M. Haines, R. Clauberg, R. Feder: Phys. Rev. Lett. **54**, 932 (1985)

4.24 D.M. Edwards: Can. J. Phys. **52**, 704 (1974)

4.25 J. Hertz, D.M. Edwards: Phys. Rev. Lett. **28**, 1334 (1972)

4.26 L.M. Roth: Phys. Rev. **186**, 428 (1969)

4.27 A. Liebsch: Phys.Rev. **B23**, 5203 (1981)

4.28 C.S. Wang, J. Callaway: Phys. Rev. **B15**, 298 (1977);
J. Callaway, C.S. Wang: Phys. Rev. **B16**, 2095 (1977)

4.29 C.S. Wang, B.M. Klein, H. Krakauer: Phys. Rev. Lett. **54**, 1852 (1985). There are also the well known discrepancies between computed and measured band widths and splittings in nickel, most likely related to the omission of virtual spin-wave effects. See the analysis of [4.27]

4.30 P.J. Brown, J. Deportes, D. Givord, K.R.A. Ziebeck: J. Appl. Phys. **53**, 1973 (1982);
P.J. Brown, H. Capellmann, J. Deportes, D. Givord, K.R.A. Ziebeck: J. Magn. Magn. Mat. **30**, 335 (1982)

4.31 P.J. Brown, H. Capellmann, J. Deportes, D. Givord, K.R.A. Ziebeck: J. Magn. Magn. Mat. **31–34**, 295 (1983);
G. Shirane, P. Boni, J.P. Wicksted: BNL Report 36586, (1985)

157

4.32   H.A. Mook, J.W. Lynn, R.M. Nicklow: Phys. Rev. Lett. **30**, 556 (1973);
        H.A. Mook, J.W. Lynn: J. Appl. Phys. **57**, 3006 (1985)
4.33   J.W. Lynn: Phys. Rev. B**11**, 2624 (1975)
4.34   J.F. Janak, A.R. Williams: Phys. Rev. B**14**, 4199 (1976);
        U.K. Poulson, J. Kollar, O.K. Andersen: J. Phys. F**6**, L241 (1976)
4.35   Gyorffy: In vol II
4.36   R.L. Stratonovich: Sov. Phys. Dok. **2**, 416 (1958);
        J. Hubbard: Phys. Rev. Lett. **3**, 77 (1959)
4.37   A.L. Fetter, J.D. Walecka, *Quantum Theory of Many-Particle Systems* (McGraw-Hill, New York 1971)
4.38   J. Hubbard: Phys. Rev. B**19**, 2626 (1979)
4.39   A.J. Holden, V. Heine, J.H. Samson: J. Phys. F**14**, 1005 (1984)
4.40   J.C. Gustafson, T.G. Phillips: Phys. Lett. **29A**, 273 (1971)
4.41   V. Korenman: unpublished
4.42   J.S. Kouvel, J.B. Comly: Phys. Rev. Lett. **20**, 1237 (1968)
4.43   P.C. Riedi: Phys. Rev. B**15**, 5197 (1977)
4.44   D.L. Connelly, J.S. Loomis, D.E. Mapother: Phys. Rev. B**3**, 924 (1971)
4.45   R.E. Pawel, E.E. Stansbury: J. Phys. Chem. Solids **26**, 757 (1965)
4.46   Since SARPES is surface sensitive, the correct magnetization and magnetic energy density to use in evaluating the parameters in D are those for the surface region. Having no information about the surface energy density, we compare the experiments with the *bulk* theory at the same value of magnetization. This circumstance is also discussed in [4.20 and 21]
4.47   C.J. Maetz, U. Gerhardt, E. Dietz, A. Ziegler, R.J. Jelitto: Phys. Rev. Lett. **48**, 1686 (1982)
4.48   I.S. Gergis: Int. I. Magn. **3**, 273 (1972)

5. Electron Correlations in Transition Metals

P. Fulde, Y. Kakehashi, and G. Stollhoff
With 14 Figures

A treatment is given of electron correlations in the ground state and in single-particle excited states of transition metals. With respect to the ground state two different approaches are discussed. One is the density functional method and the local spin density approximation to it. The other is one in which one attempts to calculate the correlated or many-body ground state wave function of the system. Emphasis is placed on a physical understanding of the different contributions to the correlation energy which result from that approach. We also discuss a generalization of the latter to finite temperatures. This is done by using the functional integral method but going beyond the static approximation to it. As a consequence one obtains at zero temperature the correlated instead of the Hartree-Fock ground state wave function. Finally an analysis is presented of the effect of electron correlations on the photoemission spectra of inner-core states.

5.1 Background

The magnetism of transition metals has intrigued and fascinated physicists for many years and, despite of tremendous progress, is still doing so. There has been a long way from the contrary points of views taken by *Van Vleck* [5.1] and *Slater* [5.2] concerning local *vs* itinerant features of d electrons to modern theories like the unified theory of *Moriya* and *Takahashi* [5.3]. The central problem has always been that of electron correlations and their influence on various properties of transition metal systems. There have never been any doubts that electron correlations are important. After all, we are dealing with metals for which it is well known that a true independent electron, i.e. Hartree-Fock (HF) type of calculation would give completely unsatisfactory results. But the question has always been whether or not the correlation effects can be incorporated into the theory by just renormalizing properly a number of characteristic quantities such as the effective interactions. For sufficiently low temperatures, this should be always the case as we know from Landau's Fermi-liquid theory [5.4]. However, this theory does apply only as long as the excitation energy and $k_B T$ are lower than those characteristic energies at which there is structure in the frequency-dependent single particle self-energy. In transition metals one is interested in temperatures of the order of the Curie or Néel temperature. The latter are typically of the order of 0.1 eV. Therefore one important question is whether or not there do exist characteristic energies in the sense described before which are less than $\cong 0.1\,\text{eV}$ and which invalidate a simple Fermi-liquid description in that temperature range.

Another important point to realize is that one can introduce correlations into a theory in two different ways. One way is by calculating the correlated ground state and excited states or at least their energies. For example, one can think of the correlated ground state as consisting of the HF ground state superimposed with a large number of HF excited states. This is the conventional configuration interaction approach in quantum chemistry [5.5]. The same holds true for the excited states. A rather different way of introducing correlation effects is by finite-temperature techniques. In order to understand the difference let us consider the functional integral method [5.3]. In that approach the interaction of an electron with the other electrons is replaced by the motion of an electron in fictitious spatially and time-dependent external fields. They appear with Gaussian weights. One may also think of these fields as resulting from a coupling of the electron to a heat bath which contains all the many-body degrees of freedom. Now let us neglect, as it is usually done, the time dependence of these external fields so that they are only space dependent. This is called the static approximation. In that case one can show that at $T = 0$ only a uniform field is left. It equals the Stoner exchange field when the ground state is ferromagnetic and vanishes when it is paramagnetic. Therefore at $T = 0$ there are no correlations left and that is due to the negligence of the time dependence of the external field. But at $T \neq 0$ there are correlations present even when the static approximation is made.

These are the correlations which are considered in conventional theories in order to explain, e.g., the large entropy changes near T_c or the Curie-Weiss susceptibility above T_c in Fe and Ni. How do these correlations come about? Apparently when an electron is moving through the system at $T \neq 0$ it experiences a variety of spatially fluctuating fields, the relative weights of which are given by Gaussian distributions. They result in correlation effects similar to those of a time fluctuating field at $T = 0$. Clearly, in a complete theory one must have both, spatially *and* time fluctuating external fields acting on an electron in order to replace the degrees of freedoms of the other electrons. Only in the high temperature limit do time fluctuations become neglegible while at T_c they are still of considerable importance.

In the present chapter we will first discuss recent developments which have taken place with respect to calculations of the correlated ground state. Thereby we will distinguish between the local spin density (LSD) approximation [5.6] to the density functional approach [5.7,8] and calculations of the correlated ground state wave function. It is well known that the density functional approach avoids calculations of wave functions and instead computes directly the ground state energy or other ground state quantities. Correlation effects are taken into account by the appropriate choice of the effective single-particle potential. This potential results in a weaker electron repulsion than the bare potential. It changes the single particle densities so that they approximate better those of the correlated system and it also lowers the total energy as correlations do.

One may also try to calculate instead the correlated ground state wave function. At present this is not possible on an ab initio level. Rather one has to start from model Hamiltonians. However, the Hamiltonians can come rather close in applying to actual transition metal systems because, e.g., the forms of the canonical d bands can be properly taken into account. After having gained insight into the correlated ground state wave function we consider the influence of correlations on the cohesive energy, the criterion for ferromagnetic order, the size of the local moment, etc..

Correlation effects play also an important role when single-particle excitations are considered. They are described by a one-electron (hole) Green's function and are measured, e.g., in photoemission experiments. For example, the existence of many-body or shake-up peaks is solely due to correlations.

In a next step we go over to finite temperatures and consider the thermodynamics of transition metals. This is done by using the functional integral method but going beyond the static approximation. At $T = 0$ the present approach reduces to the energy of the correlated ground state instead to the one of the Hartree-Fock ground state. It is a variational method which we are applying when we treat the corrections to the static approximation. It is shown that they result in considerable reductions of the Curie temperature for given Coulomb interaction and d band width compared with results of the static approximation.

It should be also pointed out that a high orbital degeneracy D improves on the quality of the static approximation. When D is considered as a free parameter, then in the limit $D \rightarrow \infty$ the mean-field theory and therefore the static approximation become exact. This suggests expansion procedures in $1/D$ [5.9]. Correct treatment of the terms of order $1/D$ requires going beyond the static approximation.

Last but not least, we consider the influence of correlations between the d electrons on the photoelectron spectra from inner core states. As will be demonstrated, there is quite a strong influence of them not only on the multiplicity of the lines but also on the line shape.

5.2 The Correlated Ground State

We will start out in this section by giving a simple and intuitive picture of electron correlations in the ground state of a d electron system. This is followed by a discussion of two different methods which treat electron correlations

1. the density-functional method and the local spin-density approximation to it, and
2. a variational calculation of the ground state wave function.

5.2.1 Physical Interpretation of Electron Correlations

In order to be able to deal with systems with large electron numbers, one often makes the assumption that the electrons move independently of each other. That does not mean that the electron interaction effects are neglected altogether. Instead they are taken into account in an averaged manner. The motion of an electron is influenced by the presence of the other electrons, but it depends only on where the other electrons are on *average* and not where they are *actually* at a given instant of time. The main deficiency is therefore that in the independent-electron approximation the electrons can come too close with respect to each other. This holds particularly true for electrons with opposite spins, because electrons with equal spins are prevented by the Pauli principle to be in the same quantum state.

In order to illustrate the effects of correlations consider a particular model with a model Hamiltonian. We assume that we are dealing with a cubic lattice of atoms which have five d orbitals each. We neglect all s and p electrons. Furthermore we assume that only d electrons interact with each other which sit at the same site. The model Hamiltonian is chosen to be of the form

$$
\begin{aligned}
H &= H_0 + H_1 \\
H_0 &= \sum_{\nu\sigma k} \varepsilon_\nu(k) n_{\nu\sigma}(k) \\
H_1 &= \sum_\ell H_1(\ell) \\
H_1(\ell) &= \tfrac{1}{2} \sum_{ij\sigma\sigma'} U_{ij} a_{i\sigma}^+ a_{j\sigma'}^+(\ell) a_{j\sigma'}(\ell) a_{i\sigma}(\ell) \\
&\quad + \tfrac{1}{2} \sum_{ij\sigma\sigma'} J_{ij} [a_{i\sigma}^+(\ell) a_{j\sigma'}^+(\ell) a_{i\sigma'}(\ell) a_{j\sigma}(\ell) \\
&\quad + a_{i\sigma}^+(\ell) a_{i\sigma'}^+(\ell) a_{j\sigma'}(\ell) a_{j\sigma}(\ell)] \ .
\end{aligned}
\tag{5.1}
$$

The part H_0 describes the canonical d bands. Their momentum dependent dispersions $\varepsilon_\nu(k)$ $(\nu = 1,\ldots,5)$ can be found, e.g. in [5.10]. They contain only one free parameter which is the d band width W. The $n_{\nu\sigma}(k)$ are the number operators of the eigenstates (Bloch states) with spin σ which correspond to the canonical band eigenvalues. The Hamiltonian H_1 describes the electron interactions. U_{ij} and J_{ij} are the Coulomb and exchange matrix elements, respectively. The operators $a_{i\sigma}^+(\ell)(a_{i\sigma}(\ell))$ create (destroy) an electron in the atomic like d orbital i at site ℓ with spin σ. Orbitals are assumed to have been orthogonalized. Generally U_{ij} and J_{ij} are anisotropic matrices. The U_{ij} have the following general form [5.11]

$$
U_{ij} = U + 2J - 2J_{ij}
\tag{5.2}
$$

where U and J are the average Coulomb and exchange interaction constant and $2J_{ij}$ contains the anisotropies. For the cubic systems under consideration the d orbitals are divided into e_g and t_{2g} orbitals. Then $J = (J_{e_g e_g} + J_{t_{2g} t_{2g}})/2$ and the anisotropy can be described by a single parameter $\Delta J = J_{e_g e_g} - J_{t_{2g} t_{2g}}$. The matrix J_{ij} expressed in J and ΔJ is explicitly given in [5.26]. The Hamiltonian (5.1) does *not* comply with rotational invariance. In order to be rotationally invariant, one must add interaction terms to $H_1(\ell)$ which are proportional to ΔJ and depend on more than two different orbital indices. Those terms play no role in the following calculations because an approximation will be made later in the framework of which they do not contribute. When one neglects any anisotropies by setting $\Delta J = 0$, one obtains $U_{ij} = U$ and $J_{ij} = J$. Finally, in the case of a single orbital we recover the well known Gutzwiller-Hubbard Hamiltonian [5.12,13].

Let us start out from the HF ground state of the Hamiltonian (5.1) which we denote by $|\phi_0\rangle$. It is of the form

$$|\phi_0\rangle = \prod_{\substack{k\nu\sigma \\ \epsilon_{k\nu} \leq \epsilon_F}} c_{\nu\sigma}^+(k)|0\rangle \tag{5.3}$$

where $c_{\nu\sigma}^+(k)$ creates an electron with momentum k and spin σ in the canonical band ν. We expand the Bloch states $c_{\nu\sigma}^+(k)$ in terms of the orbital states $a_{i\sigma}^+(\ell)$ as

$$c_{\nu\sigma}^+(k) = \sum_{i\ell} \alpha_i(\nu,k) a_{i\sigma}^+(\ell) e^{ikR_\ell} . \tag{5.4}$$

In order to visualize and study the effect of correlations we decompose $|\phi_0\rangle$ with the help of (5.4) into a sum of products of $a_{i\sigma}^+(\ell)$ operators. Each of these terms will be called a configuration. In Fig. 5.1 we have drawn two configurations which are quite different with respect to their respective interaction

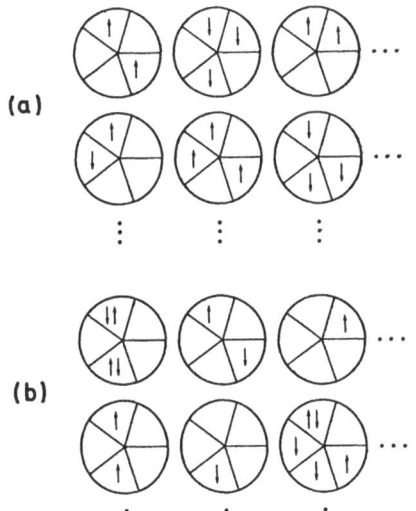

(a)

(b)

Fig. 5.1. (a) "favourable" and (b) "unfavourable" configuration contained in the nonmagnetic SCF ground state $|\phi_0\rangle$. The circles symbolize atoms and the five segments the different d orbitals. The d electron (hole) occupancy per atom is 2.5

163

energies. The configuration in Fig. 5.1a is a *favourable* configuration from the point of view having a small repulsion energy while the one in Fig. 5.1b is an *unfavourable* one. It is important to realize that the essential deficiency of the independent electron approximation is that the unfavourable configurations have too large weights in $|\phi_0\rangle$. The role of the electron correlations is to partially suppress them. In the independent-electron approximation the interaction of an electron with the other ones is taken into account only with respect to their *averaged* distributions, while in a more accurate theory it should depend on their *actual* positions. Therefore the charge fluctuations at an atomic site and connected with them the Coulomb repulsions come out too large when the independent-electron approximation is made. Electron correlations suppress considerably charge fluctuations. A measure of charge fluctuations is the quantity

$$\Delta n^2 = \langle \psi_0 | n^2(\ell) | \psi_0 \rangle - \langle \psi_0 | n(\ell) | \psi_0 \rangle^2 \quad \text{where} \tag{5.5}$$

$$n(\ell) = \sum_{i\sigma} a_{i\sigma}^+(\ell) a_{i\sigma}(\ell)$$

is the number operator for electrons at an arbitrary site ℓ and $|\psi_0\rangle$ is the ground state wave function. Therefore we expect Δn^2 to be considerably larger for $|\phi_0\rangle$ than for the correlated ground state $|\psi_0\rangle$.

Another important effect of correlations is related to Hund's rules. In an atom with an unfilled p, d or f shell, the repulsive energy is smallest when the total spin takes its maximum value. We expect also in a solid a similar relationship to hold. When there is a given number of d electrons at an atomic site at a given instant of time then those configurations will be favoured in which the electron spins are partially aligned. The gain in kinetic energy due to electron hopping on and off a site works, of course, against a perfect Hund's rule spin alignment. A measure of the degree of alignment is the quantity

$$S^2 = \langle \psi_0 | S^2(\ell) | \psi_0 \rangle \ ,$$

$$S(\ell) = \sum_i s_i(\ell) \tag{5.6}$$

where $s_i(\ell)$ is the electron spin operator for orbital i on site ℓ. S^2 is different from zero even for a nonmagnetic ground state to which the Hartree-Fock approximation is applied ($S^2 \to S_{HF}^2$). For example, also Na metal has a $S^2 \neq 0$. When we assume that the electrons do not move from site to site (localized limit) S^2 goes over into S_{loc}^2. It corresponds to fully aligned electron spins at an atomic site. The quantity

$$\Delta S^2 = \frac{\langle \psi_0 | S^2(\ell) | \psi_0 \rangle - S_{HF}^2}{S_{loc}^2 - S_{HF}^2} \tag{5.7}$$

is a measure to what extend the Hund's rule correlations are effective when the correlated ground state wave function is used. One may use ΔS^2 in order to characterize a local moment. When $\Delta S^2 > 1/2$ the system is with respect to its local moment closer to the localized limit than to the independent-electron limit. In summary we have shown that electron correlations reduce charge fluctuations and furthermore enhance those configurations at an atomic site in which electron spins are aligned.

We shall discuss in the following two very different approaches which aim at incorporating the electron correlations and their effects on ground state properties.

5.2.2 Density Functional Method

The density functional method aims at calculating ground state properties such as the energy, the density or the magnetization without having to calculate the wave function. The theoretical basis of density functional calculations has been provided by the work of *Hohenberg* and *Kohn* [5.7], and *Kohn* and *Sham* [5.8], who could build on ideas developed by J.C. Slater. Starting point is hereby the proof that the energy E_0 of the ground state can be always written as a functional of the density $n(r)$ and the spin-density $\sigma(r)$

$$E_0[n,\sigma] = \int d^3r V(r)n(r) + \frac{e^2}{2} \int d^3r d^3r' \frac{n(r)n(r')}{|r-r'|}$$

$$- \mu_B \int d^3r \sigma(r)H(r) + G[n,\sigma] \ . \tag{5.8}$$

$V(r)$ is the external potential in which the electrons are moving and $H(r)$ is the local magnetic field. $G[n,\sigma]$ is an unknown but universal functional of $n(r)$ and $\sigma(r)$. This means that the functional form of $G[n,\sigma]$ is independent of $V(r)$ so that $V(r)$ enters into it only indirectly through $n(r)$ and $\sigma(r)$. Furthermore $E_0[n,\sigma]$ is minimized by the ground state density distribution $n(r)$ and magnetization distribution $\sigma(r)$.

The functional $G[n,\sigma]$ can be decomposed into the kinetic energy of a *non*interacting electron system $T_s[n,\sigma]$ and a remaining part which is denoted by $E_{xc}[n,\sigma]$ and is defined as the exchange-correlation energy,

$$G[n,\sigma] = T_s[n,\sigma] + E_{xc}[n,\sigma] \ . \tag{5.9}$$

In order to derive from these general considerations a practicable computational scheme the assumption is made that $E_{xc}[n,\sigma]$ is of the form

$$E_{xc}[n_+, n_-] = \int d^3r \varepsilon_{xc}(n_+(r), n_-(r))n(r) \tag{5.10}$$

where $\varepsilon_{xc}(n_+, n_-)$ is the exchange-correlation energy density of a homogeneous electron gas with spin-dependent densities n_+, n_-, respectively. Equation (5.10) is the local spin-density approximation (LSD), which is crucial for practical applications of the method.

The energy E_0 is minimized by the density

$$n_\sigma(\mathbf{r}) = \sum_i^{\text{occ}} |\psi_{i\sigma}(\mathbf{r})|^2 \tag{5.11}$$

where the $\psi_{i\sigma}(\mathbf{r})$ fulfill the equations

$$\left[-\frac{1}{2m}\nabla^2 - \mu_B\boldsymbol{\sigma}\cdot\mathbf{H} + V_\sigma^{\text{eff}}(\mathbf{r}) \right]\psi_{i\sigma}(\mathbf{r}) = \varepsilon_{i\sigma}\psi_{i\sigma}(\mathbf{r}) \ . \tag{5.12}$$

The effective potential $V_\sigma^{\text{eff}}(\mathbf{r})$ is given by

$$V_\sigma^{\text{eff}}(\mathbf{r}) = V(\mathbf{r}) + e^2\int d^3r'\frac{n(\mathbf{r}')}{|\mathbf{r} - \mathbf{r}'|} + \mu_\sigma^{\text{xc}}(\mathbf{r}) \tag{5.13}$$

with the exchange-correlation potential

$$\mu_\sigma^{\text{xc}} = \frac{d}{dn_\sigma}[(n_\sigma + n_{-\sigma})\varepsilon_{\text{xc}}(n_+, n_-)] \ . \tag{5.14}$$

Solving (5.12) is a band structure problem for the solution of which sophisticated methods have been developed [5.14]. The energy $\varepsilon_{\text{xc}}(n_+, n_-)$ is thereby often chosen to be of the form (in atomic units) [5.6]

$$\mu_\sigma^{\text{xc}} = -\frac{0.611}{r_s}\left[\beta(r_s) + \frac{\sigma}{3}\frac{\delta(r_s)m}{(1 + \sigma m \cdot 0.297)}\right] \tag{5.15}$$

where $\sigma = \pm$, $m = (n_+ - n_-)/n$ and $r_s = 3(4\pi a_0^3 n)^{-1/3}$ (a_0: Bohr's radius). The functions $\beta(r_s)$ and $\delta(r_s)$ are approximated by

$$\beta(r_s) = 1 + 0.0545 r_s \ell n(1 + 11.4/r_s)$$

$$\delta(r_s) = 1 - 0.036 r_s + 1.36 r_s/(1 + 10 r_s) \ . \tag{5.16}$$

The density functional methods with the LDA have one great advantage which can not be overrated and that is their simplicity. Equation (5.12) which has to be solved, is essentially a Hartree-type of equation with a *local* self-consistent potential $V_\sigma^{\text{eff}}(\mathbf{r})$. With present days computer facilities this poses no problems even when there are several atoms per unit cell.

Before discussing the accuracy of results based on the LSD approximation we want to make some more remarks concerning the correlation which are contained in $\mu_\sigma^{\text{xc}}(\mathbf{r})$. Since the form of the latter is based on results for the homogeneous electron gas it will contain correlation effects which are present in a spin-polarized homogeneous system. This implies that, e.g., Hund's rule correlations are partially contained in it when $m \neq 0$ but are not taken into account when $m = 0$ since there are no such correlations in a nonmagnetic homogeneous system. Also charge fluctuations are treated with different accuracy in the nonmagnetic and magnetic phase of a system. This can be seen by computing the H_2 molecule for different bond lengths within the LSD approximation. Thereby

166

one finds that for an atomic separation larger than a critical value the state of lowest energy is a symmetry broken one with $m \neq 0$[6]. This is an artifact which is due to a stronger suppression of charge fluctuations in the symmetry broken case. Similar features are known from unrestricted Hartree-Fock calculations. Due to an insufficient treatment of charge fluctuations and the lack of Hund's rule correlations in the nonmagnetic phase one expects that the LSD approximation overestimates the energy difference between the ordered and non-ordered phases, favoring the former.

The calculations which have been performed within the LSD approximation have been very successful [5.15–17]. Many experimental findings like trends in binding energies, magnetic ordering and results of photoemission experiments have been described or predicted very well. They have confirmed the suggestion that d electrons in transition metals are delocalized despite of the fact that correlations among them are strong. Any improved theory must therefore start out from a delocalized description of d electrons.

Beside the successes just described the theory has also shown deviations from experimental results which demonstrate the limitations of a local ansatz for exchange and correlations. We want to discuss them briefly, and start out with the ground state. One significant deviation is in the binding energies which are usually overestimated by 1–2 eV [5.18]. This can be explained by an inaccurate evaluation of the energy changes which are connected with the charge transfer from s to d electrons as one goes over from separated atoms to a solid. However, it results, at least partially, as well from an insufficient treatment of d-electron correlations. Another deviation concerns the calculated lattice constants. Usually they come out approximately 2 % too short for systems with a magnetic ground state and partially filled d bands [5.18]. This error is due to an improper treatment of d-electron correlations. In the LSD approximation the main contribution to the pressure dependence of the total energy comes from that of the kinetic energy. Contributions from exchange and correlations are much smaller. However, when the correlations are strong one expects a noticable decrease of the kinetic energy effects. This leads to an increase in the lattice constant, as compared to the case where electron correlations are not sufficiently taken into account. Apparently that is the case when a local correlation potential is used.

Connected with the problem of the lattice constant is that of the size of the magnetic moments at $T = 0$. Of special interest is Fe in which the spin-majority bands are not completely filled. The calculations within the LSD approximation have either used the atomic sphere approximation (ASA) or have been performed within a finite basis set of Gaussian orbitals. In both cases the results were slightly different. In the first case the results for the moment was best when the *calculated* lattice constant was used, which is 2.5 % too short. In that case the error was merely 1–2 % [5.15,16]. For the *experimental* lattice constants the deviations from the experimental magnetization are approximately 5 % [5.15]. In the calculations using Gaussian orbitals the moment was found to be too large by only 2 % for the experimental lattice constant (and too small by 4 % for the calculated lattice constant [5.19]). These errors should be compared with an error of 20 % which would result if one would fill completely the majority spin band (strong ferromagnetims). Apparently partial cancellation of errors takes place between an underrating of atomic distances and an overestimation of magnetic moments.

Small discrepancies also occur with respect to the relative populations of different d subbands. The assumption of a local exchange underestimates anisotropies in the subband populations and in the exchange splitting of e_g and t_{2g} states in Ni. These effects are amplified when one goes over to the surface. The biggest problem with the LSD approximation arises when one calculates transition temperatures T_c between the magnetically ordered and the paramagnetic state. The conventional generalization of the LSD theory to finite temperatures is via a Stoner theory. Within such a theory the transition temperatures, e.g., for Fe, Co and Ni come out much too large [5.20]. As pointed out above, the energy difference between the ordered and nonmagnetic state at $T = 0$ is too large when a local approximation for exchange and correlations is made. This is one source of error when T_c is calculated. Another source of error is that spin fluctuations are neglected in a Stoner theory as in any other mean-field theory. This refers to fluctuations in the direction of magnetization as well as in their amplitude.

There have been attempts to improve on the last point. For that purpose the nonmagnetic state has been replaced by one with locally broken symmetry. This is possible by allowing for the formation of disordered local moments (DLM) at each site [5.21,22]. This way local correlations are better taken into account in the nonmagnetic state. At least they are treated with comparable accuracy as in the magnetically ordered state. Despite of considerable improvements a nonmagnetic state with local symmetry breaking is unsatisfactory, at least at $T = 0$, in the same manner as an unrestricted Hartree-Fock description is of the H_2 molecule. The local moments which are generated by a DLM calculation are decoupled from the motions of the electrons which generate them.

5.2.3 Correlated Ground State Wave Function

In distinction to the density functional method we aim in the following at calculating the ground state wave function of a d-electron system [5.23–27]. This can be done only for model systems and we shall use for the calculations the model Hamiltonian (5.1). It is evident that this Hamiltonian does not contain all of the features of realistic systems. First of all there is the restriction to d-electrons only. The 4s and 4p electrons are taken into account only indirectly by using screened d–d interactions and noninteger d-electron occupancies. Furthermore the d-bandwidth contained in the Hamiltonian is fitted to the bandwidth as it is obtained from LSD calculations. This way some of the hybridization effects with the 4s and 4p electrons are indirectly included. All other effects of the s and p electrons on magnetism and the band structures are neglected.

The three interaction matrix elements U, J and ΔJ are interrelated to each other in a further approximation by assuming $J = 0.2\,U$ and $\Delta J = 0.15\,J$. This leaves U as the only adjustable parameter. For ferromagnetic metals it is chosen so as to reproduce the experimental values for the magnetic moments when the ground state is calculated. For Fe, Co and Ni this implies the following choice of parameters. The d-band occupations are $n_d = 7.4,\ 8.4$ and 9.4, the d-bandwidths are [5.10] $W = 5.43,\ 4.84$ and $4.35\,eV$ and the interaction matrix element is $U = 2.4,\ 3.1$ and $3.3\,eV$, respectively [5.26]. Clearly, because of the electron-electron interactions the correlated ground state wave function can not be written in the form of a single Slater determinant as in the independent electron approximation. However, one can start out from the latter, i.e., $|\phi_0\rangle$ as given by (5.3) and correct it for electron correlations. By starting from $|\phi_0\rangle$ one accounts for the dominating effect of d-electron delocalization. That is the route taken by the Local Approach (LA) to the correlation problem. The latter has been successfully applied to molecules [5.28,29] and to solids [5.30]. We shall describe in the following its application to the Hamiltonian (5.3). Starting point is the following ansatz for the ground state wave function

$$|\psi_0\rangle = e^S|\phi_0\rangle \tag{5.17}$$

where S is of the form

$$S = -\sum_{ij\ell\ell'} \eta_{ij}(\ell, \ell') O_{ij}(\ell, \ell') \ . \tag{5.18}$$

We assume that the operators $O(\ell, \ell')$ are site diagonal. They are related through

$$O_{ij}(\ell) = \tilde{O}_{ij}(\ell) - \langle\phi_0|\tilde{O}_{ij}|\phi_0\rangle \tag{5.19}$$

to operators $\tilde{O}_{ij}(\ell)$. This ensures that $O_{ij}(\ell)|\phi_0\rangle$ generates a state which is orthogonal to $|\phi_0\rangle$. The $\tilde{O}_{ij}(\ell)$ take the following three forms

$$\tilde{O}_{ij}(\ell) = \begin{cases} n_{i\uparrow}(\ell)n_{i\downarrow}(\ell)\delta_{ij} \\ n_i(\ell)n_j(\ell) \\ s_i(\ell){\cdot}s_j(\ell) \ . \end{cases} \tag{5.20}$$

The spin-dependent density operators $n_{i\sigma}(\ell) = a^+_{i\sigma}(\ell)a_{i\sigma}(\ell)$. The $s_i(\ell)$ are the spin operators for orbitals i on site ℓ. The η_{ij} are parameters to be determined. The ansatz (5.17) has the form of a coupled-cluster ansatz [5.31] as it has been used in nuclear physics and quantum chemistry [5.32]. Together with (5.18,19) it can be also considered as a generalization of the *Gutzwiller* [5.12,33] or *Jastrow* [5.34] ansatz. The operator $O_{ij}(\ell)$ when applied on $|\phi_0\rangle$ picks out of it all configurations in which electrons are simultaneously in orbitals i and j on site ℓ. By choosing η_{ij} properly, the operator e^S can therefore suppress $(\eta_{ij}>0)$ or enhance $(\eta_{ij}<0)$ the amplitude of these configurations in $|\phi_0\rangle$. This way one can suppress unfavorable configurations (see Fig. 5.1b) and enhance favorable configurations (Fig. 5.1a) contained in $|\phi_0\rangle$. The operator e^S can therefore generate a correlation hole around each electron. The operator $s_i \cdot s_j$ in (5.20) ensures that the correlation hole can be spin dependent. Acutally, the operators $O_{ij}(\ell)$ contain, in addition to two-particle excitations, also one-particle excitations out of the ground state. The latter result in density changes only and do not directly contribute to the correlation hole. Since the densities are already relatively well described within the Hartree-Fock or the independent-electron approximation we want to discard the one-particle parts contained in $O_{ij}(\ell)$. This is done by excluding contractions within $O_{ij}(\ell)$ operators when expectation values containing the latter are evaluated. In order to avoid excessive notation we shall not introduce a new one for this particular procedure.

By limiting the $O_{ij}(\ell, \ell')$ to $O_{ij}(\ell)$ one excludes correlations between electrons on different atoms. This limitation is not necessary but it simplifies considerably the actual calculations. We shall restrict ourselves to it for the main body of the present discussion but we will discuss at the end the more general case.

Until now we have not yet specified the parameters η_{ij}. We will choose them so that the expectation value

$$E = \frac{\langle \psi_0 | H | \psi_0 \rangle}{\langle \psi_0 | \psi_0 \rangle} = \langle e^S H e^S \rangle_c \qquad (5.21)$$

is minimized. Here and in the following we will use the abbreviation $\langle \dots \rangle = \langle \phi_0 | \dots | \phi_0 \rangle$. Furthermore the subscript c indicates that only connected contractions must be taken when the expectation value is evaluated [5.30]. The latter cannot be evaluated without further approximation though. We shall approximate $e^S = 1 + S$ so that

$$\begin{aligned} E = &\langle H \rangle - 2 \sum_{ij\ell} \eta_{ij}(\ell) \langle O_{ij}(\ell) H \rangle_c \\ &+ \sum_{ijmn} \sum_{\ell\ell'} \eta_{ij}(\ell) \langle O_{ij}(\ell) H O_{mn}(\ell') \rangle_c \eta_{mn}(\ell') \ . \end{aligned} \qquad (5.22)$$

The conditions $\partial E / \partial \eta_{ij}(\ell) = 0$ yield a system of coupled linear equations for $\eta_{ij}(\ell)$. Solving them and setting $\eta_{ij}(\ell)$ back into (5.21) gives the correlation energy $E_{corr} = E - \langle H \rangle$. In evaluating the matrix elements $\langle O_{ij}(\ell) H \rangle_c$ and $\langle O_{ij}(\ell) H O_{mn}(\ell') \rangle_c$ another approximation is made. It consists in treating correlations on different sites as independent of each other. Therefore only those terms are kept in the matrix elements in which $O_{ij}(\ell)$, $H_1(\ell')$ and $O_{mn}(\ell'')$ refer to the same site, i.e. $\ell = \ell' = \ell''$. This corresponds to a local cluster approximation and has been termed $R = 0$ approximation by *Friedel* and coworkers [5.35,36]. This approximation is rather poor for a 1-dimensional half-filled Hubbard system where one obtains with it only 75 % of the correlation energy [5.37]. However, for a 3-dimensional system with the five-fold orbital degeneracy taken into account the $R = 0$ approximation leads to errors of less than 5 % [5.23]. This is in agreement with corresponding findings by second-order perturbation calculations [5.36]. The $R = 0$ approximation simplifies the calculations considerably. The only inputs needed are the partial occupancies of the e_g and t_{2g} states as well as the mean kinetic energies for these bands and their occupied parts.

a) Nonmagnetic State

We consider the nonmagnetic state of a transition metal or more precisely the nonmagnetic solutions of the Hamiltonian (5.1). In particular, we are interested in the ground state energy and the cohesive energy, respectively. The prevailing opinion is that the s electron contribute in approximately the same way to these energies, independently of the d band filling, and that the observed variation of the cohesive energy through the transition metal series results from the d electrons only. The cohesive energy of transition metals has been first discussed and analysed in detail by *Friedel* [5.38], and *Friedel* and *Sayers* [5.39]. We will follow, in large parts, their discussions. Thereby it is convenient to restrict oneself to a bcc structure and furthermore to set $\Delta J = 0$ in (5.1). The interaction parameters U and J are chosen as $U = W/2$ and $J = U/5$, respectively. Furthermore we shall subtract from all interaction energies an amount of

$$\tilde{E}_0 = (U - 2J/9)n_d(n_d - 1)/2 \ . \tag{5.23}$$

This is the interaction energy of n_d electrons sitting on an atom when they are distributed with equal probabilities among the different d orbitals. In that case the average interaction energy of one electron with a second one is

$$E_{12} = \frac{1}{9}[(U + 2J) + 4U + 4(U - J)] = U - \frac{2J}{9} \ . \tag{5.24}$$

On an isolated atom with n_d electrons the latter are not equally distributed among all orbitals. Instead they prefer parallel spin alignments (first Hund's rule). The residual interaction energy gain as compared with $\tilde{E}_0$ is then

$$E_{atom} = -\frac{1}{2}\tilde{n}_d(\tilde{n}_d - 1) \cdot \frac{7}{9}J \quad \text{where} \tag{5.25}$$

$$\tilde{n}_d = \begin{cases} n_d & \text{for } n_d \leq 5 \\ 10 - n_d \ ; & n_d \geq 5 \ . \end{cases} \tag{5.26}$$

This energy in units of W is shown in Fig. 5.2 as the upper solid line. It has a minimum at $n_d = 5$ where maximal spin alignment is obtained. In a solid the d electrons gain kinetic energy due to delocalization. This is shown in Fig. 5.2 for the bcc case by the lower solid line. Due to the delocalization the d-electron number on an atom fluctuates around the mean value n_d. This results in an increase in the interaction energy as compared with $\tilde{E}_0$. The Hartree-Fock interaction energy for the case of equally populated orbitals is

$$\tilde{E}_{HF}^{int} = \frac{n_d^2}{2} \frac{9}{10} \left(U - \frac{2}{9}J \right) \tag{5.27}$$

171

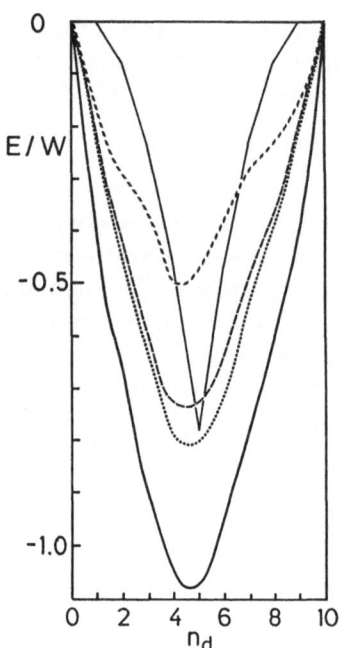

0

E / W

-0.5

-1.0

0 2 4 6 8 10
n_d

Fig. 5.2. Energies for various approximations of the ground state for materials with a bbc structure as function fo the d electron number. For comparison also atomic energies are shown (*Upper solid line:* E_{atom}). A ratio $U/W = 0.5$ was chosen. *Lower solid line:* kinetic energy; broken line: energy of uncorrelated electrons ($\langle H \rangle - \tilde{E}_0$); (- - -): including density correlations; (....): total energy ($E - \tilde{E}_0$) of the correlated ground state

and its residual part $E_{HF}^{int} = \tilde{E}_{HF}^{int} - \tilde{E}_0$

$$E_{HF}^{int} = \frac{n_d}{2}\left(1 - \frac{n_d}{10}\right)\left(U - \frac{2}{9}J\right) = \frac{1}{2}E_{12}\Delta n^2 . \qquad (5.28)$$

The latter is thus proportional to the electron-number fluctuation. The sum of the kinetic energy and E_{HF}^{int} for the bcc case is given by the broken line in Fig. 5.2, thereby allowing for different partial occupations of e_g and t_{2g} oribtals. It is seen that for most of the n_d region the completely localized atomic state is more stable than the nonmagnetic Hartree-Fock ground state. Therefore in the independent-electron approximation any symmetry breaking solution which reduces the charge fluctuations and allows for partial local spin alignment will be favored in that range of n_d values.

Charge fluctuations are also suppressed by electron correlations. When density correlations are taken into account between electrons on the same site the residual energy takes a form shown by the dashed line in Fig. 5.2. Additional inclusion of on-site spin correlations results in the dotted curve shown in the same figure. It is seen that the energy is now always lower than in the localized atomic limit. This is not unexpected because the ansatz (5.17) contains the atomic case as a particular limit. The approximation (5.22) prevents us, however, from treating correctly strongly correlated systems. Also the $R = 0$ approximation leads to a small underestimation of the correlation energy. We have found that both approximations lead to unreliable results for $U/W > 0.6$ when $n_d = 5$ and for $U/W > 0.8$ when $n = 3$ or 7.

The cohesive energy is given by the difference between the dotted and the upper solid curve. It shows the well known double peak structure with maximum values of 0.3 W or approximately 1.5 eV. This is in qualitative agreement with the known d-electron contributions to experimental binding energies. A more quantitative discussion would necessitate the explicit incorporation of s electrons and, in particular, of s − d charge transfers.

From Fig. 5.2 it is seen that the contribution of spin correlations to the cohesive energy is significant. It can be as large as 0.4 eV. Contributions of that form are partially taken into account in LSD approximation calculations. They are not contained in unpolarized calculations, though, and therefore the energy differences between the magnetic and nonmagnetic states are overestimated within that scheme. Therefore, one should be aware that a LSD approximation calculation can result in a spin-polarized ground state when the nonmagnetic state has still lower energy.

Furthermore, the exchange and correlation energy contributions depend, in our model calculations, on the partial e_g- and t_{2g}- -orbital occupancies, and not only on n_d. If one compares the fcc and bcc ground state energies for fixed parameter values W, U, J, and n_d, one finds that they differ mainly because of the different kinetic energies. But there is also a contribution from the interaction energy which depends on differences in the e_g and t_{2g} oribtal occupancies. This contribution has been calculated for the model Hamiltonian (5.1) to be 0.02 W or 1200 K for Fe [5.40]. It cannot be accounted for by local exchange-correlation potentials either and therefore the latter favor fcc structures. Elaborate calculations within the LSD approximation for the fcc and bcc structure of Fe find a nonmagnetic ground state for the fcc and a magnetic ground state for the bcc structure [5.41]. Furthermore the nonmagnetic fcc state is found to be lower by 800 K per atom than the ferromagnetic bcc state. When the nonlocal correction of 0.02 W is taken into account the ferromagnetic bcc state should become lower in energy.

We want to investigate briefly the validity of second-order perturbation expansions. They can be simulated by replacing the matrix elements $\langle O_{mn}(\ell)HO_{ij}(\ell)\rangle_c$ in (5.22) by $\langle O_{mn}(\ell)H_0O_{ij}(\ell)\rangle_c$. The two-particle excitations are limited to a small subset of all possible ones due to the choices (5.19,20) for the O_{mn} operators. When this approximation is made the correlation energy increases by a factor of two to three for parameters U and J as chosen above [5.23]. The correlation energy is then found to be larger than the decrease in interaction energy when one goes to the atomic limit and this should not be the case. This indicates that third-order corrections are of importance. They ensure, e.g., that one cannot profit more than once from a reduction of charge fluctuations as it is the case when one treats all atomic two-particle excitations as independent processes. Alternatively one can consider the third-order corrections as screening effects.

It is instructive to study the influence of electron correlations on various correlation functions. In the following we state the results for the local charge fluctuations Δn^2, see (5.5), and the local spin correlations S^2, see (5.6), or

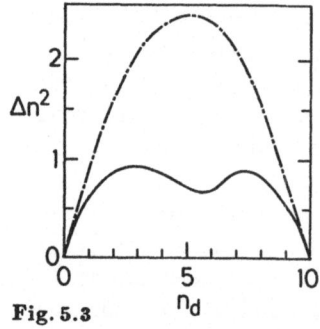
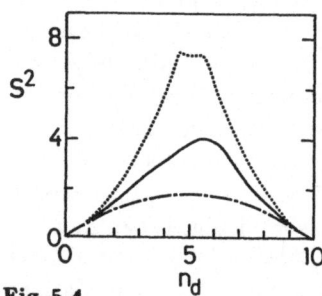

Fig. 5.3. Charge fluctuations as function of d-band filling n_d (bcc structure). Upper curve without and lower curve with electron correlations included. $U/W = 0.5$

Fig. 5.4. Local spin correlations S^2 as function of d-band filling n_d (bcc structure). *Upper curve:* atomic limit; *lower curve:* independent electron approximation. The solid line represents the results for $|\psi_0\rangle$. $U/W = 0.5$

ΔS^2 (5.7), respectively. The reduction of charge fluctuations due to correlations is shown in Fig. 5.3 for a bcc structure. The parameters U and J were chosen as follows, $U/W = 0.5$ and $J/U = 0.2$. One notices that the reduction in Δn^2 is appreciable. For $n_d = 5$ it is a factor of three. From (5.28) we expect a corresponding strong reduction of the interaction energy. However, from Fig. 5.2 it is seen that the reduction of the difference between total energy and kinetic energy of non-interacting electrons is considerably less than expected on the grounds of (5.28). This demonstrates that the introduction of correlations causes a considerable loss of kinetic energy. For the above parameters it is roughly 50 % of the energy gain which results from the reduction of the Coulomb repulsion. It is worth noticing that approaches like the renormalized atom method [5.42] work with suppressed charge fluctuations by construction. But it is not possible to write down the ground state wave function within such an approach.

Next we consider the amplitude S^2 of the local spin. The paramters U and J are the same as above. The numerical results are shown in Fig. 5.4. Also shown for comparison are the results for the HF and the completely localized case. One notices that for this parameter choice the ground state $|\psi_0\rangle$ is roughly halfways between the localized and delocalized limits. Indeed one finds that the quantitiy ΔS^2, as given by (5.7) is approximately 0.45, independent of the d electron number. This result is very different from the one which one would obtain from an unrestricted or symmetry broken SCF calculation. This is the approach taken in the "disordered local moment" (DLM) calculations [5.21,22]. Thereby one is searching for solutions for the wave functions within the independent electron approximation but allowing for local exchange fields to appear. The exchange fields are treated within the single-site approximation, i.e. the exchange fields at different sites are independent of each other. They average to zero within the sample. In that case ΔS^2 as function of U for

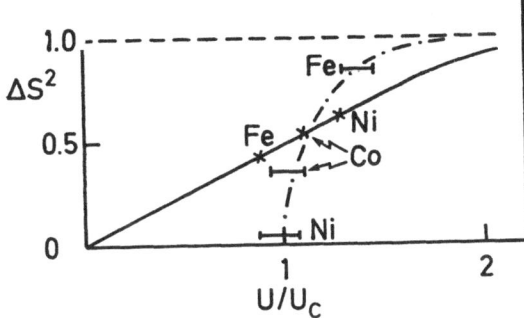

Fig. 5.5. Schematic representation of the relative changes of local spin correlations ΔS^2 as function of U/U_c, as obtained within the present approach (—) and within the disordered local moment approach (-.-.-). Crosses and bars denote the respective positions of Fe, Co, Ni in the two approaches [5.40]

fixed value of n_d takes the form shown in Fig. 5.5. Depending on the d-band occupancy there is a threshold value U_c below which $\Delta S^2 = 0$. For $U > U_c$, ΔS^2 takes values close to $\Delta S^2 = 1$ with a weak additional U dependence only. In that scheme Fe has a value $U > U_c$ while Co and Ni have values $U < U_c$ [5.21] or just above U_c [5.22]. The corresponding results of the present theory are also shown in that figure. It is important to realize that a nonmagnetic state with locally broken symmetry of the form just described has a higher energy than the nonmagnetic ground state $|\psi_0\rangle$ as given by (5.17).

b) Ferro- and Antiferromagnetic State

In the last subsection we considered the nonmagnetic state of the d electrons in transition metals and the effects of correlations on them. This was necessary in order to understand the changes in the various correlation contributions which occur when the system is magnetically ordered. We turn now to the latter case and consider first the ferromagnetic state [5.26]. The correlation calculations are performed as before and the only difference is that the ground state in the independent-electron approximation $|\phi_0\rangle$ is one with a broken spin symmetry. For a given d-electron number this state is optimized self-consistently with respect to the total magnetization and the partial magnetizations of the e_g and t_{2g} subbands. The requirement is thereby that the correlated ground state $|\psi_0\rangle = e^S |\phi_0\rangle$ has the lowest possible energy.

Correlations are of importance even in the case of strong ferromagnetism where the d bands of one spin direction are completely filled. Even in that case the charge fluctuations in $|\phi_0\rangle$ are too large and therefore are reduced by correlations. Of particular interest are the changes due to correlations of the criterion for the occurrence of magnetic order. In the independent-electron approximation it is given by the Stoner-Wohlfarth [5.43] condition. In the case of different subbands with densities of states $\varrho_e(\varepsilon_F)$ (e_g bands) and $\varrho_t(\varepsilon_F)$ (t_{2g} bands) it takes the form

$$3 \left(U + J + \frac{19}{2} \Delta J \right) \varrho_t^2(\varepsilon_F) + 2 \left(U + J + \frac{7}{2} \Delta J \right) \varrho_e^2(\varepsilon_F)$$
$$+ \left(J - \frac{3}{2} \Delta J \right) \varrho^2(\varepsilon_F) = \varrho(\varepsilon_F) \tag{5.29}$$

175

where $\varrho(\varepsilon_F) = 3\varrho_t(\varepsilon_F) + 2\varrho_e(\varepsilon_F)$ is the total density of states at the Fermi energy. For identical subbands with a rectangular density of states this condition reduces to the simple form [5.24]

$$W = U + 6J \; . \tag{5.30}$$

When correlations are included the critical values of $U(n_d)$ are changed as compared with those which follow from (5.29). The results are shown in Fig. 5.6a for a fcc structure. Thereby it is assumed that $J = 0.2\,U$ and $\Delta J = 0.15\,J$ so that U is the only free parameter. The same figure shows also the condition for ferromagnetism when only density correlations are taken into account, i.e. when spin correlations are neglected. This is the most which the LDA may achieve. It treats density correlations like in the homogeneous electron gas but it can not provide for Hund's rule correlations in a nonmagnetic state. It is seen that correlations shift the critical values of U to considerably larger val-

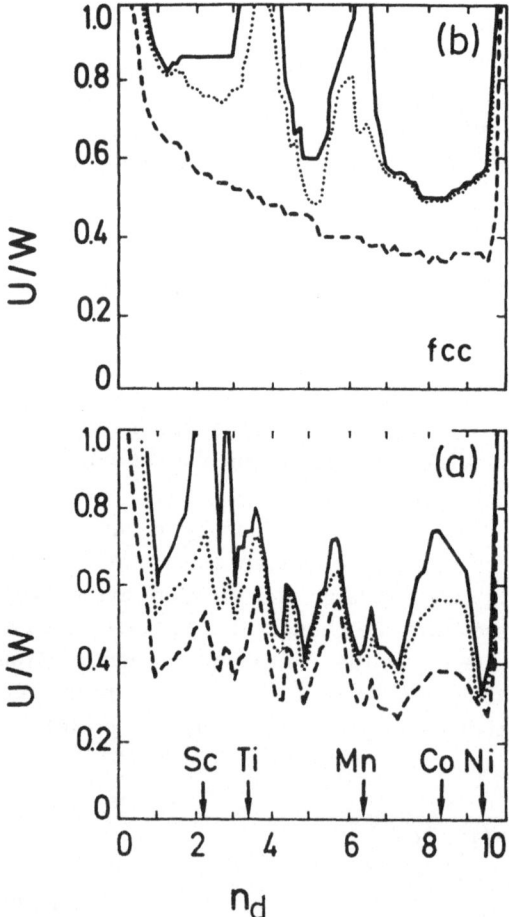

Fig. 5.6a,b. Ferromagnetic instabilities as function of d-electron band fillings assuming a fcc structure (a) Onset of ferromagnetic order (b) Complete or strong ferromagnetism. (- - -): independent electron approximation; (....): inclusion of density correlations; (—): full inclusion of correlations [5.26]

ues. However, the shift is not uniform and depends on band structure effects. It is also seen that spin correlations contribute significantly to that shift.

Another criterion of interest is that for the occurence of strong ferromagnetism. The results for a fcc structure are shown in Fig. 5.6b. In the independent-electron approximation this criterion depends apparently only weakly on the density of states. When correlations are included there are three domains of strong ferromagnetism. The middle one around $n_d = 5$ is an artifact. It is caused by a breakdown of the present approximation which occurs first for the half-filled band case as U/W becomes large (Sect. 5.2.3a). One can compare these findings with the ones which are obtained when also provision is made for the occurrence of antiferromagnetic order. In that case one finds an *anti*ferromagnetic transition around $n_d = 5$ [5.25].

Next we want to consider in more detail bcc Fe and fcc Co and Ni. The magnetic moments of the ferromagnetic ground state of the three systems are known from experiments. The values are 2.1, 1.6 and 0.6 μ_B, respectively, after subtraction of the small orbital contributions [5.17]. This implies "strong ferromagnetic" behavior for Co and Ni while Fe has d holes in both spin bands. The interaction parameter U is adjusted so that it reproduces the correct magnetic moment for each substance. This implies $U = 2.4$ eV for Fe and $U \geq 2.6$ eV and $U \geq 3.1$ eV for Co and Ni. We have actually chosen $U = 3.1$ eV for Co and $U = 3.3$ eV for Ni. The ratios U/W are 0.44, 0.65 and 0.75, respectively, and agree well with those found in [5.44]. When we use the above parameters for Fe but neglect the spin correlations, the magnetic moment increases by 5 % to $2.3\mu_B$. We list in Table 5.1 the energy gains in eV per atom which are obtained due to the appearance of ferromagnetic order. The different values refer to the HF approximation (ΔE_{HF}), to the inclusion of density correlations only (ΔE_{dc}) and to the case where also spin correlations are included (ΔE_c). It is seen that density correlations reduce drastically, i.e. by a factor of three, the energy difference between the nonmagnetic and ferromagnetic state, as compared with the independent-electron approximation. There is a further reduction of this energy when spin correlations are taken into account. In Fe and Co it amounts to 30 % while in Ni it is insignificant due to the reduced probability of finding two d holes simultaneously on a Ni site. The energy gain due to ferromagnetic order is 0.17 eV/atom in Fe and larger than the change in internal energy when the temperature changes from zero to T_c. The latter quantity is known

Table 5.1. Energy gains in eV due to ferromagnetic order in the HF approximation (ΔE_{HF}) and when density correlations (ΔE_{dc}) as well as spin correlations (ΔE_c) are included. For the choices of W and U see text.

	ΔE_{HF}	ΔE_{dc}	ΔE_c
Fe	0.564	0.221	0.150
Co	0.430	0.161	0.133
Ni	0.115	0.039	0.035

from experiments to be approximately 0.06 eV/atom [5.45]. This suggests that ferromagnetic domains or droplets exist even at $T > T_c$.

We want to use the above information in order to take up the problem of equilibrium distances and their computation within the LSD approximation. It is also related to the magnetovolume effect and its computational determination.

Within the LSD approximation the computed equilibrium lattice distances are different for a ferromagnetic and nonmagnetic state. No appreciable difference is observed experimentally when the temperature T is raised above the Curie temperature. This apparent contradiction is usually resolved by assuming disordered local moments to exist in the paramagnetic state above T_c. They have to be almost as large as the magnetic moments in the ferromagnetically ordered ground state at $T = 0$ [5.45] in order to explain why the changes in the lattice parameter are so small. However, it was found before that within the metastable nonmagnetic state the spin correlations at an atomic site are only half as strong as in the ferromagnetic state at $T = 0$. A similar result is expected to hold for the paramagnetic state at $T > T_c$ except when a sizeable fraction of sites are parts of magnetic droplets. In the following we want to discuss the consequences for the expected magnetovolume effect when the local electron correlations are included more precisely than by a local exchange-correlation potential.

The magnetovolume effect can be described as the difference in volume change of the energy in the nonmagnetic and ferromagnetic state, i.e. $\Delta(\partial E/\partial V)$. In the LSD approximation one distinguishes usually between two different contributions to the magnetovolume effect. The first is proportional to the difference in kinetic energy $\Delta E_{kin}^{(0)}$ of the noninteracting d electrons in the two states. The second contribution is proportional to the difference ΔE_{xc} in the local exchange-correlation potential. Therefore

$$\Delta\left(\frac{\partial E}{\partial V}\right) = \alpha[\Delta E_{kin}^{(0)} - f(\Delta E_{xc})] \ . \tag{5.31}$$

The first term accounts for the smaller cohesion in the ferromagnetic state and leads to a volume expansion in the latter. The second term describes a volume contraction of the ordered state due to the decreased electronic interactions in the ordered state. It also includes correlation corrections. The functional form of $f(\Delta E_{xc})$ is not of interest here. Altogether the second term reduces the first one, for example, in Fe by approximately 20 % [5.15]. When one wants to start out instead from the model Hamiltonian (5.1) one must investigate its dependence on the bandwidth W. Therefore one considers

$$\Delta\left(\frac{\partial E}{\partial W}\right) = \frac{1}{W}\Delta E_{kin} \ .$$

In distinction to (5.31) E_{kin} is here the kinetic energy of the interacting and correlated d-electron system. Other interaction contributions to $\partial E/\partial W$ are neglected. The difference $\Delta E_{kin}^{(0)}$ is 0.099 W for Fe and results from $E_{kin}^{(0)} = -0.692\,W$ and $-0.593\,W$ for the paramagnetic and ferromagnetic state, respectively. When correlation effects are included this changes to $E_{kin} = -0.613\,W$ and $-0.542\,W$ and therefore $\Delta E_{kin} = 0.071\,W$, respectively. Correlations reduce thus the kinetic energy difference by 30 % [5.37]. This change is apparently missing within the LSD approximation because the small correlation contribution contained in ΔE_{xc} can not account for it. If we add to $W^{-1}\Delta E_{kin}$ an estimated direct contribution of the interaction of the form of the second term on the right-hand side of (5.31) then the magnetovolume effect for Fe should approximately be 50 % of the term $\alpha\cdot\Delta E_{kin}^{(0)}$. In order to reduce the remaining contribution to the observed size one must presumably invoke electron correlations which extend over several atoms.

The above results for ΔE_{kin} of Fe are based on a ratio of $U/W = 0.43$. If we would have chosen instead a ratio of $U/W = 0.6$, the magnetovolume effect would have been only 25 % of the value which results from (5.31). For Co and Ni the correlation corrections to $\Delta E_{kin}^{(0)}$ are only approximately 50 % of what they are for Fe.

The considerable correlation corrections to the kinetic energy contribute to the equilibrium distances as well. As discussed above, these distances are underestimated by the LSD approximation. One can estimate the expected improvement due to correlations as follows. Correlations change the kinetic energy $E_{kin}^{(0)}$ of the ferromagnetic ground state by approximately $E_{kin}^{(0)}/2$. With exchange corrections included the equilibrium lattice constant of the ferromagnetic ground state should therefore increase by roughly 65 % of the calculated difference in the lattice constants between the ferromagnetic and nonmagnetic states. This effect should be larger for Co and even more so for Ni.

In summary, it can be stated that the proper inclusion of on-site correlations reduces the magnetovolume effect and corrects for the lattice constants as compared with the values which result from a LSD approximation.

Next we turn to the problem of the distribution of the d electrons among the different d orbitals [5.26]. In the ferromagnetic state the occupation of the minority and majority e_g and t_{2g} subbands is determined by minimization of the energy. This leads to charge transfers between different subbands, as compared with the nonmagnetic solution. The exchange potential which results from the model Hamiltonian is nonlocal. It is different for e_g and t_{2g} states. For an e_g electron it depends primarily on the filled e_g states while for a t_{2g} electron it depends mainly on the filled t_{2g} states. The exchange has a tendency to fill up those subbands which have already the highest occupancy.

Also the exchange splitting is different for the two types of subbands. The larger splitting correspond to the bands with the larger filling provided that $n_d > 5$. The effect of correlations is to reduce the anisotropies caused by the exchange.

A comment is in order, on how the exchange splitting is actually determined. Since it is a property of the quasiparticle excitations it can, of course, not be obtained rigorously from a ground state calculation. But when one assumes that the shapes of the quasiparticle excitation bands are essentially the same as those which follow from H_0, see (5.1), the splitting can be calculated. Therefore, we use the canonical bands as described by H_0 and add to it an external anisotropic field $\Delta(e_g)$ and $\Delta(t_{2g})$ which acts on the e_g and t_{2g} electrons, respectively. It is determined by requiring that the resulting partial occupational numbers $n_\sigma(e_g)$, $n_\sigma(t_{2g})$ agree with those which minimize the full Hamiltonian (5.1).

The anisotropy in the exchange splitting is largest for Ni where one finds $\Delta(e_g) = 0.15\,\text{eV}$ and $\Delta(t_{2g}) = 0.57\,\text{eV}$ and an anisotropy ratio $A = \Delta(t_{2g})/\Delta(e_g) = 3.5$. For Fe and Co it is smaller, as discussed below. In Ni part of the anisotropy comes from ΔJ, see (5.1). When one puts $\Delta J = 0$ the values for $\Delta(e_g)$ and $\Delta(t_{2g})$ reduce to $0.27\,\text{eV}$ and $0.50\,\text{eV}$, respectively, with a resulting $A = 2$. This should be compared with an empirical fit of measured partial spin densities, from which $\Delta(e_g) = 0.1\,\text{eV}$ and $\Delta(t_{2g}) = 0.4\,\text{eV}$ was calculated [5.46]. Photoelectron spectroscopy experiments yield an anisotropic exchange splitting for Ni of $\Delta(e_g) = 0.22\,\text{eV}$, $\Delta(t_{2g}) = 0.33\,\text{eV}$ [5.47], respectively (see also [5.48] for that point). Within the LSD approximation an isotropic exchange splitting is found of the order of $\Delta = 0.6\,\text{eV}$ [5.17]. That in our model the isotropic part of the exchange splitting is smaller than that of the LSD calculations is very likely accidental. However, we expect the calculated anisotropy to be reliable. The isotropic part of the exchange splitting depends on the density of states at the upper band edge only. Since this density of states may decrease when hybridization effects are taken into account, the present calculations represent an upper limit to that part. The anisotropies are much less affected by this effect. For Co the exchange splitting is found to be much more isotropic, i.e. $\Delta(e_g) = 1.28\,\text{eV}$ and $\Delta(t_{2g}) = 1.06$. The result of the LSD approximation is $\Delta = 1.5\,\text{eV}$ [5.17] while experiments [5.49] give $\Delta = 1.1\,\text{eV}$ with no noticeable anisotropy. For Fe the calculated values are $\Delta(e_g) = 1.74\,\text{eV}$ and $\Delta(t_{2g}) = 1.30\,\text{eV}$, respectively. They compare reasonably well with the results of LSD calculations [5.17] ($\Delta = 1.5 - 2.0\,\text{eV}$) and experiments [5.49] ($\Delta = 1.45\,\text{eV}$). None the less they will be influenced by $s - p - d$ hybridizations, which can not be treated within the present model. Their importance is seen when one tries to understand the Fermi surface in detail.

The above discussion shows that the anisotropies are largest for Ni, for which the occupation of the e_g and t_{2g} orbitals is considerably different. The anisotropies are further enlarged at the surface. In particular, the [001] surface of Ni has been studied in detail by very careful and complete LSD calculations [5.50]. An analysis reveals a significant charge redistribution among different d orbitals and a small increase in the total d-hole number. The e_g orbitals, which are energy split at the surface contain more holes than inside the bulk and the orbitals which stick out of the surface have fewer holes than those within the surface. It is then instructive to apply a model Hamiltonian, similar to that described by (5.1), to the surface and to perform calculations for the correlated ground state within the $R = 0$ approximation [5.27]. In that case the only

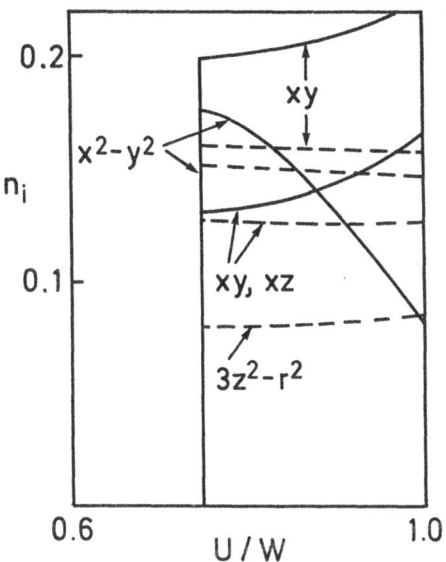

Fig. 5.7. Redistributions of d electron minority holes n_i for different orbitals i on a Ni [001] surface as function of U/W. (—): nonlocal exchange and correlations (see text); (- - -): local approximation for exchange and correlations [5.27]

information which is required are the partial densities of states for the surface layer and for the bulk. These functions are known from numerical calculations and can be approximated by simpler analytic forms. The LSD approximation is simulated in the model calculations by replacing the exchange by an isotropic one, i.e. one which is calculated by assuming the same occupation for all d orbitals. The surface potentials which appear in the Hamiltonian are determined by fitting the simulated LSD calculations to the true ones by *Jepsen* et al. [5.50]. When the isotropic exchange is replaced by the true nonlocal one, drastic charge rearrangements take place. The holes within the $(3z^2 - r^2)$ orbital disappear completely while those in the $(x^2 - y^2)$ orbital are decreased by a factor of 4-5 (Fig. 5.7). Practically all holes are then in the t_{2g} orbitals. As a consequence one finds a large anisotropy in the exchange splitting between t_{2g} and e_g type of orbitals, namely in the ratio of 3:1. The conclusion which we draw from these model calculations is that strong anisotropies exist at the surface of Ni and presumably of the other transition metals, too. Their treatment requires the inclusion of nonlocal exchange effects.

A treatment of correlations in antiferromagnetically (AF) ordered states is somewhat more complicated. First there is the problem that different types of antiferromagnetic order may exist. In addition, due to the changes in the Brillouin zone, one must repeat the SCF calculations for the Bloch states at each stage of the calculations. In distinction to the ferromagnetic case their form depends explicitly on the order parameter. Due to these complications the influence of correlations on an antiferromagnetically ordered ground state has been investigated only by making a number of drastic approximations [5.25]. Instead of extracting the single-particle properties from a microscopic Hamiltonian such as (5.1) they are taken into account by assuming a particular form for the density of states.

For the nonmagnetic system a five-fold degenerate band with a constant density of states is assumed. With respect to the AF order it is assumed that it leads to a doubling of the unit cell. For the density of states in the ordered state a very special form is assumed, namely

$$\varrho_{i\sigma}(\omega) = \begin{cases} \dfrac{1}{W} \dfrac{|\omega|}{(\omega^2 - \Delta^2)^{1/2}} \; ; & \Delta \le |\omega| \le \dfrac{1}{2}(\Delta^2 + \omega^2) \\ 0 \quad \text{otherwise} . \end{cases} \tag{5.32}$$

For the half-filled band case $(n_d = 5)$ this corresponds to the Fermi surface of a 1-dimensional system. For $n_d = 5$ there will then be always an AF-ordered state irrespective of how small U is. In Fig. 5.8 is shown the phase diagram when correlations are neglected (broken line) and when they are taken into account (solid line). We have chosen again $J = 0.2\,U$. The phase diagram is symmetric around $n_d = 5$. The inclusion of correlations stabilizes the nonmagnetic state as expected. As pointed out before, the details of the phase diagram near $n_d = 5$ depend considerably on the particular choice of the density of states.

Finally we want to discuss spin correlations betwen neighboring sites and the additional information which they provide [5.51].

The nonmagnetic state $|\psi_0\rangle$ at $T = 0$ is a singlet. This implies $S_{\text{tot}}|\psi_0\rangle = 0$ where $S_{\text{tot}} = \sum_\ell S(\ell)$ and furthermore

$$\begin{aligned} C_{q=0} &= \frac{1}{N} \sum_{\ell'} \langle \psi_0 | S(\ell) S(\ell + \ell') | \psi_0 \rangle \\ &= \frac{1}{N} \sum_{\ell'} C(\ell') = 0 \; . \end{aligned} \tag{5.33a}$$

As before $S(\ell) = \sum_i s_i(\ell)$, see (5.6).

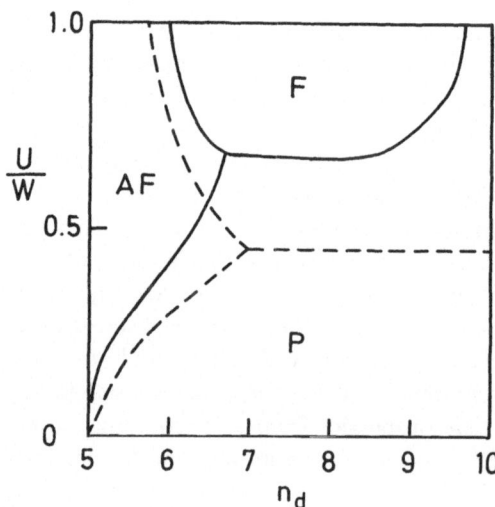

Fig. 5.8. Phase diagram for different magnetic phases in the U/W $vs.$ n_d plane. Solid lines: including electron correlations and dashed lines: within the independent electron approximation [5.25]

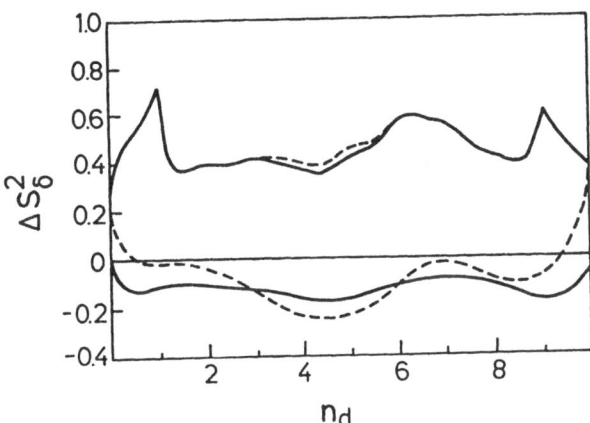

Fig. 5.9. Relative change of spin-correlation functions as function of d-band fillings. *Upper curves:* $\Delta S^2_{\delta=0}$, see (5.34). *Lower curves:* ΔS^2_δ (δ denotes nearest neighbor sites). Solid lines when on site correlation operators are taken into account and dashed lines when operators $O(\ell, \ell + \delta)$ are included

From, e.g., Fig. 5.4 we know that there is an appreciable local spin alignment $C(0)$ with values 1.85 for Fe and 0.44 for Ni. Therefore, in order that $C_{q=0} = 0$, there must be a corresponding strong antiferromagnetic correlation between different sites. In order to study their behavior we extend the form of the operator S in (5.17) by including spin correlations between neighboring sites. This is achieved by the operators

$$O(\ell, \ell + \delta) = S(\ell)S(\ell + \delta) \tag{5.33b}$$

where δ are nearest neighbors. When the calculations are repeated under the inclusion of these operators the results for $C(0)$ change. These changes are shown in Fig. 5.9 which also includes the results for $C(\delta)$. A convenient plot is in terms of ΔS^2, see (5.7), and a similar quantity

$$\Delta S^2_\delta = \frac{C(\delta) - C_{HF}(\delta)}{S^2_{loc} - S^2_{HF}} . \tag{5.34}$$

$C_{HF}(\delta)$ is the value of the correlation function when it is evaluated with respect to the HF state $|\phi_0\rangle$ instead of $|\psi_0\rangle$. The calculations were performed for the bcc lattice with $U/W = 0.5$, $\Delta J = 0.2\,U$, and $\Delta J = 0.15\,J$. One notices that ΔS^2 changes noticeably only for nearly half-filled d bands when $O(\ell, \ell + \delta)$ is included. When ΔS^2_δ is evaluated without taking the operator $O(\ell, \ell + \delta)$ into account, one finds antiferromagnetic nearest neighbor correlations for all d band fillings. When $O(\ell, \ell + \delta)$ is included one finds a decrease of the antiferromagnetic correlations except in the regime $3.5 < n_d < 6.5$ where they are stronger than before. This trend agrees with the phase diagram shown in Fig. 5.8. With the present choice of parameters the nonmagnetic state is metastable in the regime $7 < n_d < 8$. There are no anomalies seen in that regime as far as ΔS^2_δ is concerned. The energy gain due to the additional operators (5.33) is very small (less than 100 K per atom for Fe). This shows that when one starts out from the correlation operators (5.20) and includes, in addition, those of the form

(5.33) no appreciable short-range order results. However, when one includes spin-correlation operators over much larger distances than nearest neighbors (extended correlations) one expects to find energy depletions indicating the ferromagnetic instability.

One may speculate that these extended correlations are likely to exist also in the paramagnetic state at $T > T_c$. But it is not clear whether they are the ones which dominate the large magnetic scattering in neutron scattering experiments. There it is found that the quantitiy $C_{q=0}$ when generalized to finite temperatures becomes 13 for Fe and 3.5 for Ni at $T = 1.25\,T_c$ [5.52,53]. Atomic or on-site correlations alone are not sufficient to explain this strong magnetic scattering. Spin correlations above T_c which extend over an appreciable range have been proposed by *Korenman* and *Prange* [5.54], and by *Capellmann* [5.55].

5.3 Correlation Effects in Single-Particle Excitations

After having discussed the effects of correlations on the ground state we consider here their influence on single-particle excitations. This is of considerable interest due to the experimental data which is available on photoemission spectroscopy. Also we can establish some connections to the ground state correlation calculations.

Conventionally, single-particle excitations are described by the single-particle Green's function $G^{\sigma}_{\mu\nu}(k, \omega)$ matrix, where μ, ν are band indices. It is defined by

$$G^{\sigma}_{\mu\nu}(\boldsymbol{k}, \omega) = \left[\omega - \varepsilon_{\mu}(\boldsymbol{k}) - \Sigma^{\sigma}_{\mu\nu}(\boldsymbol{k}, \omega)\right]^{-1} . \tag{5.35}$$

The self-energy $\Sigma^{\sigma}_{\mu\nu}(k, \omega)$ contains HF and correlation energy contributions. The HF contribution, which shows up to order U, results generally in a redistribution of electrons between different orbitals. This effect was discussed at length in the preceeding section. The correlation contributions to $\Sigma^{\sigma}_{\mu\nu}(\boldsymbol{k}, \omega)$, which are at least of order U^2, represent themselves in the spectral densities

$$A^{\sigma}_{\mu\nu}(\boldsymbol{k}, \omega) = \frac{1}{\pi} \, \mathrm{Im}\left\{G^{\sigma}_{\mu\nu}(\boldsymbol{k}, \omega)\right\} . \tag{5.36}$$

There are three different effects which correlations have on them:

a) there are broadened lines instead of δ-functions as in the absence of correlations,

b) there are line shifts as compared with the HF case, and

c) new structures show up, when U/W is sufficiently large (shake-up or many-body processes).

When the spectral density is integrated over k, one obtains the partial densities of states. They show corresponding changes, i.e., band-narrowing effects connected with distortions in the density of states and additional structure.

We want to concentrate in the following on the photoemission from band states of Ni and also Fe. In a photoemission experiment a hole is generated in the conduction band and correlation effects result from additional electron-hole excitations accompanying such a process. For Ni the essential observations are a d-bandwidth of the order of 4 eV and a satellite peak, which appears approximately 6 eV below the Fermi energy [5.56]. When the results of energy-bandwidth calculations within the LSD approximation are compared with those from photoemission experiments then it is found that the former are somewhat too wide, i.e. by $\leq 10\%$ for Fe and 25% for Ni [5.17]. It is well known that the eigenvalues of the single-particle Schrödinger equation, as obtained in a density functional calculation should not be interpreted as quasiparticle excitation energies. Nevertheless this is commonly done with great success. The above discrepancies indicate the limitations of such a procedure. Also lifetimes of the quasi-particles become of importance, and need being taken into account.

Starting point of a correlation energy calculation are the energy bands $\varepsilon_\nu(k)$, as obtained within the LDA. The exchange splittings are taken into account as a Hartree-Fock contribution to the mass operator $\Sigma_{\mu\mu}^\sigma(k,\omega)$. The simplest way of taking correlation effects into account is by second-order perturbation theory. Energy shifts, lifetime effects and even shake-up processes can be treated this way to lowest order. Calculations based on a model Hamiltonian similar to the one in (5.1) have been performed by *Treglia, Ducastelle* and *Spanjaard* (TDS) [5.36,57], and by *Kleinman* and *Mednick* (KM) [5.11]. The former made, in addition, the $R = 0$ approximation, which was discussed in Sect. 5.2.3. Due to this single-site approximation the self-energy $\Sigma_{\mu\nu}^\sigma(k,\omega)$ becomes k independent and diagonal in the indices $\mu\nu$. Both investigations include s bands and hybridization effects. They do not consider those first-order or HF contributions which result in charge redistributions and in the case of Ni in anisotropic exchange splittings. For Ni one can try to fit the effective interaction U of the model Hamiltonian in order to correct for the differences between the band calculations and the experiments. A value of U = 2 eV was found by TDS who applied on overall d-band fit, while KM obtained a value U = 1.2 eV. Their fit was with respect to features close to the Fermi energy ε_F. The value of U obtained by TDS seems to be better founded because their fitting procedure is less sensitive with respect to the left out first-order contributions than the fitting procedure of KM. But even a value of U = 2 eV is considerably smaller than the value of U = 3.3 eV which follows from the ground state calculations. This discrepancy is apparently due to an overestimation of the correlation energy by a second-order perturbative treatment (Sect. 5.2.3a). From the ground state calculations one expects that this is larger by a factor of two. When this is taken into account U increases by a factor $> 2^{1/2}$ and the discrepancy is removed.

For Fe TDS obtained a value of U < 1 eV while the ground state calculations require U = 2.4 eV. This change is too large in order to be understood by a simple renormalization of U as it was done in the case of Ni. It requires a more detailed study of higher-order effects, in particular of d − d electron screening, which dominates for Fe.

Shake-up processes are qualitatively described by second-order perturbation theory. However, for a quantitative treatment of the peak positions and widths higher-order contributions are important. In a second-order treatment the generated electron and hole are noninteracting. The shake-up process, which is local in character, should therefore have a characteristic energy of the order of W (i.e., $2 \cdot W/2$), and a width which is also of the order of W. On the other hand, one would expect a characteristic energy of the order of U when a local charge fluctuation is generated and a more atomistic point of view is taken. The position of the observed shake-up peak in Ni is approximately 6 eV below ε_F, as compared with W = 4.35 eV. The halfwidth of the peak is much smaller than W indicating an excitation with considerable coherence.

For Ni the correlation problem has been studied also by other methods, in particular by *Liebsch* [5.58,59], *Penn* [5.60], and *Davis* and *Feldkamp* [5.61]. Thereby use was made of the Kanamori t-matrix approach [5.62]. This approach is particularly suited for almost empty or filled bands. In fact it becomes exact in the dilute hole (or electron) limit. In that case it can be shown that the dominant scattering processes are the ladder diagrams in the hole-hole or particle-particle channel. The calculations in [5.58] were done for a model Hamiltonian similar to the one in (5.1) while in [5.60,61] the interaction part of the Hamiltonian was simplified further. By taking the t-matrix between the two holes into account one can accomplish an overall interpretation of the photoemission spectra. However, it turns out that one needs two different values of U in order to explain the observed band narrowing and the peak position at the same time. We return to this problem later. From the self-energy $\Sigma_n^\sigma(\omega)$ the anisotropic exchange splittings Δ_i are obtained through

$$\Delta_i = \Sigma_i^\downarrow(\varepsilon_F) - \Sigma_i^\uparrow(\varepsilon_F) \ , \quad i = e_g, t_{2g} \ . \tag{5.37}$$

Those splittings are very insensitive with respect to possible improvements of the Kanamori t-matrix scheme. With the above value of U the values $\Delta(e_g) = 0.21$ eV and $\Delta(t_{2g}) = 0.37$ eV were found.

They agree well with the experimental values of $\Delta(e_g) = 0.2$ eV, $\Delta(t_{2g}) = 0.33$ eV [5.47] and are smaller than the values found in Sect. 5.2.3b. This difference can be understood qualitatively. Due to the reduction of the quasiparticle bandwidth by approximately 25 % the value for the average exchange splitting found in Sect. 5.2.3b has to be reduced by the same amount. Furthermore in distinction to [5.58] the exchange has been treated self-consistently in Sect. 5.2.3b which also results in an increase of the anisotropy.

Let us return to the problem of the satellite-peak position. There are two possible ways of improving the Kanamori t-matrix type of approach. The first consists in going beyond the low-density limit in which the Kanamori approach

186

is exact. The second is to study the influence of s − p hybridization which has been left out until now.

In [5.59] the first way was persued, i.e. the Kanamori t-matrix was extended by including, in addition, hole-electron scattering processes (ladder diagrams). This scattering channel is closely related to the dynamic susceptibility $\chi(\boldsymbol{q}, \omega)$. As such it contains also contributions from the magnon pole. The calculations beyond the Kanamori t-matrix were done within a one-band model and for $\chi(\boldsymbol{q}, \omega)$ the RPA expression was used, as given in [5.46]. The corrections are similar to the ones which follow from the hole-magnon coupling, as introduced by *Roth* [5.63], and *Hertz* and *Edwards* [5.64]. A generalization to a five-band model was treated in [5.65] but with the interaction restricted to electrons within the same orbital. A recent, careful study of the one-band model by *Igarashi* [5.66] calls into question the importance of contributions which go beyond the Kanamori t-matrix. He did not find significant corrections in the case of Ni when he included self-consistently the next-higher corrections. It is interesting to compare these findings with those of the ground state calculations with almost complete d-band filling (Ni). There it is found that higher-order corrections are negligible as long as one is dealing with a one-band model. This changes when the fivefold degeneracy of the d orbitals is taken into account. Then mutual screening of d electrons in different states at a given site becomes important. In the ground state the correlation energy is reduced between 40 % (nonmagnetic state with all orbitals equally populated) and 20 % (ferromagnetic state with the t_{2g} states equally populated) depending on circumstances [5.37]. The inclusion of d − d electron screening may perhaps resolve some of the difficulties in interpreting the Ni spectra mentioned above.

It would be also of interest to understand better the role of s − d and p − d screening for the hole self-energy. In any case there is still considerable room for improvements on the theoretical treatment of shake-up satellite peaks.

5.4 Finite Temperature Calculations

With increasing temperatures transition metals show more and more local moment characteristics while the itinerant or delocalized features seem to become less important. For example, in Fe, Co and Ni the magnetization curves as function of temperature resemble closely Brillouin curves which would follow from a localized electron description. Also the observed entropy and the effective Bohr magneton number for Fe are close to what is expected from a local moment model. It has therefore been a main problem in the theory of magnetism to understand these features by going beyond a Stoner type mean-field theory. Due to the work of *Schrieffer* [5.67], *Cyrot* [5.68], *Hubbard* [5.69], *Hasegawa* [5.70], and others [5.3] one may claim that this problem is qualitatively understood at present. A basis for this has been provided by the functional integral method [5.71,72]. This method replaces the effects on the motion of an electron due to the Coulomb interactions with all the other electrons by the introduction of external fields. They fluctuate in space and time and at finite temperatures a Gaussian average must be taken with respect to their amplitudes. However, until now all calculations have been limited to the static approximation [5.68–70,73–81]. In it the fluctuations in time of the external fields are neglected. Because of the central role played by the static approximation we shall discuss it in the following subsection. Thereby the discussion will be concentrating on the thermodynamics evolving from that approximation because our aim is constructing afterwards an improved form of the free energy. This is done in the second subsection. There we present a variational theory which goes beyond the static approximation [5.82]. In particular it reduces at T = 0 to the one discussed in Sect. 5.2.

5.4.1 Static Approximation

The functional integral method transforms an interacting electron system into a one-electron problem with fictitious time- and space-dependent external fields (Stratonovich-Hubbard transformation) [5.71,72]. In the static approximation the time dependence of those fields is neglected [5.83]. In order to demonstrate the consequences of this assumption we shall consider in the following a single-band Hubbard model Hamiltonian for N sites

$$H = \sum_{i\sigma}(\varepsilon_i^0 - \mu)n_{i\sigma} + \sum_{ij\sigma}t_{ij}a_{i\sigma}^+a_{j\sigma} + \frac{U}{4}\sum_i(n_i^2 - m_i^2) \ . \tag{5.38}$$

Here ε_i^0, μ and U are the energy of the atomic-like orbital, the chemical potential and the Coulomb integral at a site, respectively.

The transfer integral between sites i and j is denoted by t_{ij}. The $a_{i\sigma}^+(a_{i\sigma})$ are creation (annihilation) operators for d electrons with spin σ at site i. The n_i and m_i are local charge and spin operators.

The functional integral method introduces exchange and charge fields $\{\xi_i(\tau), \varsigma_i(\tau)\}$ which are conjugate to the spin and charge operators on site i. They replace the electron-electron interactions in (5.38). This is called the two-fields method [5.83]. We want to determine the thermodynamic potential Ω which is related to the free energy F through $\Omega = F - \mu N_{el}$. N_{el} is the electron number. For Ω the following exact expression holds

$$e^{-\beta\Omega} = \int \mathcal{D}z(\tau)e^{-\beta\tilde{\Omega}(z(\tau))}$$

$$\beta\tilde{\Omega}(z(\tau)) = \frac{U}{4}\sum_i^N\int_0^\beta d\tau|z_i(\tau)|^2 - \ell nY(z(\tau))$$

$$Y(z(\tau)) = TrT_\tau \exp\{-\beta H_0$$

$$+ \frac{U}{2}\sum_i^N\int_0^\beta d\tau[n_{i\uparrow}(\tau)z_i(\tau) - n_{i\downarrow}(\tau)z_i^*(\tau)]\} \ . \tag{5.39}$$

Here $\mathcal{D}z(\tau)$ is a functional differential defined through

$$\mathcal{D}z_i(\tau) = \lim_{M\to\infty} \prod_{n=1}^{M} \left[\frac{\Delta\tau}{\beta} \frac{\beta U}{4\pi} dz_i(\tau_n) \right] \; ; \quad \mathcal{D}z(\tau) = \prod_{i=1}^{N} \mathcal{D}z_i(\tau) \qquad (5.40)$$

where $\Delta\tau$ denotes a segment of size β/M of the "time" interval $(\beta,0)$. As usual $\beta = (k_B T)^{-1}$. The complex field $z_i(\tau)$ is related to the two fields $\xi_i(\tau)$ and $\varsigma_i(\tau)$ by

$$\xi_i(\tau) = \frac{1}{2}[z_i(\tau) + z_i^*(\tau)] \; ,$$

$$i\varsigma_i(\tau) = \frac{1}{2}[z_i(\tau) - z_i^*(\tau)] \; . \qquad (5.41)$$

The Hamiltonian H_0 is obtained from (5.38) by setting $U = 0$. Furthermore in (5.39) Tr implies taking the trace and T_τ denotes the "time" ordering operator. In the static approximation the τ dependence of $z_i(\tau)$ is neglected. Equation (5.39) simplifies then to

$$\exp(-\beta\Omega_{st}) = \int \prod_i \left(\frac{\beta U}{4\pi} \right)^{1/2} d\xi_i \exp[-\beta E_{st}(\xi, T)] \qquad (5.42)$$

with

$$E_{st}(\xi, T) = -\beta^{-1} \ell n \mathrm{Tr}\{ \exp[-\beta H^0(\xi, \varsigma(\xi))]\} + \frac{U}{4} \sum_i [\xi_i^2 - \varsigma_i^2(\xi)], \qquad (5.43)$$

ξ_i being a static-field variable at site i which is defined by $\xi_i = \beta^{-1} \int_0^\beta d\tau \xi_i(\tau)$.

For the charge field a saddle-point approximation has been made. The Hamiltonian $H^0(\xi, \varsigma(\xi))$ is that of an electron moving in random fields ξ_i

$$H^0(\xi, \varsigma(\xi)) = \sum_i \left\{ \varepsilon_i^0 - \mu + \frac{1}{2}U[\varsigma_i(\xi) - \xi_i\sigma] \right\} n_{i\sigma} + \sum_{ij\sigma} t_{ij} a_{i\sigma}^+ a_{j\sigma} \; . \qquad (5.44)$$

The second term on the right-hand side of (5.43) gives the random field with a Gaussian weight.

The static approximation describes within the same framework local-moment features as well as itinerant featurs of a system [5.70]. At $T = 0$ Ω_{st} reduces to the Hartree-Fock energy. Therefore itinerant features of transition metals at low temperatures such as a non-integer local magnetization, the existence of a Fermi surface and the large Sommerfeld specific heat coefficient γ are contained in the static approximation. When the temperature is increased the field variables $\{\xi_i\}$ in (5.43,44) give rise to thermal spin fluctuations. It is not difficult to show that, within that theory, local exchange splittings can persist even above T_c as it has been observed in Fe. The Curie temperature is therefore much lower than in a Stoner theory. Curie-Weiss susceptibilities are obtained for weak as well as strong interactions. In the case of strong interactions the functional $E_{st}(\xi)$ has a double-minimum structure. This yields the magnetic entropy near the magnetic phase transition which is required to explain the experiments.

Since the random exchange fields are multiplied by the Coulomb inter-actions U, see (5.44), they can produce a gap in the density of states of the paramagnetic state provided that U is sufficiently large. This results in a metal-insulator transition for a half-filled band [5.68,70].

The static approximation is essentially a high-temperature approximation. This is seen by using

$$\beta^{-1} \int\limits_0^\beta d\tau \, \xi_i(\tau) m_i(\tau) \rightarrow \xi_i \beta^{-1} \int\limits_0^\beta d\tau \, m_i(\tau) \tag{5.45}$$

and realizing that this relation becomes exact for $k_B T \rightarrow \infty$ or $\beta \rightarrow 0$. Therefore it is no surprise that it suffers from severe shortcomings at low temperatures. One is that it reduces to the Hartree-Fock energy at $T = 0$ and therefore contains the same problems as the Hartree-Fock approximation for the ground state [5.1,12,13,62]. In particular, the energy gain due to magnetic order is overestimated. This results in a overestimation of the Curie temperature. One way to avoid this problem is by introducing an effective Coulomb interaction $U_{eff} < U$. The latter can be adjusted so that the Curie temperature has the observed size. It is not possible, however, to improve on the missing suppression of charge fluctuations or the amplitude of the local moment because a reduced U_{eff} increases the delocalization of the d electrons (Sect. 5.2).

Another shortcoming of the static approximation is that it does not include low energy spin-wave excitations. Spin wave excitations are responsible for the $T^{3/2}$ behavior of the magnetization M at low temperatures. In the static approximation $M \sim T$. This implies that certain thermodynamic relations are violated for $T = 0$. One example is $(\partial M/\partial T)_{T=0} \neq 0$, but similar difficulties appear for the specific heat and the thermal expansion coefficient [5.75,84–86].

5.4.2 Beyond the Static Approximation

In the static approximation, which reduces to the Hartree-Fock theory at $T = 0$, the local correlation effects discussed in Sect. 5.2 are missing. These effects are also present at finite temperatures, provided that T is not too high. We want to obtain an estimate of the temperature T_s beyond which the correlations are severely affected by thermal fluctuations. Thereby we must take into account that the correlation energy gain due to density or charge correlations is larger than the one due to spin correlations. For a study of the effect of T on the former consider a single half-filled Hubbard band. Assume that the on-site Coulomb repulsion U is much larger than the kinetic energy of the electrons. Then at $T = 0$ one finds $\langle H \rangle = 0$ since there is just one electron at each site due to corre-lations. At $T > T_s$ the correlations are assumed to be destroyed. Therefore there are double occupancies of sites with probability $1/4$. Hence $\langle H \rangle = 1/4$ U. We expect therefore that $k_B T_s \cong 1/4$ U. This argument can be refined by defining $k_B T_s$ as the energy difference between the nonmagnetic Hartree-Fock ground state and the correlated magnetic ground state. This definition can be rewritten as

$$k_B T_s = -E_{corr}(\xi = 0) + E_0(\xi = 0) - E_0(\xi) \ . \tag{5.46}$$

Here $E_{corr}(\xi = 0)$ is the correlation energy of the nonmagnetic ground state and $E_0(\xi)$ is the exact energy in the presence of an exchange field ξ. By using the results of Sect. 5.2 for $E_{corr}(\xi = 0)$ for Fe and Ni one finds that T_s is much larger than the Curie temperature T_c in both cases. Therefore one expects that the density correlations which are present at $T = 0$ are also present at $T > T_c$. When, in addition to U, the exchange J is included the additional energy gain due to spin correlations is of the order of T_c in Fe and somewhat less than T_c in Ni [5.26]. Therefore these correlations are modified to some extent when the temperature increases up to T_c. We try to take account of these correlations adiabatically at finite temperatures on the basis of the variational principle.

The following theory is based on a single-band description and therefore allows for charge but not for the Hund's-rule correlations. The case of degenerate orbitals is discussed at the end. We proceed in constructing a theory which reduces to the correlated ground state at $T = 0$ and to the static approximation in the high temperature limit.

Let us start out from the Feynman inequality for the thermodynamic potential

$$\Omega \leq \Omega_t + \langle E(\xi, T) - E_t(\xi, T) \rangle \ , \tag{5.47}$$

$$\exp(-\beta \Omega_t) = \int \left[\prod_i \left(\frac{\beta U}{4\pi} \right)^{1/2} d\xi_i \right] \exp\left[-\beta E_t(\xi, T) \right] \ , \tag{5.48}$$

$$\langle A \rangle = \frac{\int \left[\prod_i d\xi_i \right] e^{-\beta E_t} A}{\int \left[\prod_i d\xi_i \right] e^{-\beta E_t}} \ . \tag{5.49}$$

A denotes an arbitrary functional of ξ. $E_t(\xi, T)$ is a trial energy functional. We assume that E_t depends on a set of variational parameters $\eta_i(\xi, T)$ which describe adiabatically the correlated electron motion, i.e. $E_t = E_t(\xi, \eta(\xi, T), T)$.

By minimizing the right hand side of (5.47) one obtains the following set of equations for the determination of the η_i

$$\frac{\partial E_t(\xi, \eta(\xi, T), T)}{\partial \eta_i(\xi, T)} = 0 \ . \tag{5.50}$$

An appropriate ansatz for E_t is of great importance in the present approach. Thereby we shall use that at $T = 0$, E_t reduces to the energy of the correlated ground state $|\psi_0\rangle$. The latter is related to the uncorrelated ground state $|\phi_0(\xi, \varsigma)\rangle$ of the Hamiltonian $H^0(\xi, \varsigma)$ through

$$|\psi_0\rangle = Q|\phi_0(\xi, \varsigma)\rangle \ . \tag{5.51}$$

This should be compared with (5.17). We have chosen here the notation Q instead of e^S because it will turn out later in connection with the single-site approximation that a product ansatz for Q is more convenient than an equivalent exponential ansatz. However, the following considerations are very general and Q need not to be specified. We assume that Q is norm conserving.

We shall make the following ansatz

$$E_t(\xi, \varsigma(\xi, T), \eta(\xi, T), T) = E_{st}(\xi, \varsigma(\xi, T), T)$$
$$+ \langle Q\tilde{H}Q \rangle_0(\xi, \varsigma(\xi, T), \eta(\xi, T), T) . \qquad (5.52)$$

Here $\tilde{H} = H - \langle H \rangle_0$ and $\langle \ldots \rangle_0$ is defined through

$$\langle Q H Q \rangle_0 = \frac{\text{Tr}\{ \exp[-\beta^* H^0(\xi, \varsigma)] Q \tilde{H} Q \}}{\text{Tr}\{ \exp[-\beta^* H^0(\xi, \varsigma)] \}} . \qquad (5.53)$$

$(\beta^*)^{-1} = k_B T^*$ is an effective temperature which we shall assume to be zero. This is necessary in order to ensure that for $T \to \infty$ the thermodynamic potential Ω_t approaches that of the static approximation Ω_{st} (see below). This point has been discussed in more detail in [5.82]. In (5.52) the $\varsigma_i(\xi, T)$ and $\eta_i(\xi, T)$ are regarded as variational parameters. The former build up the best charge potentials while the latter describe best the correlated motions of the electrons in the static exchange fields ξ_i. Equation (5.50) is then replaced by

$$\frac{\partial}{\partial \eta_i} \langle Q\tilde{H}Q \rangle_0 = 0 , \qquad (5.54)$$

$$\frac{\partial}{\partial \varsigma_i} (E_{st} + \langle Q\tilde{H}Q \rangle_0) = 0 . \qquad (5.55)$$

Later, when we apply the theory, we will replace the ς_i by those of the static approximation

$$\varsigma_i(\xi) \cong n_i^0(\xi) = \int d\omega \, f(\omega) \varrho_i(\omega, \xi) . \qquad (5.56)$$

Here $n_i^0(\xi)$ is the electron charge at site i which follows from the Hamiltonian $H^0(\xi, \varsigma)$ and $\varrho_i(\omega, \xi)$ is the local density of states belonging to it. This approximation facilitates the proof of the following statements.

1. The thermodynamic potential Ω_t is lower than that of the static approximation

   $$\Omega_t \leq \Omega_{st} . \qquad (5.57)$$

2. At $T = 0$, Ω_t gives an upper bound of the exact ground state energy E_0. It is also instructive to compare it with the variational energy $\langle H \rangle_t = \langle \phi_{HF} | Q H Q | \phi_{HF} \rangle$ which has been discussed in Sect. 5.2. Here $| \phi_{HF} \rangle$ is the Hartree-Fock ground state corresponding to the Hamiltonian (5.38). The following inequalities hold at $T = 0$

$$E_0 \leq \langle H \rangle_t(\xi_t, \eta_t(\xi_t)) \leq \Omega_t(T = 0) \leq \langle H \rangle_t(\xi_{HF}, \eta_t(\xi_{HF})) \ . \tag{5.58}$$

Here ξ_t is the exchange field in $|\phi_{HF}\rangle$ which minimizes $\langle H \rangle_t$ while ξ_{HF} is the one which minimizes the Hartree-Fock energy. When $\xi_t = 0$ or ξ_{HF} then $\Omega_t(T = 0) = \langle H \rangle_t$.

3. In the high-temperature limit Ω_t agrees with that of the static approximation, i.e.

$$\lim_{T \to \infty} \Omega_t = \Omega_{st} \ . \tag{5.59}$$

In order for that relation to hold it is essential that the effective temperature T^* in (5.53) is chosen to be zero [5.82].

4. The theory produces correctly the atomic limit at $T = 0$ and $T = \infty$.

Next we want to demonstrate that Ω_t also leads to improvements of the thermodynamic quantities [5.82].

The internal energy is given by

$$\langle H \rangle - \mu N_{el} = \int d\omega\, f(\omega) \omega \langle \varrho(\omega, \xi) \rangle - \frac{U}{4} \sum_i \left(\langle \varsigma_i(\xi)^2 \rangle - \langle \xi_i^2 \rangle + \frac{2}{\beta U} \right)$$
$$+ \langle \langle Q\tilde{H}Q \rangle_0 \rangle \ . \tag{5.60}$$

Here $f(\omega)$ is the Fermi distribution function and $\varrho(\omega, \xi)$ is the total density of states. The first and the second term denote the energy in the static approximation, but with a renormalization due to the energy E_t which appears in the thermal average. Here and in the following we shall leave out the index t in $\langle \ldots \rangle_t$. The last term represents the correlation energy contribution. It suggests that the correlation energy gain in the paramagnetic state at $T > T_c$ is considerably reduced as compared with that of the nonmagnetic state at $T = 0$ because of the large random fields.

For the entropy the following relation is derived

$$S = -\int d\omega \langle \varrho(\omega, \xi) \rangle \{ [1 - f(\omega)] \ln [1 - f(\omega)]$$
$$+ f(\omega) \ln f(\omega) \} + \ln \int \left[\prod_i \left(\frac{\beta U}{4\pi} \right)^{1/2} d\xi_i \right] e^{-\beta(E_t - \langle E_t \rangle)} - \frac{1}{2} N \ . \tag{5.61}$$

The first term is the well-known entropy of a system of independent electrons with a T-dependent density of states. The second term is the renormalized (due to correlations) magnetic entropy. The renormalization generally reduces the entropy change at intermediate temperatures because the correlated motion of electrons suppresses some of the degrees of freedom. The third term is a consequence of the prefactor $(\beta U/4\pi)^{1/2}$ [5.82]. The local charge $\langle n_i \rangle$ and magnetization $\langle m_i \rangle$ are given by

$$\langle n_i \rangle = \langle \varsigma_i(\xi) \rangle + \langle \langle Q\tilde{n}_i Q \rangle_0 \rangle \ ,$$

$$\langle m_i \rangle = \langle \xi_i \rangle + \langle\langle Q\tilde{m}_i Q\rangle_0\rangle \; , \qquad (5.62)$$

where $\tilde{n}_i = n_i - \langle n_i\rangle_0$ and $\tilde{m}_i = m_i - \langle m_i\rangle_0$. The second term on the right-hand side of these equations is a correlation effect. The term $\langle\langle Q\tilde{m}_i Q\rangle_0\rangle$ reduces generally the magnetization, as discussed in Sect. 5.2.

The thermal averages of the square of the local charge and the squared amplitude of the local moment are obtained as follows [5.82]

$$\left[\begin{matrix} \langle n_i^2 \rangle \\ \langle m_i^2 \rangle \end{matrix} \right] = \langle \varsigma_i(\xi)\rangle \pm \frac{1}{2}\left[\langle \varsigma_i(\xi)^2\rangle - \langle \xi_i^2\rangle + \frac{2}{\beta U} \right]$$
$$+ \langle\langle Q[\begin{smallmatrix} \tilde{n}_i^2 \\ \tilde{m}_i^2 \end{smallmatrix}]Q\rangle_0\rangle \qquad (5.63)$$

where $\tilde{n}_i^2 = n_i^2 - \langle n_i^2\rangle_0$ and $\tilde{m}_i^2 = m_i^2 - \langle m_i^2\rangle_0$.

The forms of the first terms result from the static approximation. They are different, however, from the formulae used before in the literature, i.e.,

$$\left[\begin{matrix} \langle n_i^2 \rangle \\ \langle m_i^2 \rangle \end{matrix} \right] = \left[\begin{matrix} \langle \varsigma_i^2(\xi)\rangle \\ \langle \xi_i^2\rangle - \frac{2}{\beta U} \end{matrix} \right] \; . \qquad (5.64)$$

These expressions do not take account of the quantum fluctuations, (i.e., at $T = 0$ it is $\langle m\rangle = \langle m^2\rangle^{1/2}$) and therefore do not yield the correct results in the delocalized or independent-electron limit [5.82]. The last term on the right-hand side of (5.63) which results from the electron correlations has the effect of partially suppressing charge fluctuations and enhancing the amplitude of the moments even above T_c.

When one applies the present theory to transition metals one is facing two difficulties. One is the evaluation of the energy functional $\langle Q\tilde{H}Q\rangle_0$, and the other is the evaluation of the integrals $\int [\Pi d\xi_i]$, e.g., for Ω_t. The simplest way of making a numerical calculation possible is by applying the single-site approximation (SSA). This approximation is well established within the static approximation [5.69,70]. An effective medium $\mathcal{L}_{i\sigma}^{-1}$ is introduced into the diagonal part of the Hamiltonian (5.44) which describes thermally averaged one-electron states. The deviations in the energy functional $E_{st}(\xi)$ with respect to the effective medium are expanded with respect to the sites [5.76]. The zeroth-order term does not depend on the field variables ξ_i. The first-order correction term is expressed as a sum of energy functionals $E_{st}^{(i)}(\xi_i)$ of impurities embedded in the effective medium. Higher-order terms are neglected in SSA.

For the evaluation of the functional $\langle Q\tilde{H}Q\rangle_0$ we adopt the following form for Q [5.23–29]

$$Q = \langle \prod_i (1 - \eta_i O_i)^2 \rangle_0^{-1/2} \prod_i (1 - \eta_i O_i) \quad \text{where} \qquad (5.65)$$

$$O_i = (n_{i\uparrow} - \langle n_{i\uparrow}\rangle_0)(n_{i\downarrow} - \langle n_{i\downarrow}\rangle_0) \; . \qquad (5.66)$$

Except for a slight change in the definition of the operators O_i this form of Q corresponds to the *Gutzwiller* ansatz for the correlated wave function [5.12,33]. It implies that we correlate electrons only when they sit in the same orbital (remember that there is only one orbital per site). We could have started out instead as well with an equivalent form $Q = \exp(S)$, as in Sect. 5.2. However, when SSA is made it turns out that the simplifications which follow from it are more easily derived when Q has the form of (5.65). In order to distinguish the following considerations based on the special form of (5.65) from the general ones valid for all forms of Q we shall replace the index "t" (trial) by the index "G" (*Gutzwiller*). By making use of Wick's theorem one can express $\langle Q\tilde{H}Q\rangle_0$ as a sum of products of different contractions of one-electron operators.

When all contractions between operators on different sites are neglected one obtains for the thermodynamic potential in SSA

$$\Omega_G(\text{SSA}) = \Omega_{st}(\text{SSA}) - \beta^{-1} \sum_i \ln\langle \exp[-\beta E_c^{(i)}(\xi_i)]\rangle_{st} \qquad (5.67)$$

where the single-site correlation energies $E_c^{(i)}(\xi_i)$ are given by

$$E_c^{(i)}(\xi_i) = \frac{-2\eta_i\langle O_i\tilde{H}\rangle_0 + \eta_i^2\langle O_i\tilde{H}O_i\rangle_0}{1 + \eta_i^2\langle O_i^2\rangle_0} \quad . \qquad (5.68)$$

$\Omega_{st}(\text{SSA})$ is the thermodynamic potential in the SSA and $\langle...\rangle_{st}$ is defined as in (5.49) but with $E_{st}^{(i)}(\xi_i)$ replacing $E_t(\xi)$. The parameters η_i are obtained in analytical form from $\partial E_c^{(i)}/\partial\eta_i = 0$. The great advantage of SSA is that $\langle O_i^2\rangle_0$, $\langle O_i\tilde{H}\rangle_0$, $\langle O_i\tilde{H}O_i\rangle_0$ and therefore $E_c^{(i)}(\xi_i)$ can be expressed in terms of the local density of states $\varrho_i(\omega, \xi)$ for any configuration of exchange fields ξ_i. Thus one can approximate $\varrho_i(\omega, \xi)$ by the CPA form. The same holds true for the different correlation terms in (5.60–63). The effective medium in (5.67,68) is determined self-consistently by the CPA equations.

In the following we present and discuss numerical results obtained within SSA of the present theory. First we consider the case of a half-filled band [5.87]. Thereby the Fermi distribution function is replaced by a step function whereever it appears. It turns out that this approximation results in minor changes only. For the density of states of the noninteracting system a bell-shaped form is adopted. The ground state may become antiferromagnetic and the magnetic phase diagram is shown in Fig. 5.10. The Néel temperature T_N has a maximum near $U \cong W$. The local electron correlations slightly shift the boundary between the paramagnetic metallic state and the paramagnetic insulator to larger values of U/W. They also reduce T_N considerably in the regime $0 < U/W < 1$. For $U/W > 1$ the theory reduces to the molecular field approximation of a Heisenberg Hamiltonian. The susceptibility in the paramagnetic state follows a Curie-Weiss law. The Curie constant remains approximately the same when the electron correlations are taken into account.

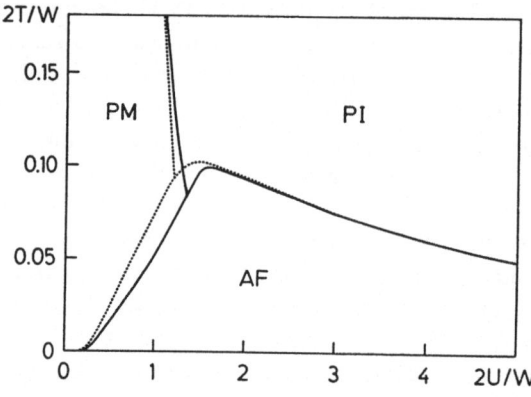

Fig. 5.10. Magnetic phase diagram for a half-filled bell shaped band, showing the paramagnetic metal (PM), the paramagnetic insulator (PI), and the antiferromagnetic state (AF). Full lines (dotted lines) show the phase boundaries in the variational approach (in the static approximation [5.70])

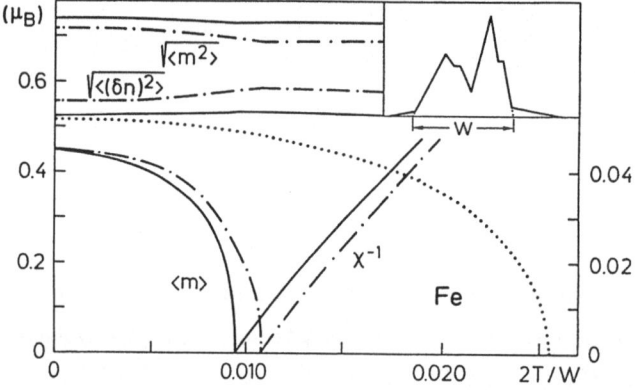

Fig. 5.11. Magnetization $\langle m \rangle$, inverse susceptibility χ^{-1}, amplitude of local moment $(\langle m^2 \rangle)^{1/2}$, and charge fluctuation $((\langle (\delta n)^2 \rangle))^{1/2}$ for Fe as a function of temperature.
(—): the different quantities within the variational approach. (...): $\langle m \rangle$ in the static approximation. (-.-.-): the different quantities in the static approximation but with $U_{eff} = 0.694\,U$. The scale on the r.h.s. refers to χ^{-1}. The model DOS is shown in the inset together with the definition of the d-bandwidth W

Next we consider bcc Fe where correlation effects can be appreciable [5.82]. The magnetization vs. T curves $\langle m(T) \rangle$ are shown in Fig. 5.11. Correlations reduce the Curie temperature by a factor of three. This is due to the fact that the energy difference between the paramagnetic state above T_c and the ferromagnetic state at T = 0 is strongly reduced by local-electron correlations. The effects of correlations on the magnetization and susceptibility can be simulated, however, by employing the static approximation with an effective Coulomb interaction $U_{eff}(\cong 0.7\,U)$. It is chosen to be consistent with the observed ground state magnetization. The reduction of T_c is about 10 % when the results of the present theory are compared with those of the static approximation using U_{eff}. The reduced magnetization is close to a Brillouin curve when correlations are included. Near T_c it is larger than a Brillouin curve with S = 1/2 which is is in agreement with experiments. The paramagnetic susceptibility $\chi(T)$ follows a Curie-Weiss law with a $m_{eff} \cong 1.75\langle m(T = 0)\rangle$, a value which is close to that in the static approximation. The experimental prefactor is 1.44. The quantum fluctuations enhance $\langle m^2\rangle^{1/2}$ by 70 %. Local-electron correlations yield an additional increase by a few percent. The value $\langle m^2(0)\rangle^{1/2} = 0.73\,\mu_B$ is close to the atomic value $0.75\,\mu_B$. The correlations also reduce the T dependence of $\langle m^2\rangle^{1/2}$. It is found that $\langle m^2(T_c)\rangle^{1/2}/\langle m^2(0)\rangle^{1/2} = 0.99$, as compared with 0.96 in the static approximation. This change may seem insignificant but has in fact consequences for the magnetovolume effect [5.88].

Let us consider next Ni. Here the correlation-energy effects are smaller than in Fe since the probability is small to find more than one hole at a Ni site. The T dependence of $\langle m(T)\rangle$, $\langle m^2(T)\rangle^{1/2}$ and $\langle(\delta n)^2(T)\rangle$ are shown in Fig. 5.12. There is a reduction in T_c by about 20 % as compared with the static approximation. The functions $\langle m(T)\rangle$ and $\chi(T)$ are well described by the static approximation provided that U is replaced by $U_{eff} \cong 0.8\,U$. This is so because the field dependence of $E_c(\xi)$ is very small in Ni and influences little the functional dependence of the total energy. $\chi(T)$ follows again a Curie-Weiss law with $m_{eff} \cong 3.5\cdot\langle m(T = 0)\rangle$. The experimental prefactor is 2.68. Values between 2.4 and 3.0 have been reported in previous calculations based on the two-fields method [5.70,76,77,81]. The amplitude $\langle m^2\rangle^{1/2}$ calculated from (5.63) is by more than a factor 3.5 larger than that obtained from (5.64). Inclusion of correlations results in a further increase by a few percent. The calculated value is close to the atomic value of $0.45\,\mu_B$. There is no anomaly in $\langle m^2(T)\rangle$ and $\langle(\delta n)^2(T)\rangle$ at T_c. The amplitude is found to be practically independent of T. This is consistent with the theoretical analysis of the thermal expansion coefficient of Ni [5.75].

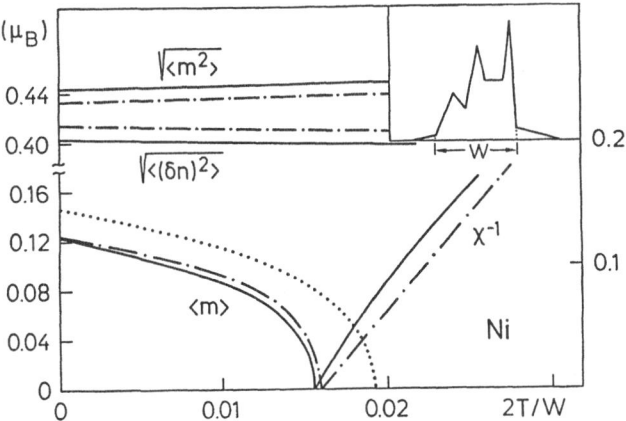

Fig. 5.12. Magnetization, inverse susceptibility, amplitude of local moment and charge fluctuation for Ni as a function of temperature. The notation is the same as in Fig. 5.11

The above numerical results are based on the single-band model. Therefore it is of interest to discuss the changes which one might expect when the 5-fold degeneracy of the d bands is included. The effects of the quantum fluctuations on the local moments are suppressed by the inclusion of the 5-fold degeneracy (d = 5). Therefore decreases the ratio $\langle m^2 \rangle^{1/2}/\langle m(0) \rangle$ decreases. On the other hand, the Hund's rule correlations between electrons in different orbitals at the same site increase the size of the local moment. However, the Hund's rule coupling is partially destroyed by thermal fluctuations when T becomes of the order of T_c. This is so because the correlation energy gain due to spin correlations is only of the order of 700 K per atom in Fe. Therefore one expects a gradual decrease of $\langle m^2 \rangle^{1/2}$ with increasing T at temperatures of interest. The charge fluctuations $\langle (\delta n)^2 \rangle$ should be temperature independent because the correlation energy gain due to partial suppressions of charge fluctuations is much larger than $k_B T_c$.

In the static approximation a scaling relation $T_c(d) = dT_c(1)$ was found when the Fermi distribution function is neglected [5.8]. $T_c(d)$ is hereby the Curie temperature in the presence of a band degeneracy d. This relation does not hold anymore when electron correlations are taken into account.

Recently *Hasegawa* extended his theory of magnetism to the case of degenerate orbitals [5.81]. For Fe he obtained a T_c of 3000 K by using a U_{eff} which fits the observed magnetization of $2.2\,\mu_B$ at T = 0. Local correlations reduce T_c by 10 % (Fig. 5.11). A further reduction by 25 % is obtained when the Fermi distribution functions are used instead of step functions [5.75]. Nearest-neighbor spin correlations of the Heisenberg type reduce T_c by another 35 %. An additional decrease of T_c by 10 % is expected when transverse quantum degrees of freedom are taken into account. They are not contained in [5.81] and therefore in the atomic limit of that theory the entropy is always ln 2 instead of ln 3 as it should be for S = 1 [5.86]. When these effects are all added up one ends up with a T_c of the order of 1200 K which comes close to the experimental value of 1044 K.

In Ni the situation is somewhat different. There is no reduction of T_c as calculated in the static approximation with U_{eff} replacing U (Fig. 5.6). Also the entropy is not underestimated as in Fe [5.86]. After inclusion of the Fermi functions *Hasegawa* obtained $T_c = 700$ K. Reducing this value by 30 % due to short-range spin correlations [5.89] results in $T_c = 500$ K. This is not too far from the experimental value of 630 K.

We have considered here the effects of electron correlations on the thermodynamic properties of transition metals. Apparently they are responsible to a large extent that the phenomenological Heisenberg model describes rather well the thermodynamic properties of these systems [5.90,91]. Another effect of the electron localization is seen in the inner-core spectra of transition metals. We shall study it in detail in the next section and show that it reveals new aspects of electron correlations.

5.5 Influence of Correlations
on the Photoelectron Spectra from Inner-Core States

The correlated motion of electrons strongly influences the inner-core photoelectron spectra in transition metals. In the following we concentrate our discussion onto the 3s spectra of 3d transition metals because in that case there is no spin-orbit coupling.

An inner-core hole which is created by the incident photons interacts with the conduction eletrons on the same site via the Coulomb (V_c) and exchange (J_c) interactions, see (5.71). Therefore the cross section for inner-core photoelectron processes contains local information about conduction electrons. This is easily seen from a moment analysis of the cross-section [5.92]. It is

$$\int d\omega(\omega - \omega_c)I_s(\omega) = \tfrac{1}{2}J_c\langle S_z\rangle \qquad (5.69)$$

$$\int d\omega(\omega - \omega_c)^2 I_c(\omega) = \Gamma^2 + V_c^2\langle(\delta n)^2\rangle + \tfrac{1}{4}J_c^2\langle S^2\rangle \ . \qquad (5.70)$$

Here

$$\omega_c = \int d\omega\,\omega I_c(\omega) \ , \quad I_c \equiv I_+ + I_- \ , \quad I_s \equiv (I_+ - I_-)/2 \ ,$$

and $I_\pm$ are the cross-sections for photoelectrons with spin $\pm$. Γ is a phenomenologically introduced width which is caused by the life time of the core hole, the convolution effects, and other processes of nonmagnetic origin. S and $\delta n = n - \langle n\rangle$ are the spin and fluctuating number operator of the d electrons at the site of the core hole. The average $\langle \dots\rangle$ is with respect to temperature and different d-electron configurations. It is seen that the first moment is proportional to the magnetization, i.e. $\langle S_z\rangle$ of the d electrons. The second moment of the unpolarized spectra depends on the charge fluctuations $\langle(\delta n)^2\rangle$ and the local moment $\langle S^2\rangle$ which were discussed in Sect. 5.2 and 3. Electron correlations suppress charge fluctuations and enlarge the local moment due to the Hund-rule correlations. Both effects influence the width of the spectrum in opposite ways, see (5.70). Large changes with temperature of the amplitude of the local moment as e.g. in invar alloys [5.75,93,94] lead also to changes in the width.

In addition to changing the width of the spectra, correlations of the d-electrons cause also changes in the line shape of the 3s spectra. In order to study these effects, we shall consider, for simplicity, the case of a single Hubbard band coupled to a core hole. The local interaction H_I is given by

$$H_I = -V_c n n_c - J_c S \cdot S_c \qquad (5.71)$$

where $n(n_c)$ are the number operators for the 3d electrons (and the 3s-core electron) at site 0 and $S(S_c)$ are the corresponding spin operators. In this case the core spectra depend on the ratios V_c/W, J_c/W, and U/W when the temperature T and the d electron number are given. W is the d band width as before. The inner-core spectra of atoms with an unfilled shell were investigated theoretically by *Van Vleck* many years ago [5.95]. He found multiplet structures for the spectra which depend on J_c only. On the other hand, the core spectra of simple metals were extensively investigated because of the appearence of an infrared divergency in them which is due to the existence of a Fermi surface [5.96–98]. In that case the line shape is determined by V_c/W only. In both cases the line shape of the core spectra does not depend on the Hubbard U explicitly (U/W is approximately ∞ and 0 in the two cases, respectively). This is not the case any longer in 3d transition metals where U/W is of order one. Here the line form shows features which reflect the delocalized as well as the localized character of the 3d electrons. In order to demonstrate this one must develop an interpolating theory for the cross section which is correct in both, the atomic as well as the delocalized limit. Such a theory has been developed with the help of the Mori-Zwanzig method [5.99].

The cross section for photoelectron emission from inner-core states is proportional to the Fourier transform $I_\alpha(\omega)$ of the time-correlation function of core-electron creation and annihilation operators b_α^+, b_α. In terms of the Liouville operator formalism [5.100–102] it is written as

$$I_\alpha(\omega) = -\frac{1}{\pi} \operatorname{Im} \phi_\alpha(z) \ , \tag{5.72}$$

$$\phi_\alpha(z) = \left(b_\alpha \left| \frac{1}{z - K} b_\alpha \right. \right) \ , \tag{5.73}$$

$$K = L + H_I \ . \tag{5.74}$$

Furthermore $z = \omega + \varepsilon_c + i\delta$ where ε_c is the core hole energy ε_{co} measured from the Fermi surface plus the Coulomb repulsion U_c between core electrons, i.e. $\varepsilon_c = \varepsilon_{co} + U_c$. The Liouville operator L is with respect to the d-electron system and is defined through

$$LA = [H, A]_- \tag{5.75}$$

for any arbitrary operator A. H is given by (5.38). The scalar product $(A|B)$ between any operators A and B is defined by the thermal average

$$(A|B) = \langle A^+ B \rangle \ . \tag{5.76}$$

The thermal average in (5.73) can be treated independently of the time evolution of the core-hole propagator. It is evaluated by means of the functional integral method within the static approximation. The correlation function (5.73) is then given by

$$\phi_\alpha(z) = \langle \tilde{\phi}_\alpha(z, \xi) \rangle \tag{5.77}$$

where $\langle \ldots \rangle$, which acts on a scalar, is defined by (5.49). The adiabatic correlation function $\tilde{\phi}_\alpha(z, \xi)$ has the same form as (5.73) but now with the thermal average $\langle \ldots \rangle$ replaced by $\langle \ldots \rangle_0$ where the latter is taken with respect to the d electron Hamiltonian (5.44).

The time evolution of $\tilde{\phi}_\alpha(z, \xi)$ is treated by applying the Mori-Zwanzig technique. The basic idea for the construction of a theory with the required interpolating properties is to start out from a properly chosen space of dynamic variables. We choose the operators $|p_\mu b_\alpha)$ as such. They span a space which contains the atomic as well as the delocalized limits. The p_μ are operators which project on the eigenstates μ of the Hamiltonian H_I. There are four eigenstates of H_I within the one-core hole space. They are the zero d electron state (0), the singlet (1s) and triplet (1t) state formed between one d electron and the core hole, and the state with two d electrons (2) at site 0. By inserting the identity $\sum_\mu p_\mu = 1$ into $\tilde{\phi}_\alpha(z, \xi)$ one obtains

$$\tilde{\phi}_\alpha(z, \xi) = \sum_{\mu\nu} \phi_{\mu\nu\alpha}(z, \xi) , \qquad (5.78)$$

$$\phi_{\mu\nu\alpha}(z, \xi) = \left(p_\mu b_\alpha \left| \frac{1}{z - K} p_\nu b_\alpha \right. \right)_0 . \qquad (5.79)$$

The scalar product $(A|B)_0 = \langle A^+ B \rangle_0$. The equation of motion for $\phi_{\mu\nu\alpha}$ is given by [5.102].

$$\sum_\lambda (z\delta_{\mu\lambda} - K_{\mu\lambda\alpha} - M_{\mu\lambda\alpha}) \phi_{\lambda\nu\alpha} = \langle p_\mu \rangle_\alpha \delta_{\mu\nu} . \qquad (5.80)$$

The relative intensities $\langle p_\mu \rangle_\alpha$ are defined by $\langle b_\alpha^+ p_\mu b_\alpha \rangle_0$. They have been determined in [5.99]. $K_{\mu\lambda\alpha}$ and $M_{\mu\lambda\alpha}$ are the frequency matrix and memory function, respectively. The frequency matrix leads to line shifts and describes both, the delocalized as well as the atomic limits. The memory function gives rise to a line broadening and additional line shifts. A simple approximation scheme for calculating both matrices was proposed in [5.99]. It enabled one to perform numerical calculations, the results of which we want to describe in the following. All quantities can be expressed again in terms of the local densities of states $\varrho(\omega, \xi)$. This enables one to make single-site approximations. The theory is more reliable in the atomic regime than in the delocalized limit because the equation of motion (5.80) starts out from the atomic states.

We want to study the effects of electron correlations on the line shape by changing U/W along the line $V_c/W = J/W = 1$ (Fig. 5.13) [5.103]. The numerical work to be presented is limited to the half-filled band case. The correlation correction is then relatively small and we can discuss it within the static approximation. That implies that only those correlations are considered explicitly which result from the random fields ξ_i at $T \neq 0$. A single-site-approximation is applied in the calculations. Furthermore a bell-shaped model density of states is used. The temperature T is fixed at $2T/W = 0.12$, because the static approximation and the SSA work best at high T. This implies that one is in the paramagnetic state.

The relative intensities $\langle p_\mu \rangle = \sum_\alpha \langle\langle b_\alpha^+ p_\mu b_\alpha \rangle_0\rangle$ of the four components (0, 1s, 1t, 2) are 1/2:1/4:3/4:1/2 for a noninteracting system. Note that $\sum_\mu \langle p_\mu \rangle$

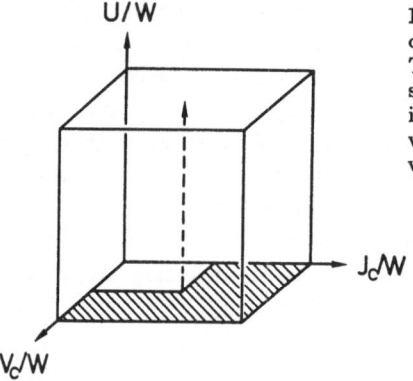

Fig. 5.13. The space $(V_c/W,\ J_c/W,\ U/W)$ which determines basic features of inner-core spectra. The hatched area in the space $(V_c/W,\ J_c/W)$ shows the region where the present theory [5.99] is available. Dashed line shows a trajectory along which the numerical results represented in Fig. 5.14 were calculated

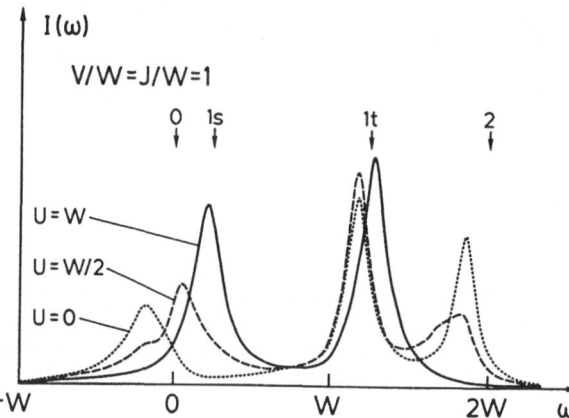

Fig. 5.14. Calculated spectra for the case $V_c/W = J_c/W = 1$ as a function of Coulomb interaction U when $2T/W = 0.12$. The arrows in the figure show the position of peaks in the atomic limit

$= 2$. With increasing value of U the charge fluctuations are suppressed so that $\langle p_0 \rangle$ and $\langle p_2 \rangle$ decrease rapidly. They become neglegible for $U/W > 1$. Note that the intensity ratio between $\langle p_{1t} \rangle$ and $\langle p_{1s} \rangle$ is 3:1, independent of U/W.

In transition metals the Coulomb interaction U drastically changes the line shape of core spectra. Therefore their proper inclusion in the theory is very important. Fig. 5.14 shows the change of spectra when U changes along the line $V_c = J_c = W$ in Fig. 5.13. For $U = 0$ there are three peaks. The one at $\omega \cong -W/4$ contains 0 and 1s components. When U increases it splits into two peaks, a smaller one of component 0 and a larger one of component 1s. For $U/W = 1$ the former has vanished because of the suppression of charge fluctuations. The same holds true for the peak at $\omega \cong 2W$ which contains the component 2. Therefore for $U/W = 1$ only two peaks remain which have 1s and 1t character, respectively. This is just the atomic multiplet structure and it is brought out by the Coulomb interaction U. It is likely that the observed 3s splitting in $\alpha - Mn$ [5.104] is the multiplet which is caused by the d-electron correlations. The observed splitting in Ni [5.105] must result from charge fluctuations, i.e. the formation of a two-hole bound state [5.106]. The spin amplitude on a site is too small in order to produce a distinct multiplet. The 3s core spectra of Fe also show splittings [5.105]. The origin of the splitting is not clear because the situation is between the ones for α – Mn and Ni. However, it should be pointed out that experiments coupled with a spin polarization analysis of the emitted core electron should be able to conclusively determine the origin of the splitting [5.103]. In a ferromagnetic metal one expects the peaks strongly to depend on the spin of the electron if they are the atomic like multiplets discussed before. On the other hand, one does not expect a spin dependence of the spectra if the splitting is caused by charge fluctuations.

It is also worth pointing out that we can estimate the temperature dependence of the amplitude of the local magnetic moment by analysing the experimental data with respect to different moments, see (5.69,70) [5.92]. The data on the 3s inner-core spectra of Fe show hardly any changes in the line forms as the temperature varies from below to above T_c. This implies that the amplitude of the magnetic moment remains essentially unchanged in that temperature regime. This is consistent with the theoretical findings in Sect. 5.4 [5.74–81] and with the interpretation of the experimental data on the magnetovolume effect [5.45].

Unfortunately there are no systematic experiments available of inner-core spectra of materials which are expected to have strongly T dependent local moments.

Finally we draw attention that the numerical results presented here were limited to a regime V_c, $J_c \geq W$ which is close to the atomic one. There is another important effect of electron correlations on the core spectra if one is in the regime V_c, $J_c \leq W$. It is well known that there exists a distinct edge singularity in that case [5.96–98]. It is a Fermi-surface effect. When U increases and a Mott transition takes place, the singularity should be replaced by an atomic-like multiplet structure, as discussed before. The disappearance of the absorption peak asymmetry (i.e., the edge singularity) has been observed in vanadium-oxide compounds [5.107]. The present theory which is based on the Mori-Zwanzig method does not describe the edge singularity in the regime V_c, $J_c \leq W$. *Kakehashi* and *Kotani* have developed an alternative interpolation theory for inner-core spectra [5.108]. It reproduces the edge singularity in the delocalized regime but does not describe well the atomic regime. A theory which describes both, the delocalized and the atomic regime equally well remains to be developed.

Acknowledgements. The authors would like to thank Dr. A.M. Oleś for numerous discussions on the present subject.

References

5.1   J.H. Van Vleck, *The Theory of Electric and Magnetic Susceptibilities* (Oxford U. Press, Oxford 1932)
5.2   J.C. Slater: Phys. Rev. **49**, 537 and 931 (1936)
5.3   For an introduction see T. Moriya: *Spin Fluctuations in Itinerant Eelctron Magnetism*, Springer Ser. Solid-State Sci., Vol. 56 (Springer, Berlin, Heidelberg 1985)

5.4   For a discussion see D. Pines, P. Nozières: *The Theory of Quantum Liquids* (Benjamin, New York 1966)
5.5   H.F. Schaefer III: "Methods in Electronic Structure Theory", *Modern Theoretical Chemistry*, Vol. 3 (Plenum, New York 1977)
5.6   O. Gunnarsson, B.I. Lundqvist: Phys. Rev. B13, 4274 (1976)
5.7   P. Hohenberg, W. Kohn: Phys. Rev. 136, 864 (1964)
5.8   W. Kohn, L.J. Sham: Phys. Rev. 140, 1133 (1965)
5.9   J.H. Samson: Phys. Rev. B30, 1437 (1984)
5.10   O.K. Andersen, O. Jepsen: Physica B91, 313 (1977)
5.11   L. Kleinmann, K. Mednick: Phys. Rev. B24, 6880 (1981)
5.12   M.C. Gutzwiller: Phys. Rev. Lett. 10, 159 (1963)
5.13   J. Hubbard: Proc. Roy. Soc. A276, 238 (1963)
5.14   O.K. Andersen: Phys. Rev. B12, 3060 (1975)
5.15   U.K. Poulson, J. Kollar, O.K. Andersen: J. Phys. F6, L 241 (1976)
5.16   J.F. Janak, A.R. Williams: Phys. Rev. B14, 4199 (1976)
5.17   J. Callaway: In *Physics of Transition Metals*, 1980, ed. by P. Rhodes, Conf. Series Nr. 55 (Inst. of Physics, Bristol 1981)
5.18   V.L. Moruzzi, J.F. Janak, A.R. Williams: *Calculated Electronic Properties of Metals* (Pergamon, New York 1978)
5.19   D. Bagayoko, J. Callaway: Phys. Rev. B28, 5420 (1983)
5.20   O. Gunnarsson: J. Phys. F6, 587 (1976)
5.21   V. Heine, J.H. Samson, G.M.M. Nex: J. Phys. F11, 2645 (1981)
5.22   J. Staunton, B.L. Gyorffy, A.J. Pindor, G.M. Stocks, H. Winter: J. Mag. Mag. Mat. 45, 15 (1984)
5.23   G. Stollhoff, P. Thalmeier: Z. Physik B43, 13 (1981)
5.24   A.M. Oleś: Phys. Rev. B23, 271 (1981)
5.25   A.M. Oleś: Phys. Rev. B28, 327 (1983)
5.26   A.M. Oleś, G. Stollhoff: Phys. Rev. B29, 314 (1984)
5.27   A.M. Oleś, P. Fulde: Phys. Rev. B30, 4259 (1984)
5.28   G. Stollhoff, P. Fulde: Z. Physik B26, 257 (1977); B29, 231 (1978)
5.29   G. Stollhoff, P. Fulde: J. Chem. Phys. 73, 4548 (1980)
5.30   B. Kiel, G. Stollhoff, C. Weigel, P. Fulde: Z. Physik B46, 1 (1982); P. Horsch, P. Fulde: Z. Physik B36, 23 (1979)
5.31   H. Kümmel, K.H. Lührmann, J.G. Zabolitzky: Phys. Rept. 36C, 1 (1978)
5.32   W. Kutzelnigg: In [5.5]
5.33   M.C. Gutzwiller: Phys. Rev. 137, 1726 (1965)
5.34   R. Jastrow: Phys. Rev. 98, 1479 (1955)
5.35   F. Kajzar, J. Friedel: J. Physique 39, 379 (1978)
5.36   G. Treglia, F. Ducastelle, D. Spanjaard: J. Physique 41, 281 (1980)
5.37   G. Stollhoff: unpublished
5.38   J. Friedel: In *The Physics of Metals* ed. by J.H. Ziman (Cambridge U. Press, Cambridge 1969)
5.39   J. Friedel, C.M. Sayers: J. Physique 38, 697 (1977)
5.40   G. Stollhoff: In Proc. Workshop on 3d Metallic Magnetism, Grenoble (1983)
5.41   C.S. Wang, B.M. Klein, H. Krakauer: Phys. Rev. Lett. 54, 1852 (1985)
5.42   L. Hodges, R.E. Watson, H. Ehrenreich: Phys. Rev. B5, 3955 (1972)
5.43   E.P. Wohlfarth: Proc. Roy. Soc. A195, 434 (1949)
5.44   B.N. Cox, M.A.Coulthard, P. Lloyd: J. Phys. F4, 807 (1974)
5.45   A.J. Holden, V. Heine, J.H. Samson: J. Phys. F14, 1005 (1984)
5.46   J.F. Cooke, J.W. Lynn, H.L. Davis: Phys. Rev. B21, 4118 (1980)
5.47   P. Heimann, F.J. Himpsel, D.E. Eastman: Solid State Commun. 39, 219 (1981)
5.48   J. Kirschner, D. Rebenstorff, H. Ibach: Phys. Rev. Lett. 53, 698 (1984)
5.49   D.E. Eastman, F.J. Himpsel, J.A. Knapp: Phys. Rev. Lett. 44, 95 (1980)
5.50   O. Jepsen, J. Madsen, O.K. Andersen: Phys. Rev. B26, 2790 (1982)
5.51   G. Stollhoff: to be published
5.52   P.J. Brown, D. Deportes, D. Givord, K.R.A. Ziebeck: J. Appl. Phys. 53, 1973 (1982).
5.53   P.J. Brown, D. Deportes, D. Givord, K.R.A. Ziebeck: unpublished
5.54   V. Korenman, J.L. Murray, R.E. Prange: Phys. Rev. B16, 4032 (1977)
5.55   H. Capellmann: J. Phys. F4, 1966 (1974); Z. Physik B34, 29 (1979)

5.56   S. Hüfner, G.K. Wertheim, N.V. Smith, M.M. Traum: Solid State Commun. **11**, 323 (1972)
5.57   G. Treglia, F. Ducastelle, D. Spanjaard: J. Physique **43**, 341 (1982)
5.58   A. Liebsch: Phys. Rev. Lett. **43**, 1431 (1979)
5.59   A. Liebsch: Phys. Rev. B**23**, 5203 (1981)
5.60   D. Penn: Phys. Rev. Lett. **42**, 921 (1979)
5.61   L.C. Davis, L.A. Feldkamp: Solid State Commun. **34**, 141 (1980)
5.62   J. Kanamori: Progr. Theor. Phys. **30**, 235 (1963)
5.63   L.M. Roth: Phys. Rev. **186**, 426 (1969)
5.64   J.A. Hertz, D.M. Edwards: J. Phys. F**3**, 2174 (1973)
5.65   S.R. Allan, D.M. Edwards: J. Phys. F**12**, 1203 (1982)
5.66   J. Igarashi: J. Phys. Soc. Jpn. **52**, 2827 (1983); **54**, 260 (1985)
5.67   J.R. Schrieffer, W.E. Evanson, S.Q. Wang: J. Physique **32**, CI, Suppl. **2-2**, 1 (1971)
5.68   M. Cyrot: Phys. Rev.Lett. **25**, 871 (1970); J. Physique **33**, 125 (1972)
5.69   J. Hubbard: Phys. Rev. B**19**, 2626 (1979); B**20**, 4584 (1979); B**23**, 5970 (1981)
5.70   H. Hasegawa: J. Phys. Soc. Jpn. **46**, 1504 (1979); **49**, 178 (1980)
5.71   R.L. Stratonovich: Sov. Phys. Doklady **2**, 416 (1957)
5.72   J. Hubbard: Phys. Rev. Lett. **3** , 77 (1959)
5.73   T. Moriya, H. Hasegawa: J. Phys. Soc. Jpn. **48**, 1490 (1980)
5.74   K. Usami, T. Moriya: J. Mag. Mag. Mat. **20**, 171 (1980)
5.75   Y. Kakehashi: J. Phys. Soc. Jpn. **49**, 2421 (1980); **50**, 1925 (1981); **50**, 2236 (1981); **51**, 3183 (1982)
5.76   Y. Kakehashi: J. Phys. Soc. Jpn. **50**, 1505 (1981); **50**, 3620 (1981)
5.77   Y. Kakehashi: J. Phys. Soc. Jpn. **50**, 2251 (1981)
5.78   Y. Kakehashi: J. Phys. Soc. Jpn. **50**, 3177 (1981); **51**, 94 (1982); **52**, 637 (1983)
5.79   Y. Kakehashi: J. Mag. Mag. Mat. **37**, 189 (1983); **43**, 79 (1984)
5.80   E. Evangelou, H. Hasegawa, D.M. Edwards: J. Phys. F**12**, 2035 (1982)
5.81   H. Hasegawa: J. Phys. F**13**, 1915 (1983)
5.82   Y. Kakehashi, P. Fulde: Phys. Rev. B**32**, 1595 (1985)
5.83   For example, G. Morandi, E. Galleani, D'Agliano, F. Napoli, C.F. Ratto: Adv. Phys. **23**, 867 (1974)
5.84   K.K. Murata: Phys. Rev. B**12**, 282 (1975)
5.85   J.H. Samson: J. Physique **45**, 1675 (1984)
5.86   Y. Kakehashi: Phys. Rev. B**31**, 3104 (1985)
5.87   Y. Kakehashi, J.H. Samson: Phys. Rev. B**33**, 298 (1986)
5.88   Y. Kakehashi, J.H. Samson: Phys. Rev. B**34** August (1986)
5.89   C. Domb, M.F. Sykes: Proc. Roy. Soc. (London) A**240**, 214 (1957)
5.90   B.S. Shastry, D.M. Edwards, A.P. Young: J. Phys. C**14**, 1665 (1981)
5.91   T. Oguchi, K. Terakura, N. Hamada: J. Phys. F**13**, 145 (1983)
5.92   Y. Kakehashi: Phys. Rev. B**31**, 7482 (1985)
5.93   H. Hasegawa: J. Phys. C**14**, 2793 (1981)
5.94   T. Moriya, K. Usami: Solid State Commun. **34**, 95 (1980)
5.95   J.H. Van Vleck: Phys. Rev. **15**, 405 (1934)
5.96   G.D. Mahan: Phys. Rev. **163**, 612 (1967)
5.97   D. Nozières, C.T. De Dominicis: Phys. Rev. **178**, 1097 (1969)
5.98   S. Doniach, M. Sunjić: J. Phys. C**3**, 285 (1970)
5.99   Y. Kakehashi, K. Becker, P. Fulde: Phys. Rev. B**29**, 16 (1984)
5.100   H. Mori: Prog. Theor. Phys. **33**, 423 (1965)
5.101   R. Zwanzig: *Lectures in Theoretical Physics* (Interscience, New York 1961)
5.102   For example, D. Forster: *Hydrodynamic Fluctuations, Broken Symmetry and Correlation Functions* (Benjamin, London 1975)
5.103   Y. Kakehashi: Phys. Rev. B**32**, 1607 (1985)
5.104   F.R. McFeely, S.P. Kowalczyk, L. Ley, D.A. Shirley: Solid State Commun. **15**, 1051 (1974)
5.105   C.S. Fadley, D.A. Shirley: Phys. Rev. A**2**, 1109 (1970)
5.106   S. Hüfner, G.K. Wertheim: Phys. Rev. B**7**, 2333 (1973)
5.107   F. Werfel, O. Brümmer: Report Series, TURKU-FTL-R 42, (1983)
5.108   Y. Kakehashi, A. Kotani: Phys.Rev. B**29**, 4292 (1984)

6. Magnetovolume Effects

I.A. Campbell and G. Creuzet

With 11 Figures

It has been known for almost a century that there are strong magnetovolume effects in certain magnetic alloys, and this has had great technical importance with the development of invar. The composition of invar was chosen such that the magnetic thermal contraction exactly compensated the normal positive thermal expansion term around room temperature.

Numerous explanations of the invar effect have been proposed over the years; something of a consensus seems to be appearing at recent meetings held in Nagoya and devoted to invar and to magnetoeleasticity [6.1,2]. Here we will give a brief summary of physical mechanisms leading to magnetovolume effects, and we will discuss a selection of experimental results on systems where these effects have been studied. We will not discuss anisotropic magnetoelastic effects, which are also of great fundamental and technical importance.

6.1 Physical Mechanisms

6.1.1 Non-Magnetic Effects

We should recall that in the absence of magnetic terms all usual materials have a positive thermal expansion. This can be expressed by

$$\frac{1}{\ell}\frac{d\ell}{dT} = \alpha_L(T) = \frac{KC_L}{3V}\gamma_L \tag{6.1}$$

where K is the compressibility, C_L/V is the lattice specific heat per unit volume, and $\gamma_L = -d\ln\theta/d\ln V$ is a Gruneisen coefficient expressing the volume dependence of the Debye temperature.

There is also a progressive softening of the elastic moduli with increasing temperature. Both these effects are basically due to nonlinear terms in the atomic vibrations, or in other words to phonon-phonon coupling.

Magnetic Gruneisen coefficients can be defined in a similar way, to relate the magnetic specific heat and the magnetic contribution to the thermal expansion.

6.1.2 Crystal Field Effects

Suppose we have an isolated well defined local moment with an intrinsically non-spherical charge distribution; typically this would be a non-S rare earth

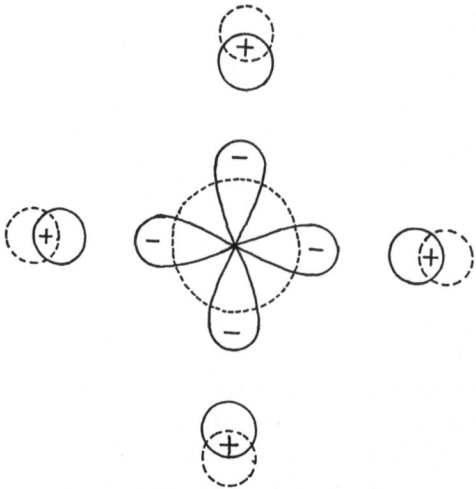

Fig. 6.1. Schematic crystal field magneto-volume effect. Full lines indicate the low temperature charge distribution and neighbour positions. Dashed lines are the high temperature equivalent

(RE) ion. At low temperature in the crystal field ground state the negative charge lobes of the ion will point towards positive charges on the neighbour sites so the neighbour atoms will tend to move in towards the RE site. At high temperature the RE ion will occupy all crystal-field states, including those which correspond to the lobes pointing away from the neighbour sites. The neighbour sites will be less attracted towards the RE sites, and so will occupy positions further from it (Fig. 6.1). This corresponds to a positive thermal expansion term whose magnitude and temperature dependence can be readily estimated if we know the crystal field parameters of the ion and the elastic moduli of the matrix. This will generally be a rather small term changing gradually with temperature and so difficult to isolate in the presence of the normal positive thermal expansion. This type of term has been discussed by *Lee* in the context of RE Al$_2$ compounds [6.3]. There will be a related volume magnetostriction term in applied fields [6.4]. In this type of system anisotropic effects are generally much larger than the isotropic magnetovolume terms.

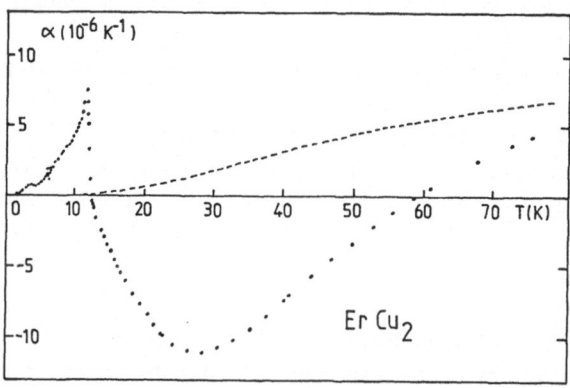

Fig. 6.2. Thermal expansion coefficient of ErCu$_2$, after *Luong* et al [6.5]. The peak near 12 K is an ordering effect; the broad minimum around 25 K is a crystal field effect. The dashed line indicates the phonon contribution

For a RE system in non-cubic symmetry the axial crystal field term can be volume dependent, which can give rise to a strong effect of this sort, anisotropic and of either sign. Striking results on RE Cu$_2$ samples (Fig. 6.2) show both the crystal field effect and the interaction effect to be discussed in Sect. 6.2 [6.5]; by comparing with specific heat data the Grüneisen parameters for the different terms can be separated out. In the case shown in Fig. 6.2, the crystal field thermal expansion is actually negative.

In a metal with band magnetism, a term of this type can, in principle, also exist if the relative occupation of d states of different symmetry changes as a function of temperature.

6.1.3 Local Moment Interactions

Suppose now that we have a system of local moments undergoing exchange interactions, with a magnetic Hamiltonian of the form

$$\mathcal{H} = \sum J_{ij}(r_{ij}) S_i \cdot S_j \ . \tag{6.2}$$

If the ground state is ferromagnetic, for instance, and near neighbour interactions J_{nn} dominate, then the system can gain magnetic energy by contracting the lattice if (dJ_{nn}/dr) happens to be negative. At temperatures above the Curie temperature T_c where the spins are disordered there will be as many anti-parallel near neighbour pairs as parallel, so the magnetovolume interaction will average to zero. For this particular sign of (dJ_{nn}/dr), there will be a magnetic term tending to expand the lattice in warming from the ordered to the disordered state. The equilibrium $r(T)$ will be determined by the minimization of a magnetic term linear in δr and an elastic term in $(\delta r)^2$. If a single parameter, e.g. $J_{nn}(r)$, dominates both the total magnetic energy and the volume derivative of the magnetic energy, then the magnetic contribution to the thermal expansion coefficient $\alpha_m \equiv (V^{-1}dV/dT)_m$ will be proportional to the total magnetic specific heat $C_m(T)$.

Generally for local moments with this Hamiltonian the magnetic contribution to the thermal expansion coefficient is [6.6]

$$\alpha_m(T) = \frac{K}{V} \left(-\frac{\partial \ln J_{ij}}{\partial \ln V} \right) \frac{\partial}{\partial T} (- \sum J_{ij} \langle S_i \cdot S_j \rangle) \ . \tag{6.3}$$

This mechanism was studied by *Argyle* et al [6.7] in elegant work on the ferromagnetic insulator EuO which has a Curie temperature at 70 K and in which the near neighbour interaction dominates. They found $V^{-1}(dV/dT)$ closely proportional to $C_m(T)$ (Fig. 6.3), and deduced a temperature independent magnetic Grüneisen coefficient $\gamma_m(\equiv -\partial \ln J/\partial \ln V) = +5.3$.

This led through simple relationships to a pressure dependence of the Curie temperature $dT_c/dP = +0.34$ K/kbar which agreed with direct measurements. The integrated magnetovolume effect $(\Delta V/V)_m(T \to \infty) = 0.24\%$.

A negative volume magnetostriction proportional to m^2 was observed above T_c, in agreement with the same model.

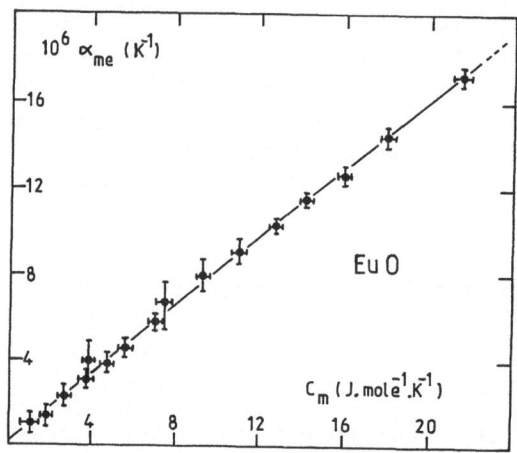

Fig. 6.3. The magnetoelastic thermal expansion α_{me} compared to the magnetic specific heat C_m (with T as an implicit parameter) in the insulating ferromagnet EuO after *Argyle* et al.[6.17]

The same mechanism also exists in metallic local moment systems or even in band magnetism, where the problem is to estimate its magnitude a priori in order to distinguish it from band terms which may also be present. To the extent that magnetic interactions between localized moments decrease with distance, this mechanism will tend to lead to an ordered state of smaller volume than the disordered state, but because interactions in metals generally oscillate with distance either sign can occur.

6.1.4 Band Mechanism

Transition metals have d bands a few eV wide. If the volume of the sample is reduced the $d - d$ overlaps will increase and the bands will tend to widen. We can look at a very simple minded picture of the d band of a transition metal in a non-magnetic or magnetic state (Fig. 6.4). When the bands are widened it is obvious that the integrated kinetic energy of the electrons is reduced in the first case and not in the second. We can deduce quite generally that for a transition metal, the magnetic state will have a larger volume than the non-magnetic state, giving magneto-volume terms $\Delta V/V \propto K\langle m_i^2\rangle$. In the Stoner model this yields directly a spontaneous magnetovolume proportional to $\overline{m}^2(T)$ which disappears above T_c [6.8].

Many model calculations are based on this physical picture. Rather than discussing the different calculations, we will follow the approach of *Friedel* and *Sayers* (FS) [6.9] because of the insight it gives concerning the problem. FS discuss the low temperature atomic volumes V and bulk moduli B (B being the inverse of the compressibility K) for the three transition element series. For the 4d and 5d elements V and B follow a very regular parabolic pattern as a function of atomic number while for the 3d series the elements in the middle of the series have larger V and weaker B than expected on a parabolic intrapolation. FS consider a d band containing five overlapping rectangular d bands, and first assume the elements non magnetic. As well as the band energy in the absence of correlations

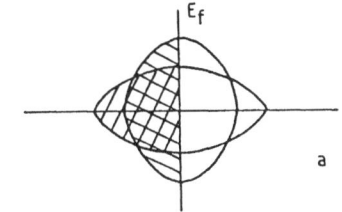

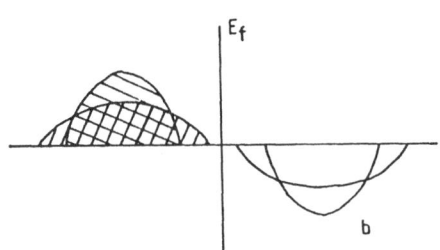

Fig. 6.4a,b. Schematic band magnetovolume mechanism. (a) is non magnetic, (b) is magnetic. The wider bands correspond in both cases to smaller volume, so more overlap. In (a) the average electron energy compared to E_F is lower when the volume is smaller

$$E_B = -z(10 - z)/20W \tag{6.4}$$

where z is the number of d electrons and W is the band width, they introduce the correlation energy

$$E_c = -45[(z/10)(1 - z/10)U]^2/W \tag{6.5}$$

which is the result of electrons being prevented from hopping onto the same atom, plus an ad hoc short distance repulsive term E_R.

The correlation term is relatively small for the 4d and 5d series, and the variation of E_B with z leads to the parabolic curves. The introduction of E_c for the 3d series leads to changes in V and B which are maximum in the middle of the series, where FS estimate changes (compared to zero correlation) of $\Delta V/V \simeq + 30\%$ and $\Delta B/B \simeq - 50\%$. Now if the elements are assumed to be magnetic, the total correlation induced effects are further enhanced by a term proportional to the moment squared (plus higher-order terms in some cases).

Thus according to FS even for non-magnetic 3d elements near the middle of the series we should expect substantial volume increases and softening compared with zero correlation values; when the elements become magnetic both effects should increase. Realistic band structures will modulate these predictions to some extent, and it is incorrect to treat an atom in an alloy (particularly a dilute alloy) on the same footing as the atom in the pure metal, but we will find that experimental data on both pure metals and alloys bear out the overall physical conclusions remarkably well.

Both the correlation term and the onset of magnetism cost the electrons kinetic energy; the system can reduce the kinetic energy cost by undergoing a lattice expansion, at the price of some elastic energy. We can note that at this level of approximation the type of magnetic order (ferromagnetic, antiferromagnetic, ...) is not important for the magnetic term; it is the change in the mean square moment $\overline{\langle m_i^2 \rangle}$ which should be the relevant parameter. This is of course, an over simplification; the type of ordering can be important, and can influence $\langle m_i^2 \rangle$.

Explicit 3d, 4d and 5d band calculations were performed by *Janak* and *Williams* [6.10]. These do not include correlation effects in the nonmagnetic state, and the magnetic polarization was introduced using an effective Stoner approach. These calculations give excellent estimates for the zero correlation (ZC) atomic volumes and moduli of the 4d and 5d series, so the non-magnetic 3d values can be used as ZC estimates with confidence. The calculated V and B for the 3d series including the Stoner polarization appear to underestimate the "3d effect". Following FS, it would appear to be important to include pure correlation as well as magnetic polarization terms. We will discuss this point in Sect. 6.2.

Phenomenologically, the increase of volume accompanying an increase of atomic moment along a ferromagnetic alloy series was discussed by *Shiga* [6.11] and by *Schlosser* [6.12] who wrote for the atomic volume $V = V_0 + k\mu_i^2$, V_0 being an estimated non-magnetic volume appropriate to that element, μ_i the atomic moment in the particular matrix considered, and k a "universal" constant for 3d elements whose value is close to $1\%/\mu_B^2$. Estimates from a wide range of alloy series [6.11,12] give excellent agreement with this simple rule.

6.1.5 Temperature Dependence

Up to now we have only been concerned with equilibrium properties at a fixed (low) temperature. What happens at finite temperature? A useful indication of what might be expected is given by the Stoner model [6.8]. If a magnetocoupling term is included in a Landau expression for the free energy of a ferromagnet, we expect once again a magnetic volume term

$$\Delta V_m / V = KC(m^2(H, T)) \tag{6.6}$$

where K is the compressibility and C is a coupling constant. In the spirit of the Stoner approach, $m \equiv \langle m \rangle$ the average overall moment which goes to zero above T_c in zero field. We would thus expect a spontaneous volume change of $KC\,m^2(0)$ between $T = T_c$ and $T = 0$, and no magnetic term at higher temperatures. The forced volume magnetostriction at each temperature should also obey the same equation, giving an independent way of estimating KC.

This approach has been applied to the invar systems and to weak itinerant ferromagnets. However, the assumption of a complete collapse of the atomic moments above T_c is generally an oversimplification. For weak ferromagnets as temperature increases order is destroyed by spin fluctuations which are not only longitudinal with q = 0 (Stoner's approximation) but both longitudinal and transverse with a range of q [6.13]. The resulting magnetovolume effect in the weak ferromagnetic limit was discussed by *Moriya* and *Usami* [6.14].

We will give a poor man's version of the Moriya-Usami reasoning for the fluctuation model. Instead of the Landau energy in the Stoner model

$$E = -mH - \frac{\alpha}{2}m^2 + \frac{\gamma}{4}m^4 \qquad (6.7)$$

with the same moment m on all atoms, we write

$$E = -\sum m_i H - \frac{\alpha}{2}\langle m_i^2 \rangle + \frac{\gamma}{4}\langle m_i^4 \rangle \qquad (6.8)$$

where m_i is the local moment which can vary in magnitude and direction. At $T = 0$, assume as in the Stoner model that all m_i are equal, $m_i = m_0$. Minimizing total E gives the well known result $m_0^2 = \alpha/\gamma$.

For the paramagnetic state $T > T_c$, there will be fluctuating moments on each site. Assume that we can represent these fluctuations by each atom having with equal probability one of 6 possible states

$$|m_i| = m_{||} \quad \text{along} \quad \pm z \quad \text{or} \quad |m_i| = m_\perp \quad \text{along} \quad \pm x \quad \text{or} \quad \pm y \ .$$

For a small applied field H along $+z$, we add an extra moment $+\delta$ along z in each case. Now, we have

$$\begin{aligned}
E = \frac{-\alpha}{2}&[(m_{||} + \delta)^2 + (m_{||} - \delta)^2 + 4(m_\perp^2 + \delta^2)] \\
&+ \frac{\gamma}{4}[(m_{||} + \delta)^4 + (m_{||} - \delta)^4 + 4(m_\perp^2 + \delta^2)^2] \\
&- 6\delta H \ .
\end{aligned} \qquad (6.9)$$

Collecting terms up to δ^2 and minimizing E with respect to δ gives

$$-6H - 6\alpha\delta + 2\gamma\delta(3m_{||}^2 + 2m_\perp^2) = 0 \quad \text{or} \qquad (6.10)$$

$$\frac{H}{\delta} = \frac{1}{\chi} = -\alpha + \frac{\gamma}{3}(3m_{||}^2 + 2m_\perp^2) \ . \qquad (6.11)$$

The fluctuating moments will increase with T, giving a susceptibility χ which is weaker at higher temperature from (6.11). As $T \to T_c$, $\chi \to \infty$, so at $T = T_c$, we have

$$\frac{\gamma}{3}(3m_{||}^2 + 2m_\perp^2) = \alpha \quad \text{or} \qquad (6.12)$$

$$\langle m_i^2 \rangle_{T_c} = \frac{3}{5}\frac{\alpha}{\gamma} = \frac{3}{5}m_0^2 \ . \qquad (6.13)$$

This is precisely the Moriya-Usami result. Now we can expect quite generally that there will be a magnetovolume term equal to $KC\langle m_i^2 \rangle_T$, so in this model we expect a drop in volume from $T = 0$ to $T = T_c$ of $\frac{2}{5}KCm_0^2$, instead of KCm_0^2 which is the Stoner model value. As T increases further above T_c, in the Moriya-Usami view the fluctuations increase, increasing $\langle m_i^2 \rangle$ and hence the magnetic volume term (Fig. 6.5). This naive description of fluctuation effects is physically almost equivalent [6.13] to a detailed model of weak ferromag-

213

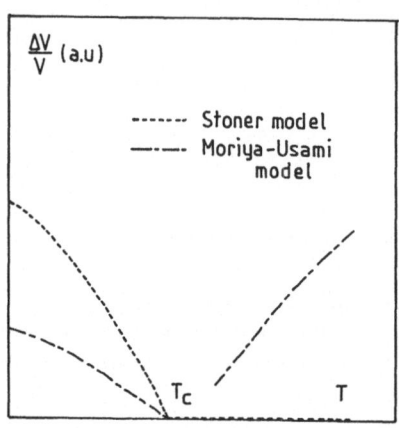

Fig. 6.5. Schematic magnetovolume effects for a weak ferromagnet according to the Stoner [6.8] and the Moriya-Usami [6.14] models

nets which provides a remarkable quantitative description of the properties of Ni$_3$Ga [6.16].

Finally, even metallic systems may have local moments which are stable when the temperature is raised. In this case we should invoke the same Heisenberg approach as for insulators, where any magnetovolume term is due to the volume dependence of the magnetic interactions.

We should also note that the relation $\Delta V/V = KC\langle m_i^2 \rangle$ should be used with precaution, as it can lead, for instance, to an apparent paradox for completely local moment systems such as certain dilute alloys. If we have atoms with spin S ordered ferromagnetically at low T and paramagnetic at high T, there is a temptation to write

$$\langle m^2 \rangle_{T>T_c} = 4S(S+1) \quad \text{and} \tag{6.14}$$

$$\langle m^2 \rangle_{T=0} = 4S^2 \ . \tag{6.15}$$

The low temperature expression is incorrect, as it leaves out terms such as $\langle S_z | S_x^2 | S_z \rangle$ which are non zero. In fact $\langle m^2 \rangle \equiv 4S(S+1)$ for a true local moment at all temperatures whatever the ordering. It is much less clear how the quantum correction should come in for itinerant systems [6.13].

A number of models have been presented which attempt to bridge the gap between the weak ferromagnetic solution and the Heisenberg model in order to explain the finite temperature behaviour of band ferromagnets [6.17]. One of the aims of these models is to relate the behaviour of particular systems to their band structure. The general conclusion is that large thermal magnetovolume effects with V dropping as T increases are associated with a partial collapse of unstable local moments but that the volume dependence of the interaction terms cannot be neglected totally. Obviously the same models (Chapt. 2 and 4), should provide a coherent explanation of finite temperature phenomena of all kinds and not just the magnetovolume. For the moment these models do not seem able to give reliable a priori predictions of just which alloys or compounds can be expected to show strong local moment collapse effects.

214

By extension of the earlier discussion, one would expect that sytems which show large-band magnetovolume changes with temperature should also have anomalous temperature dependence of their elastic moduli; this is frequently found to be the case experimentally. *Kim* [6.18] has pointed out that the inverse effect should also be important – raising the temperature of the sample will influence magnetism directly but also through the phonons acting via the magnetoelastic terms.

The strong negative thermal expansion term due to partial collapse of the local moments is generally accompanied by other effects which are characteristic of "soft" magnetic moments. These are a strong high-field susceptibility, a strong forced-volume magnetostriction, and strong variations of the spontaneous magnetization with applied pressure [6.18,19]. The latter two are thermodynamically realted through the Maxwell relation

$$\left(\frac{\partial V}{\partial H}\right)_P = -\left(\frac{\partial m}{\partial P}\right)_H \tag{6.16}$$

and if we write $(\Delta V/V)_m = Cm^2$, then

$$\frac{1}{V}\frac{\partial V}{\partial H} = 2Cm\chi \tag{6.17}$$

so it is natural that a strong susceptibility should be accompanied by the other two effects. We can note that the local moments may only become "soft" as the temperature is raised, so a sample which shows a strong drop in volume in the region of T_c does not necessarily show a strong high field susceptibility at low temperatures.

6.2 Experimental Results

6.2.1 Properties of the d Series at Fixed Temperature

We reproduce (Fig. 6.6a,b), the literature values [6.20] for the atomic volumes and bulk moduli of the 3d, 4d and 5d series. It is clear that the elements in the middle of the 3d series show deviations from the parabolic curves characteristic of the other two series. The deviations are particularly striking for the bulk moduli B, where there are strong deviations for Cr, Mn, Fe, Co and Ni.

Figure 6.7 shows the atomic volumes for the elements, zero correlation estimates of the atomic volumes from *Janak* and *Williams* [6.10] which reproduce the parabolic behaviour, and effective atomic volumes for the elements when they are in Au, Ag or Cu lattices. These we define by extrapolating alloy volumes as a function of concentration to 100 % concentration of the transition element. Composition curves of this type for Mn alloys are given in Fig. 6.8. The point of this is that Cr, Mn and Fe have strongly developed local moments

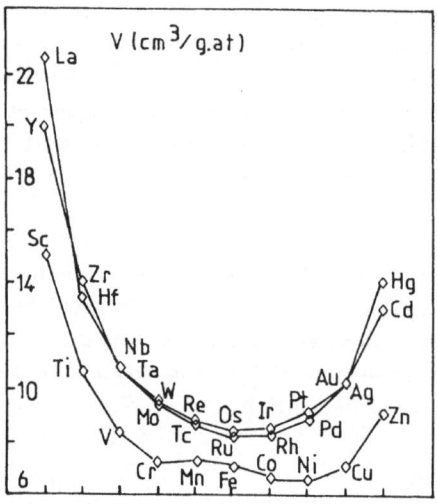

Fig. 6.6a. Atomic volumes for the 3d, 4d and 5d transition series [from 6.20]

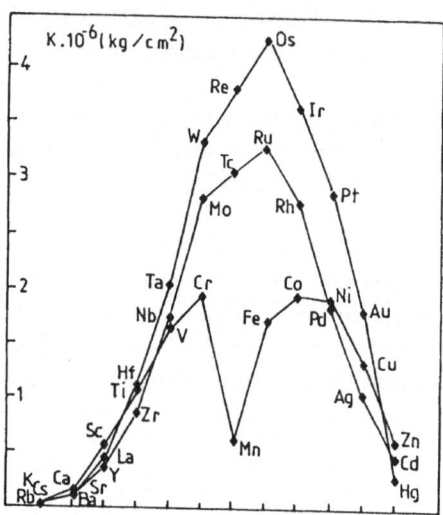

Fig. 6.6b. Bulk modulus for the 3d, 4d and 5d transition series [from 6.20]

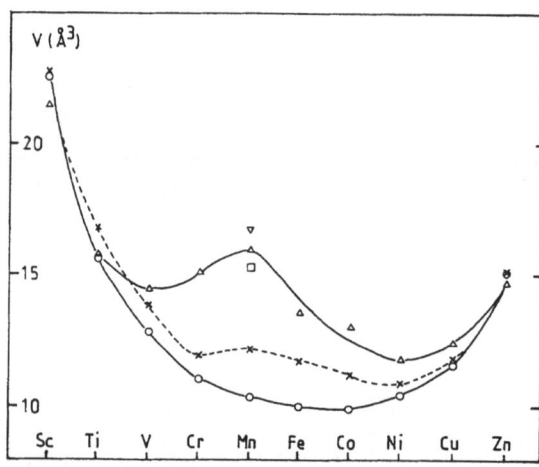

Fig. 6.7. Effective atomic volumes for the 3d elements. (○) zero correlation estimate [6.10], (×) atomic volumes in the pure metals [6.20], (△▽□) respectively estimates of effective atomic volumes of the 3d elements in Au, Ag and Cu hosts (see text)

in noble metal hosts, so the "effective volumes" defined in this way provide us with an estimate of the maximum atomic volume the element can have when it is fully magnetic. *Weiss* [6.21] and *Schlosser* [6.12] estimated the volume of "strongly magnetic γMn" in just this way.

The plot in Fig. 6.7 is instructive. First, it underlines the importance of the correlation plus magnetic volume terms for the 3d elements. Mn in Au, where it has a local moment of almost $5\mu_B$ has an effective atomic volume about 50% greater than the zero correlation Mn value. Secondly, the magnetic term does not depend crucially on the local moments being in an ordered state: all the noble metal based alloy values actually correspond to room temperature where, for the concentrations we have extrapolated from, the alloys are paramagnetic. Even at low temperature most of these alloys are spin glasses. So an assembly of disordered or fluctuating local moments gives us a large spontaneous magneto-volume effect. Finally, we can note that part of the extra volume is not strictly magnetic but is a correlation effect. βMn is non-magnetic at $T = 0$ [6.22], and it has almost the same atomic volume as the weakly magnetic αMn. The extra volume between the elementary α or β Mn and the zero correlation Mn can be ascribed essentially to correlation effects. For Cr the pure metal moment is also small, so again the extra atomic volume is principally due to correlation effects.

From the results displayed in the figure, we can give as rough estimates:
a) An extra volume due to correlations for non-magnetic state Mn of about + 15 % of the zero correlation value; rather lower terms for the neighbouring elements.
b) An extra magnetic volume term of about +1 %/μ_B^2 (in agreement with the Shiga-Schlosser value [6.11,12]).

Turning to bulk modulus values, results on the CuMn series [6.23] provide spectacular evidence for the importance of specific 3d terms. The bulk modulus B drops towards zero with increasing Mn concentration as the average local moment per atom in the alloy increases; near 70 % Mn the shear mode characterized by $(C_{11} - C_{12})/2$ softens suddenly and antiferromagnetic order with a tetragonal distortion sets in. The effective moment per Mn atom drops off when the Mn concentration goes above 50 %, as can be seen from the curvature of the V(c) plot, and from the decrease in the spin glass temperature $T_g(c)$ [6.24] (Fig. 6.8). By pure fcc γMn the local Mn moment has become quite small. The real crystal structures of pure Mn are complex, but it seems clear from these

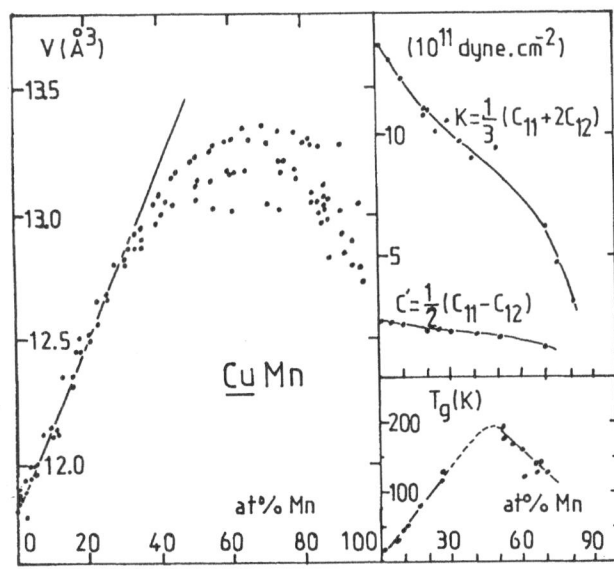

Fig. 6.8a–c. Results for CuMn alloys. (a) average atomic volume of the alloy [6.12]. (b) elastic moduli [6.23]. (c) spin glass cusp temperatures [6.24]

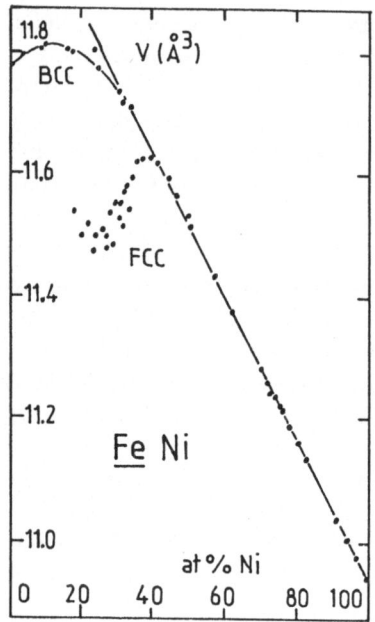

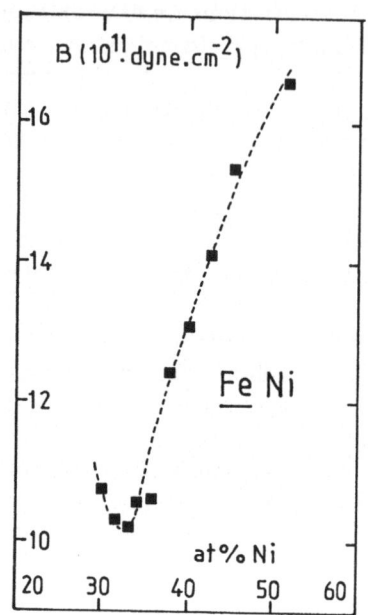

Fig. 6.9a. Average atomic volume for the FeNi alloy series [6.20]

Fig. 6.9b. Bulk modulus at low temperature for the FeNi alloy series [6.26]

results that one reason for which high moment pure Mn cannot form is because it is mechanically unstable.

Other alloy series containing Mn have been studied and show qualitatively similar results. In particular work on NiMnC alloys near the critical concentration show dramatic phonon softenings and non-linear phonon dispersion curves, again arising from very strong magnetoelastic interactions [6.25].

The low temperature properties of the celebrated NiFe series (which includes the invar alloys at about 35 % Ni) are analogous to those seen in CuMn. These alloys are ferromagnetic; up to about 50 % Fe, the average moment per atom increases linearly, the atomic volume increases and the bulk modulus drops (Fig. 6.9). Beyond 50 % Fe the average moment begins to decrease, the atomic volume decreases while the bulk modulus continues to drop [6.12,26]. There is some magnetic disorder and finally a phase change to a bcc structure at 30 % Ni. Again it appears that the fcc high Fe moment state tends to be mechanically unstable – or at least this is part of the story.

6.2.2 Temperature Dependence – Weakly Magnetic Systems

There are a number of metallic alloys or compounds which are strongly paramagnetic or weakly ferromagnetic or antiferromagnetic and whose thermal expansion and volume magnetostriction have been carefully studied. As we have seen, magnetovolume effects are proportional to the square of the local moment, and so are intrinsically small in these systems. This is not much of a hinderance for magnetostriction measurements, as accurate techniques exist, but poses problems for thermal expansion where it is essential to separate out the magnetic term from the normal phonon term which increases rapidly as the temperature is raised. As a result, thermal expansion results on weak ferromagnets do not provide as clear cut tests of the model predictions as one would hope. The behaviour expected on the simple Stoner model and on the spin fluctuation Moriya-Usami model are shown in Fig. 6.5. In both cases, the parameter C can be obtained accurately from high field magnetostriction and magnetisation measurements at $T \ll T_c$. Unfortunately, the estimate of the magnetic contribution to the volume change between $T = 0$ and $T = T_c$ depends critically on the choice of the background thermal expansion term. As a result certain experiments can be interpreted equally plausibly on either model. We will give a selection of results which have been obtained.

a) $ZrZn_2$

This weak itinerant ferromagnet has been carefully studied by *Ogawa* [6.27], who compared magnetostriction and thermal expansion. There is a positive magnetostriction proportional to $\langle m \rangle^2$ and a thermal expansion which becomes negative at low temperatures because of the magneto-volume effect. Unfortunately the lattice thermal expansion is so big compared with the magnetic term that it is difficult to decide if the Stoner or spin fluctuation models give the best agreement with experiment.

b) $MnSi$

This compound has an ordering temperature of 30 K and a moment per Mn atom of $0.4\,\mu_B$. It has a conical magnetic structure which becomes ferromagnetic in an applied field of 6 kG. The ordering temperature depends strongly on pressure, with $d\ln T_c / d\ln V = 53$ [6.28]. There is a spontaneous volume change on ordering giving a total low temperature magneto-volume $\Delta V/V \simeq 3 \times 10^{-4}$. Comparison with magnetostriction indicated agreement with the spin fluctuation model [6.29].

c) Ni_3Al

The magnetic properties of the series $Ni_{3+x}Al_{1-x}$ were intensively studied [6.30] and compared with Stoner model predictions. From a critical comparison of these results together with those on $ZrZn_2$ and MnSi [6.31], it is concluded that in Ni_3Al and $ZrZn_2$ Stoner excitations are dominant while in MnSi the spin fluctuation picture is more appropriate.

d) $TiBe_{2-x}Cu_x$

$TiBe_2$ is an enhanced paramagnet; doping with Cu drives it ferromagnetic. From a detailed study of the magnetostriction, magnetisation and thermal expansion of this alloy series it was concluded that agreement was good with the spin fluctuation model, in particular as concerns the extra positive thermal expansion above T_c for the ferromagnetic samples [6.32].

219

e) Sc₃In

e) Sc_3In

This is a very weak ferromagnetic with $T_c \sim 6\,K$ and $\mu_B \sim 0.05\,\mu_B/$at. Sc. In contrast to the other weak ferromagnets, there is an extra *positive* thermal expansion below T_c and the magnetostriction is negative [6.33]. There seems to be no obvious explanation for this anomaly.

6.2.3 Thermal Variation – Systems with Strong Moments

We group in this subsection the results on alloys and compounds where the local moments at low temperatures are relatively strong – of the order of $1\,\mu_B$ or more. From the preceding discussion we can expect two limiting forms of behaviour:

1. the local moments can remain well defined at all temperatures, so the magnetovolume effect will be due to the $J(r)$ term, will be relatively small, and will generally correspond to an extra negative magnetovolume term at $T < T_c$.
2. Alternatively, the local moments will partially collapse as the temperature is raised to T_c giving a strong positive magnetovolume effect at low T of the same qualitative form as for the fluctuation model of the weak ferromagnets (Fig. 6.5) but of much greater amplitude as the m_i^2 values involved are much higher.

a) Fe, Ni and FeNi Alloys

The experimental thermal expansion results on the ferromagnetic elements Fe and Ni show small magnetovolume effects [6.34]. For Ni there is a drop of volume on ordering and for Fe an increase of volume as temperature is lowered through T_c and then an apparent decrease. The integrated magnetic volume change is estimated as about $-0.3\,\%$ for Ni and $+0.5\,\%$ for Fe. These results have been discussed many times [6.35] and the overall conclusion is that if there is any reduction of the local moments at high temperature it is very small. A more detailed picture of the magnetic moment fluctuations in Fe above T_c has been obtained by various neutron diffraction techniques as discussed by K. Ziebeck, Chapter 7, Volume II. The results appear to confirm an effective $\langle m_i^2 \rangle$ in Fe above T_c which is close to the low temperature value.

For the NiFe alloys near the fcc phase boundary we find the invar alloys. These clearly show a strong magnetovolume contraction when the temperature is increased, which compensates the normal positive expansion over a considerable range of temperature. The negative term already appears at low temperatures with $\Delta V_m(T) \propto T^2$ as expected by the Stoner model [6.8,36]. Here the magnetic term dominates as the low temperature phonon term is small. The general form of the magnetic term as a function of temperature is shown in Fig. 6.10.

These alloys show a strong positive forced volume magnetostriction which continues to above T_c, in contrast to pure Ni and Fe which only have weak forced magnetostriction with small peaks at T_c [6.37]. Mössbauer experiments

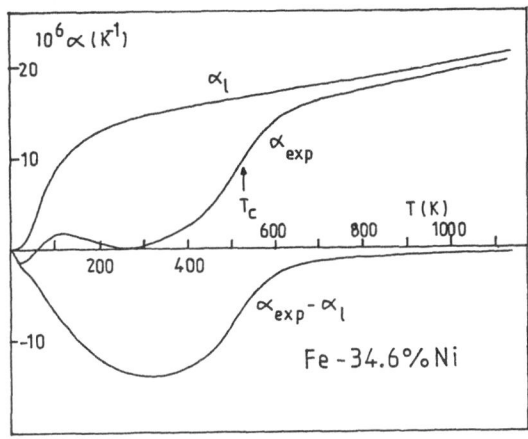

Fig. 6.10. Experimental thermal expansion coefficient for invar Fe 34.6 at. % Ni, α_{exp}; estimated lattice term α_L; and the magnetic term $\alpha_{exp} - \alpha_L$, as quoted by *Chikazumi* [6.36]

can be used to distinguish between different models for invar behaviour [6.38]. In $Fe_{65}Ni_{35}$ they are further evidence in a favour of a rather homogeneous partial collapse of local moments, with local environment effects modulating but not dominating the invar phenomenon.

The bulk modulus B and the shearing mode modulus $(C_{11}-C_{12})/2$ also show anomalous temperature variations which are clearly related to the magnetic collapse [6.26]. Near the invar concentration there is the elinvar (elastic invar) concentration where the modulus B is temperature independent around room temperature because of a compensation between phonon and magnetic effects. In pure Fe, Co or Ni, the Young's modulus (which is closely related to B) when measured in applied field only shows a small extra positive magnetic term below T_c [6.39], in contrast to the NiFe alloys.

b) Fe_3Pt and FePd Alloys

Here also there are strong invar-type magnetovolume effects, again present over the whole range of temperatures from low temperatures to above T_c [6.36,40,41]. Fe_3Pt poses an interesting test case because it shows invar behaviour even when chemically well ordered [6.40], demonstrating that atomic disorder does not play an essential role in driving the magnetovolume instability. Fe_3Pt in either the ordered or disordered state shows giant temperature anomalies in the bulk moduli [6.42]. Volume magnetostriction in Fe_3Pt [6.43] also shows a strong characteristic positive effect.

FePd alloy invar effects have also been studied at high and low temperatures [6.36,41,44].

c) Cr

The antiferromagnetic element Cr shows a sharp anomaly in thermal expansion at T_N, which may correspond to a reduction in the Cr moment or to an interaction term [6.45]. By comparison with the alloy $Cr_{95}V_5$ which does not order magnetically one can estimate a total spontaneous volume change on ordering in Cr of $\Delta V/V = +1.6 \times 10^{-3}$. There appears to be a high temperature fluctuation effect as the thermal expansion of Cr or of the CrV alloys continues to increase up to at least 700 K. The thermal expansion of Cr is negative at low temperatures [6.46].

d) Cr Based Alloys

CrFe alloys pass through antiferromagnetic, spin glass and ferromagnetic phases as a function of the Fe concentration between 0 % and 30 % Fe. Over all this concentration range there are strong positive magnetovolume terms at low temperature [6.47], indicating that the Fe local moments (which are much larger than the Cr moments) are unstable when the temperature is raised.

CrCo also show invar-like thermal expansion effects [6.48].

6.2.4 Compounds

A number of the intermetallic compounds of Fe, Co or Mn with RE or transition metals have large magnetovolume effects.

The magnetism of the $RECo_2$ compounds has been thoroughly studied [6.49]. The compound YCo_2 is a strong paramagnet, while in the compounds with magnetic rare earths a moment is induced on the Co by the RE sublattice. This Co moment is antiparallel to the RE moment for the heavy rare-earths. As one would naively expect, as the temperature is lowered below T_c and the Co moments are induced a strong positive magnetovolume term appears proportional to $m^2(Co)$ [6.50,51]. The forced volume magnetostriction at low temperatures is *negative*, as the external field counteracts the effective internal field inducing the Co moments and so reduces the Co sublattice magnetisation. At temperatures near T_c the direct effect of the external field dominates and the volume magnetostriction becomes positive. This is perhaps one of the clearest examples of band magnetovolume effects (Fig. 6.11b).

The alloy series $Gd_{1-x}Y_xCo_2$ magnetovolume effects are observed which again illustrates the influence of induced Co moments on the volume [6.52].

Fe compounds such as Y_2Fe_{17} [6.53] and $ZrFe_2$ [6.43] show strong invar type magnetovolume effects. In some cases these have been discussed in terms of volume dependent interaction terms, but from the preceding discussion a band effect would seem more likely (Fig. 6.11a).

The compound YMn_2 has a Laves structure and orders antiferromagnetically below 110 K with 2.7 μ_B on the Mn site. The transition is first order and is accompanied by a giant magnetovolume effect – about 5 % increase in volume [6.51] (Fig. 6.11c). This clearly points to a transition from a paramagnetic state in which the Mn has essentially no local moment to a low temperature ordered phase with a strong Mn local moment. Recent results on $Y(Mn_{1-x}Al_x)_2$ with x up to 0.1 show that in the Al substituted samples the Mn keeps its strong local moment even above the ordering temperature [6.54]. The experiments also demonstrate that in pure YMn_2 there is a positive fluctuation moment magnetovolume effect above T_N of the type indicated in Fig. 6.5.

The ferromagnetic compound Au_4V shows a positive spontaneous magnetovolume term at low temperature [6.55].

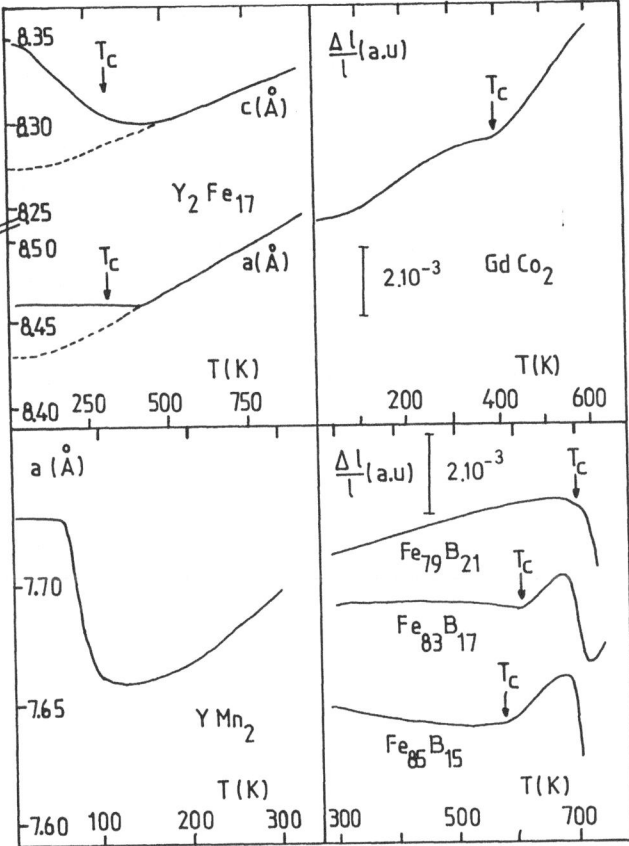

Fig. 6.11a–d. Thermal expansion for various systems showing invar behaviour. **(a)** Ferromagnetic compound Y_2Fe_{17} along a and c axes; dashed line indicates the estimated lattice term [6.53]. **(b)** Ferromagnetic compound $GdCo_2$ [6.50,51]. **(c)** Antiferromagnetic Laves structure compound YMn_2 [6.51]. **(d)** The amorphous alloys $Fe_{79}B_{21}$, $Fe_{83}B_{17}$ and $Fe_{85}B_{15}$ [6.57]

6.2.5 Heusler Alloys

In these cubic Mn compounds which are frequently ferromagnets the Mn atoms have large well localized local moments. In those cases which have been studied it appears that there are weak magnetovolume effects with a small drop in volume on ordering due to the volume dependence of interaction terms [6.56].

6.2.6 Amorphous Ferromagnets

The thermal expansion and volume magnetostriction have been measured on a number of Fe based amorphous ferromagnets. The amorphous alloys of FeB (Fig. 6.11d), FeZr and similar ferromagnetic alloys show a total spontaneous magnetovolume increase of the order of 1.5 % in the ordered state [6.57]. The paramagnetic alloy $Ni_{77}Si_{10}B_{13}$ gives a standard thermal expansion curve as a function of temperature.

Magnetostriction effects in amorphous alloys have been reviewed in [6.58].

Rather surprisingly, the elastic moduli of these alloys appear to show no strong temperature effects related to the magnetovolume effects as long as a magnetic field is applied to eliminate the "ΔE effect" related to domain wall movement [6.59]. The general physical arguments relating the bulk modulus effects to the magnetovolume effects would not seem to be closely dependent on the structure, so it is not clear why crystalline samples show the effects on the elastic coefficients while amorphous ones do not. Amorphous FeZr alloys do show softening of Young's modulus as the temperature decreases [6.60].

Magnetovolume effects in amorphous alloys have been reviewed by *Fukami-chi* [6.61].

6.2.7 Spin Glasses

In the dilute alloys such as CuMn or AuFe which have well defined local moments and which order as spin glasses at low temperature, weak magnetovolume effects have been observed from thermal expansion, magnetostriction or pressure measurements [6.62]. These are generally interpreted in terms of the volume dependence of the exchange interactions coupling together the local spins. There do not seem to be any moment collapse effects.

6.2.8 Mn Alloys

In the concentrated fcc MnCu alloys which order antiferromagnetically there is *drop* of volume at a first-order T_N [6.63], accompanied by a tetragonal distortion. It appears that the Mn is "less magnetic" in the ordered state than in the paramagnetic state.

The thermal expansion of Mn 15 % Cu is positive at low temperature [6.46].

The antiferromagnetic Mn 15 % Ni shows invar behaviour [6.64].

6.2.9 Frequency Dependence of Elastic Moduli

It is possible to compare elastic moduli values obtained from acoustic measurements corresponding to relatively low frequencies with those obtained from inelastic neutron scattering which are in a much higher frequency range. In normal metals the values are in good agreement or, in other words, the phonon dispersion remains linear over a wide range of frequencies. In certain invar-type alloys, however, there are distinct differences between the values and the temperature variations of the moduli measured at 10^7 Hz and at 10^{12} Hz [6.25]. For instance, in Fe_3Pt the neutrons do not show a bulk modulus softening as the temperature is lowered below T_c [6.65], while there is a dramatic effect of this type in the acoustic measurements [6.42]. This seems to indicate that the atoms with "soft" local moments need a certain time to change volume in response to external constraint.

6.2.10 Intermediate Valence Systems

For completeness, we will also briefly discuss intermediate valence compounds containing the rare earths Ce or Yb. In certain compounds these elements are intermediate between an effective valence of 3^+ or 4^+ (Ce), or 2^+ or 3^+ (Yb). The different valence states, corresponding to one localized f electron more or less on the rare earth site, have quite different volumes. For instance, *Harris* et al. [6.66] for $CePd_3$ estimate the lattice parameter in the magnetic Ce 3^+ limit would be 4.151 Å and in the non magnetic Ce 4^+ limit would be 4.099 Å i.e. a reduction in volume per formula unit by 3.8 %. They correlate the lattice spacing with the susceptiblity in the $CeRh_{3-x}Pd_x$ alloy series. For $CeSn_3$, *Harris* and *Raynor* [6.67] show there is an extra positive lattice expansion term as the valence drops towards 3^+ as the temperature increases.

Similar estimates for $YbAl_2$ by *Iandelli* and *Palenzona* [6.68] give a non-magnetic Yb 2^+ volume per formula unit which is 4.6 % greater than that for the magnetic Yb 3^+. There is a gradual drop of volume with increasing temperature which corresponds to an evolution from 2^+ at low temperature to 2.4^+ at 550° C. The compound CeNi has very strong anomalous thermal expansion terms which differ in sign according to the axes [6.69]. *Ribault* et al. [6.70] showed a different effect in the heavy fermion system $CeAl_3$ where the thermal expansion became negative below 1 K as coherent effects set in. The strong absolute value of the thermal expansion can be related thermodynamically to a strong dependence of the giant electronic specific heat on volume. The heavy fermion system UPt_3 has a strong positive thermal expansion at low temperatures [6.71]. Finally, the element Pu changes volume by up to 20 % as it goes through its various phase changes [6.72]. This is again an intermediate valence effect.

Acknowledgements. We warmly thank J. Friedel, V. Heine, P. Wohlfarth and P.E. Brommer for their helpful remarks on the manuscript.

References

6.1   J. Mag. Mag. Mat. **10** (1979)
6.2   Physica **119B** (1983)
6.3   EW Lee: J. Mag. Mag. Mat. **10**, 274 (1979)
6.4   G. Creuzet, I.A. Campbell: Phys. Rev. B**23**, 3375 (1981)
6.5   N.H. Luong, J.J.M. Franse, T.D. Hien: J. Mag. Mag. Mat. **50**, 153 (1985);
       J.J.M. Franse, N.H. Luong, T.D. Hien: J. Mag. Mag. Mat. **52**, 202 (1985)
6.6   E. Callen, H.B.Callen: Phys. Rev. **139A**, 455 (1965)
6.7   B.E. Argyle, N. Miyata, T.D. Schultz: Phys. Rev. **160**, 413 (1967);
       B.E. Argyle, N. Miyata, Phys. Rev. **171**, 555 (1968)
6.8   D.M. Edwards, E.P. Wohlfarth: Proc. Roy. Soc. A**303**, 127 (1968);
       E.P. Wohlfarth: Physica **91B**, 305 (1977)
6.9   J. Friedel, C.M. Sayers: J. Physique Lett. **38**, L 263 (1977)
6.10   J.F. Janak, A.R. Williams: Phys. Rev. B**14**, 4199 (1976)
6.11   M. Shiga: Sol. State Commun. **10**, 1233 (1972); J. Phys. Soc. Jap. **50**, 2573 (1981)
6.12   W.F. Schlosser: Phys. Stat. Sol. A**17**, 199 (1973); J. Mag. Mag. Mat. **1**, 102, 106, 293
       (1976)

6.13   T. Moriya: J. Mag. Mag. Mat. **45**, 79 (1984)

6.14   T. Moriya, K. Usami: Sol. State Commun. **34**, 95 (1980)

6.15   We are grateful to Dr. Lonzarich for discussions on this point

6.16   G.G. Lonzarich: J. Mag. Mag. Mat. **45**, 43 (1984)

6.17   D.G. Pettifor: J. Mag. Mag. Mat. **15–18**, 847 (1980);
Y. Kakehashi: J. Phys. Soc. Jap. **49**, 2421 (1980); **51**, 3183 (1982);
J. Kanemori, Y. Terakura: J. Mag. Mag. Mat. **10**, 217 (1979);
H. Hasegawa: J. Phys. C**14**, 2793 (1981);
V. Korenman, B. Wyman: Phys. Rev. B**24**, 5413 (1981);
M. Shimizu, J. Inoue, Y. Ohta, K. Niwa: Physica **119B**, 3 (1983);
A.J. Holden, V. Heine, J.H. Samson: J. Phys. F**14**, 1005 (1984);
A.R. Williams, V.L. Moruzzi, C.D. Gelatt. Jr., J. Kübler: J. Mag. Mag. Mat. **31–34**, 88 (1983)

6.18   D.J. Kim: Physica **119B**, 30 (1983)

6.19   J.S. Kouvel, R.H. Wilson: J. App. Phys. **32**, 435 (1961)

6.20   K.A. Gschneider: Sol. St. Phys. **16**, 275 (1964);
W.B. Pearson: *Handbook of Lattice Spacings ans Structures of Metals and Alloys* (Pergamon, London 1958 and 1967)

6.21   R.J. Weiss: Phil. Mag. **26**, 261 (1972)

6.22   T. Kohara, K. Asayama: J. Phys. Soc. Jap. **37**, 401 (1974)

6.23   Y. Tsunoda, N. Oishi, N. Kunitomi: Physica **119B**, 51 (1983); J. Phys. Soc. Jap. **53**, 359 (1984)

6.24   N. Cowlan, A.M. Shamah: J. Phys. F**11**, 27 (1981)

6.25   R.D. Lowde, R.T. Harley, G.A. Saunders, M. Sato, R. Scherm, C. Underhill: Proc. Roy. Soc. A**374**, 87 (1981)

6.26   G. Hausch, H. Warlimont: Z. Met. **63**, 547 (1972)

6.27   S. Ogawa: J. Phys. Soc. Jap. **22**, 1514 (1967); **27**, 789 (1969); Physica **119B**, 68 (1983)

6.28   D. Bloch, J. Voiron, V. Jaccarino, J.H. Wernick: Phys. Lett. **51A**, 259 (1975)

6.29   M. Matsunaga, Y. Ishikawa, T. Nakajima: J. Phys. Soc. Jap. **51**, 1153 (1982)

6.30   N. Buis, P.E. Brommer, P. Disveld, M.S. Schalkwijk, J.J.M. Franse: J. Mag. Mag. Mat. **15–18**, 291 (1980);
N. Buis, J.J.M. Franse, P.E. Brommer: Physica **106B**, 1 (1981)

6.31   P.E.Brommer, J.J.M. Franse: J. Mag. Mag. Mat. **45**, 129 (1984)

6.32   G. Creuzet, I.A. Campbell, J.L. Smith: J. Physique Lett. **44**, L 547 (1983)

6.33   E. Fawcett, P.P.M. Meincke: J. Physique **32**, C1–629 (1971)

6.34   C. Williams: Phys. Rev. **46**, 1011 (1934);
N. Ridley, H. Stuart: J. Appl. Phys. **2**, 1291 (1968);
W.J. Carr: J. Mag. Mag. Mat. **10**, 197 (1979)

6.35   See R. Joynt, V. Heine: J. Mag. Mag. Mat. **45**, 74 (1984)

6.36   F. Ono, T. Kittaka, H. Maeta: Physica **119B**, 78 (1983);
S. Chikazumi: J. Mag. Mag. Mat. **10**, 113 (1979);
Y. Tanji: J. Phys. Soc. Jap. **31**, 1366 (1971)

6.37   S. Ishio, M. Takahashi: J. Mag. Mag. Mat. **50**, 271 (1985);
O. Yamada, E. du Tremolet de Lacheisserie: J. Phys. Soc. Jap. **53**, 729 (1984)

6.38   M. Shiga, Y. Nakamura: J. Mag. Mag. Mat. **40**, 319 (1984)

6.39   J.L. Lytton: J. Appl. Phys. **35**, 2397 (1964)

6.40   K. Sumiyama, M. Shiga, Y. Nakamura: J. Phys. Soc. Jap. **40**, 996 (1976)

6.41   M. Matsui, T. Shimizu, K. Adachi: Physica **119B**, 84 (1983)

6.42   G. Hausch: J. Phys. Soc. Jap. **37**, 819 (1974)

6.43   M. Shiga, Y. Muraoka, Y. Nakamura: J. Mag. Mag. Mat. **10**, 280 (1979)

6.44   Y. Tino, Y. Iguchi: J. Mag. Mag. Mat. **31–34**, 117 (1983)

6.45   R.B. Roberts, G.K. White, E. Fawcett: Physica **119B**, 63 (1983)

6.46   G.K. White: Proc. Phys. Soc. **86**, 159 (1965)

6.47   V.E. Rode, S.A. Finkelberg, A.I. Lyalin: J. Mag. Mag. Mat. **31–34**, 293 (1983)

6.48   H.L. Alberts, J.A.J. Lourens: J. Phys. F**13**, 873 (1983)

6.49   K.N.R. Taylor: Adv. Phys. **20**. 551 (1971);
D.Bloch, R. Lemaire: Phys. Rev. B**2**, 2648 (1970);
J. Voiron, A. Berton, J. Chaussy: Phys. lett. **50A**, 17 (1974)

6.50   M. Shiga, K. Minakata, T. Tsuchida, Y. Nakamura: J. Phys. Soc. Jap. **42**, 814 (1977);
D. Gignoux, D. Givord, F. Givord, R. Lemaire: J. Mag. Mag. Mat. **10**, 288 (1979);
Y. Muraoka, M. Shiga, Y. Nakamura: J. Mag. Mag. Mat. **31–34**, 121 (1983)

6.51   Y. Nakamura: J. Mag. Mag. Mat. **31–34**, 829 (1983)

6.52   Y. Muraoka, H. Okuda, M. Shiga, Y. Nakamura: J. Phys. Soc. Jap. **53**, 331, 1453 (1984)

6.53   D. Givord, R. Lemaire: IEEE Trans MAG-**10**, 109 (1974);
M. Brouha, K.H.J. Buschow, A.R. Miedema: IEEE Trans MAG-**10**, 182 (1974);
K. Mori, A.E. Clark, O.D. McMasters: J. Mag. Mag. Mat. **31–34**, 855 (1983)

6.54   M. Shiga, H. Wada, K. Yoshimura, Y. Nakamura: J. Mag. Mag. Mat. **54–57**, 1073 (1986)

6.55   N. Kasai, S. Ogawa: J. Phys. Soc. Jap. **30**, 736 (1971)

6.56   T. Kaneko, K. Watanabe, K. Shirakawa, H. Masumoto: J. Mag. Mag. Mat. **31–34**, 79 (1983)

6.57   K. Fukamichi, M. Kikuchi, S. Arakawa, T. Masumoto: Sol. State Commun. **23**, 955 (1977);
K. Fukamichi, H. Hiroyoshi, M. Kikuchi, T. Masumoto: J. Mag. Mag. Mat. **10**, 294 (1979);
K. Shirakawa, S. Ohnuma, M. Nose, T. Masumoto: IEEE Trans. MAG-**16**, 910 (1980);
S. Ishio, M. Takahashi, Z. Xianyu, Y. Ishakawa: J. Mag. Mag. Mat **31–34**, 1491 (1983);
S. Ishio, M. Takahashi: J. Phys. Soc. Jap. **50**, 93 (1985)

6.58   E. du Tremolet de Lacheisserie: J. Mag. Mag. Mat. **25**, 251 (1982);
H.K. Lachowicz: J. Mag. Mag. Mat. **41**, 327 (1984)

6.59   K. Fukamichi, M. Hiroyoshi, M. Kikuchi, T. Masumoto: J. Mag. Mag. Mat. **10**, 294 (1979);
M. Kikuchi, K. Fukamichi, T. Masumoto: J. Mag. Mag. Mat. **10**, 300 (1979);
E. Torok, G. Hausch: J. Mag. Mag. Mat. **10**, 303 (1979);
G. Hausch, E. Torok, T. Mohri, Y. Nakamura: J. Mag. Mag. Mat. **10**, 157 (1979);
G. Hausch: J. Mag. Mag. Mat. **10**, 163 (1979);
S. Ishio, Y. Sato, T. Ikeda, M. Takahashi: J. Non Cryst. Sol. **61**, 955 (1984)

6.60   K. Fukamichi, M. Kikuchi, T. Masumoto: J. Non Cryst. Sol. **61**, 961 (1984)

6.61   K. Fukamichi: In *Amorphous Metal Alloys* ed. by F.E. Luborsky, (Butterworths, London 1983)

6.62   U. Hardebusch, W. Gerhardt, J.S. Schilling: Phys. Rev. Lett. **44**, 352 (1980);
M.A. Simpson, T.F. Smith, E. Gmelin: J. Phys. F**11**, 1655 (1981);
G. Creuzet, I.A. Campbell: J. Physique Lett. **43**, L575 (1982)

6.63   Y. Tsunoda: J. Mag. Mag. Mat. **31–34**, 67 (1983)

6.64   G. Hausch: Phys. Stat. Sol. A**41**, K 35 (1977)

6.65   K. Tajima, Y. Endoh, Y. Ishikawa, W.G. Stirling: Phys. Rev. Lett. **37**, 519 (1976);
Y. Ishikawa, S. Onodera, K. Tajima: J. Mag. Mag. Mat. **10**, 183 (1979)

6.66   I.R. Harris, M. Norman, W.E. Gardner: J. Less Com. Metals **29**, 299 (1972)

6.67   I.R. Harris, G.V. Raynor: J. Less Com. Metals **9**, 7 (1965)

6.68   A. Iandelli, A. Palenzona: J. Less Com. Metals **29**, 293 (1972)

6.69   D. Signoux, F. Givord, R. Lemaire, F. Tasset: J. Less Com. Metals **94**, 165 (1983);
G. Creuzet, A. Fert, C. Gaonach, D. Gignoux: Physica **130B**, 138 (1985)
G. Creuzet, D. Gignoux: Phys. Rev. B**33**, 515 (1986)

6.70   M. Ribault, A. Benoit, J. Flouquet, J. Palleau: J. Physique Lett. **40**, L 413 (1979);
see also K. Andres, J.E. Graebner and H.R. Ott: Phys. Rev. Lett. **35**, 1779 (1975)

6.71   A. de Visser, J.J.M. Franse: Physica **130B**, 177 (1985)

6.72   J.L. Smith, Z. Fisk, S.S. Hecker: Physica **130B**, 151 (1985)

227

Additional References with Titles

Chapter 3

Allenspach, R., Colla, E., Mauri, D., Landolt, M., Wohlfahrt, E.P.: Spin polarized threshold photoemission of amorphous $Fe_{83}B_{17}$ and $Co_{78}Mn_2B_{20}$: Weak evidence of strong itinerant ferromagnetism. Phys. Lett. **105** A, 145 (1984)

Carbone, C., Jonker, B.T., Walker, K.-H., Prinz, G.A., Kisker, E.: The epitaxial growth of Fe on GaAs(110): Development of the electronic structure and interface formation. Solid State Commun., in press

Clauberg, R., Hopster, H., Raue, R.: Reinterpretation of a controversial structure in the photoemission spectrum of Ni(100). Phys. Rev. B **29**, 4395 (1984)

Feder, R. (ed.): *Polarized Electrons in Surface Physics* (World Scientific Publishing, Singapore 1985)

Feigerle, C.S., Seiler, A., Pena, J.L., Celotta, R.J., Pierce, D.T.: CO chemisorption on Ni(110): Effect on surface magnetism

Hopster, H., Kurzawa, R., Raue, R., Schmitt, W., Güntherodt, G., Walker, K.-H., Güntherodt, G.: Spin polarized photoemission study of Fe-Ni ferromagnetic metallic glasses. J. Phys. F **15**, L11 (1985)

Jonker, B.T., Walker, K.-H., Kisker, E., Prinz, G.A., Carbone, C.: Spin-polarized photoemission study of epitaxial Fe(001) films on Ag(001). Phys. Rev. Lett. **57**, 142 (1986)

Kirschner, J.: *Polarized Electrons at Surfaces*, Springer Tracts Mod. Phys., Vol. 106 (Springer, Berlin, Heidelberg 1985)

Schmitt, W., Hopster, H., Günterodt, G.: Influence of adsorbates on surface magnetism studied by spin-resolved photoemission spectroscopy. Phys. Rev. B **31**, 4035 (1985)

Taborelli, M., Allenspach, R., Boffa, G., Landolt, M.: Magnetic coupling of surface adlayers: Gd on Fe(100). Phys. Rev. Lett. **56**, 2869 (1986)

Weller, D., Alvarado, S.F., Gudat, W., Schröder, K., Campagna, M.: Observation of surface-enhanced magnetic order and magnetic surface reconstruction on Gd(0001). Phys. Rev. Lett. **54**, 1555 (1985)

Subject Index